Computer Vision and Imaging in Intelligent Transportation Systems

Computer Vision and Imaging in Intelligent Transportation Systems

Edited by

Robert P. Loce
Conduent Labs, Webster, NY, USA

Raja Bala
Samsung Research America, Richardson, TX, USA

Mohan Trivedi
University of California, San Diego, CA, USA

Library of Congress Cataloging-in-Publication data applied for

ISBN: 9781118971604

A catalogue record for this book is available from the British Library.

Cover design: Wiley
Cover image: ©Anouar Akrouh / Eyeem / Gettyimages

Setin 10/12pt Warnock by SPi Global, Pondicherry, India
Printed and bound in Malaysia by Vivar Printing Sdn Bhd

10 9 8 7 6 5 4 3 2 1

Contents

List of Contributors

Raja Bala
Samsung Research America
Richardson, TX
USA

Edgar A. Bernal
United Technologies Research Center
East Hartford, CT
USA

Orhan Bulan
General Motors Technical Center
Warren, MI
USA

Aaron Burry
Conduent Labs
Webster, NY
USA

Yang Cai
Carnegie Mellon University
Pittsburgh, PA
USA

Gianni Cario
Department of Informatics, Modeling,
Electronics and System Engineering (DIMES)
University of Calabria
Rende
Italy

Alessandro Casavola
Department of Informatics, Modeling,
Electronics, and System Engineering (DIMES)
University of Calabria
Rende
Italy

Johan Casselgren
Luleå University of Technology
Luleå
Sweden

Zezhi Chen
Kingston University
London
UK

Shashank Deshpande
Carnegie Mellon University
Pittsburgh, PA
USA

Timothy J. Ellis
Kingston University
London
UK

Rodrigo Fernandez
Universidad de los Andes
Santiago
Chile

Hasan Fleyeh
Dalarna University
Falun
Sweden

Patrik Jonsson
Combitech AB
Östersund
Sweden

Vladimir Kozitsky
Conduent Labs
Webster, NY
USA

Matti Kutila
VTT Technical Research Centre of Finland Ltd.
Tampere
Finland

Yuriy Lipetski
SLR Engineering GmbH
Graz
Austria

Robert P. Loce
Conduent Labs
Webster, NY
USA

Marco Lupia
Department of Informatics, Modeling,
Electronics and System Engineering (DIMES)
University of Calabria
Rende
Italy

Vishal Monga
Pennsylvania State University
University Park, PA
USA

Brendan Tran Morris
University of Nevada
Las Vegas, NV
USA

Wiktor Muron
Carnegie Mellon University
Pittsburgh, PA
USA

Peter Paul
Conduent Labs
Webster, NY
USA

Pasi Pyykönen
VTT Technical Research Centre of Finland Ltd.
Tampere
Finland

Ravi Satzoda
University of California
San Diego, CA
USA

Mohammad Shokrolah Shirazi
Cleveland State University
Cleveland, OH
USA

Oliver Sidla
SLR Engineering GmbH
Graz
Austria

Mohan Trivedi
University of California
San Diego, CA
USA

Sergio A. Velastin
Universidad Carlos III de Madrid
Madrid
Spain

Wencheng Wu
Conduent Labs
Webster, NY
USA

Beilei Xu
Conduent Labs
Webster, NY
USA

Muhammad Haroon Yousaf
University of Engineering and Technology
Taxila
Pakistan

Preface

There is a worldwide effort to develop smart transportation networks that can provide travelers with enhanced safety and comfort, reduced travel time and cost, energy savings, and effective traffic law enforcement. Computer vision and imaging is playing a pivotal role in this transportation evolution. The forefront of this technological evolution can be seen in the growth of scientific publications and conferences produced through substantial university and corporate research laboratory projects. The editors of this book have assembled topics and authors that are representative of the core technologies coming out of these research projects. This text offers the reader with a broad comprehensive exposition of computer vision technologies addressing important roadway transportation problems. Each chapter is authored by world-renowned authorities discussing a specific transportation application, the practical challenges involved, a broad survey of state-of-the-art approaches, an in-depth treatment of a few exemplary techniques, and a perspective on future directions. The material is presented in a lucid tutorial style, balancing fundamental theoretical concepts and pragmatic real-world considerations. Each chapter ends with an abundant collection of references for the reader requiring additional depth.

The book is intended to benefit researchers, engineers, and practitioners of computer vision, digital imaging, automotive, and civil engineering working on intelligent transportation systems. Urban planners, government agencies, and other decision- and policy-making bodies will also benefit from an enhanced awareness of the opportunities and challenges afforded by computer vision in the transportation domain. While each chapter provides the requisite background required to learn a given problem and application, it is helpful for the reader to have some familiarity with the fundamental concepts in image processing and computer vision. For those who are entirely new to this field, appropriate background reading is recommended in Chapter 1. It is hoped that the material presented in the book will not only enhance the reader's knowledge of today's state of the art but also prompt new and yet-unconceived applications and solutions for transportation networks of the future.

The text is organized into Chapter 1 that provides a brief overview of the field and Chapters 2–15 divided into two parts. In Part I, Chapters 2–9 present applications relying upon the infrastructure, that is, cameras that are installed on roadway structures such as bridges, poles and gantries. In Part II, Chapters 10–15 discuss techniques to monitor driver and vehicle behavior from cameras and sensors placed within the vehicle.

In Chapter 2, Burry and Kozitsky present the problem of license plate recognition—a fundamental technology that underpins many transportation applications, notably ones pertaining to law enforcement. The basic computer vision pipeline and state-of-the-art solutions for plate recognition are described. Muron, Deshpande, and Cai present automatic vehicle classification (AVC) in Chapter 3. AVC is a method for automatically categorizing types of motor vehicles based on the predominant characteristics of their features such as length, height, axle count, existence of a trailer, and specific contours. AVC is also an important part of intelligent transportation system (ITS) in applications such as automatic toll collection, management of traffic density, and estimation of road usage and wear.

Chapters 2, 4, 5, and 8 present aspects of law enforcement based on imaging from the infrastructure. Detection of passenger compartment violations is presented in Chapter 4 by Bulan, Xu, Loce, and Paul. The chapter presents imaging systems capable of gathering passenger compartment images and computer vision methods of extracting the desired information from the images. The applications it presents are detection of seat belt usage, detection of mobile phone usage, and occupancy detection for high-occupancy lane tolling and violation detection. The chapter also covers several approaches, while providing depth on a classification-based method that is yielding very good results. Detection of moving violations is presented in Chapter 5 by Wu, Bulan, Bernal, and Loce. Two prominent applications—speed detection and stop light/sign enforcement—are covered in detail, while several other violations are briefly reviewed.

A major concern for urban planners is traffic flow analysis and optimization. In Chapter 6, Fernandez, Yousaf, Ellis, Chen, and Velastin present a model for traffic flow from a transportation engineer's perspective. They consider flow analysis using computer vision techniques, with emphasis given to specific challenges encountered in developing countries. Intersection modeling is taught by Morris and Shirazi in Chapter 7 for the applications of understanding capacity, delay, and safety. Intersections are planned conflict points with complex interactions of vehicles, pedestrians, and bicycles. Vision-based sensing and computer vision analysis bring a level of depth of understanding that other sensing modalities alone cannot provide. In Chapter 8, Sidla and Lipetski examine the state of the art in visual parking space monitoring. The task of the automatic parking management is becoming increasingly essential. The number of annually produced cars has grown by 55% in the past 7 years. Most large cities have a problem of insufficient availability of parking space. Automatic determination of available parking space coupled with a communication network holds great promise in alleviating this urban burden.

While computer vision algorithms can be trained to recognize common patterns in traffic, vehicle, and pedestrian behavior, it is often an unusual event such as an accident or traffic violation that warrants special attention and action. Chapter 9 by Bala and Monga is devoted to the problem of detecting anomalous traffic events from video. A broad survey of state-of-the-art anomaly detection models is followed by an in-depth treatment of a robust method based on sparse signal representations.

In Part II of the text, attention is turned to in-vehicle imaging and analysis. The focus of Chapters 10–12 are technologies that are being applied to driver assistance systems. Chapter 10 by Deshpande and Cai deals with the problem of detecting pedestrians from road-facing cameras installed on the vehicle. Pedestrian detection is critical to intelligent transportation systems, ranging from autonomous driving to infrastructure surveillance, traffic management, and transit safety and efficiency, as well as law enforcement. In Chapter 11, Casavola, Cario, and Lupia present lane detection (LD) and lane tracking (LT) problems arising in lane departure warning systems (LDWSs). LWDSs refer to specific forms of advanced driver assistant systems (ADAS) designed to help the driver to stay into the lane, by warning her/him with a sufficient advance that an imminent and possibly unintentional lane departure is going to take place so that she/he can take the necessary corrective measures. Chapter 12 by Satzoda and Trivedi teaches the technologies associated with vision-based integrated techniques for collision avoidance systems. The chapter surveys related technologies and focuses on an integrated approach called efficient lane and vehicle detection using integrated synergies (ELVIS) that incorporates the lane information to detect vehicles more efficiently in an informed manner using a novel two part–based vehicle detection technique.

Driver inattention is a major cause of traffic fatalities worldwide. Chapter 13 by Bala and Bernal presents an overview of in-vehicle technologies to proactively monitor driver behavior and provide appropriate feedback and intervention to enhance safety and comfort. A broad survey of the state of the art is complemented with a detailed treatment of a few selected driver monitoring techniques including methods to fuse video with nonvisual data such as motion and bio-signals.

Traffic sign recognition is present in Chapter 14 by Fleyeh. Sign recognition is a field-concerned detection and recognition of traffic signs in traffic scenes as acquired by a vehicle-mounted camera. Computer vision and artificial intelligence are used to extract the traffic signs from outdoor images taken in uncontrolled lighting conditions. The signs may be occluded by other objects and may suffer from various problems such as color fading, disorientation, and variations in shape and size. It is the field of study that can be used either to aid the development of an inventory system (for which real-time recognition is not required), or to aid the development of an in-car advisory system (when real-time recognition is necessary). Road condition monitoring is presented in Chapter 15 by Kutila, Pyykönen, Casselgren, and Jonsson. The chapter reviews proposed measurement principles in the road traction monitoring area and provides examples of sensor solutions that are feasible for vehicle on-board and road sensing. The chapter also reviews opportunities to improve performance with the use of sensor data fusion and discusses future opportunities. We do have an enhanced eBook available with integrated video demonstrations to further explain the concepts discussed in the book.

Robert P. Loce
Raja Bala

March 2017

Acknowledgments

We would like to take this opportunity to thank each and every author for their painstaking efforts in preparing chapters of high quality, depth, and breadth, and also for working collaboratively with the editors to assemble a coherent text in a timely manner.

About the Companion Website

Don't forget to visit the companion website for this book:

www.wiley.com/go/loce/ComputerVisionandImaginginITS

There you will find valuable material designed to enhance your learning, including:

- Videos
- Figures

1

Introduction

Raja Bala[1] and Robert P. Loce[2]

[1] *Samsung Research America, Richardson, TX, USA*
[2] *Conduent Labs, Webster, NY, USA*

With rapid advances in driver assistance features leading ultimately to autonomous vehicle technology, the automobile of the future is increasingly relying on advances in computer vision for greater safety and convenience. At the same time, providers of transportation infrastructure and services are expanding their reliance on computer vision to improve safety and efficiency in transportation and addressing a range of problems, including traffic monitoring and control, incident detection and management, road use charging, and road condition monitoring. Computer vision is thus helping to simultaneously solve critical problems at both ends of the transportation spectrum— at the consumer level and at the level of the infrastructure provider. The book aims to provide a comprehensive survey of methods and systems that use both infrastructural and in-vehicle computer vision technology to address key transportation applications in the following three broad problem domains: (i) law enforcement and security, (ii) efficiency, and (iii) driver safety and comfort. Table 1.1 lists the topics addressed in the text under each of these three domains.

This chapter introduces and motivates applications in the three problem domains and establishes a common computer vision framework for addressing problems in these domains.

1.1 Law Enforcement and Security

Law enforcement and security are critical elements to maintaining the well-being of individuals and the protection of property. Societies rely on law enforcement agencies to provide these elements. Imaging systems and computer vision are means to sense and interpret situations in a manner that can amplify the effectiveness of officers within these agencies. There are several common elements shared by computer vision law enforcement and security applications, such as the detection and identification of events of interest. On the other hand, there are also distinctions that separate a security application from law enforcement. For instance, prediction and prevention are important for security applications, while accuracy and evidence are essential for law enforcement. In many cases, modules and components of a security system serve as a front end of a law enforcement system. For example, to enforce certain traffic violations, it is necessary to detect and identify the occurrence of that event.

Table 1.1 Taxonomy of problem domains and applications described in the book.

Problem domains	Applications and methods	Imaging system employed
Law enforcement and security	License plate recognition for violations	Infrastructure
	Vehicle classification	Infrastructure
	Passenger compartment violation detection	Infrastructure, in-vehicle
	Moving violation detection	Infrastructure
	Intersection monitoring	Infrastructure
	Video anomaly detection	Infrastructure
Efficiency	Traffic flow analysis	Infrastructure
	Parking management	Infrastructure, in-vehicle
	License plate recognition for tolling	Infrastructure
	Passenger compartment occupancy detection	Infrastructure, in-vehicle
Driver safety and comfort	Lane departure warning	In-vehicle
	Collision avoidance	In-vehicle
	Pedestrian detection	In-vehicle
	Driver monitoring	In-vehicle
	Traffic sign recognition	In-vehicle
	Road condition monitoring	In-vehicle, infrastructure

Consider the impact of moving vehicle violations and examples of benefits enabled by computer vision law enforcement systems. There is a strong relationship between excessive speed and traffic accidents. In the United States in 2012, speeding was a contributing factor in 30% of all fatal crashes (10,219 lives) [1]. The economic cost of speeding-related crashes was estimated to be $52 billion in 2010 [2]. In an extensive review of international studies, automated speed enforcement was estimated to reduce injury-related crashes by 20–25% [3]. The most commonly monitored moving violations include speeding, running red lights or stop signs, wrong-way driving, and illegal turns. Most traffic law enforcement applications in roadway computer vision systems involve analyzing well-defined trajectories and speeds within those trajectories, which leads to clearly defined rules and detections. In some cases, the detections are binary, such as in red light enforcement (stopped or passed through). Other applications require increased accuracy and precision, such as detecting speed violations and applying a fine according to the estimated vehicle speed. There are other deployed applications where the violation involves less definitive criterion, such as reckless driving.

Several moving violations require observation into the passenger compartment of a vehicle. Failure to wear a seat belt and operating a handheld cell phone while driving are two common safety-related passenger compartment violations. Seat belt use in motor vehicles is the single most effective traffic safety device for preventing death and injury to persons involved in motor vehicle accidents. Cell phone usage alone accounts for roughly 18% of car accidents caused by distracted drivers [4]. In addition, the National Highway Traffic Safety Administration (NHTSA) describes other behaviors resulting in distracted driving, including occupants in the vehicle eating, drinking, smoking, adjusting radio, adjusting environmental controls, and reaching for an object in the car. The conventional approach to enforcement of passenger compartment violations has been through traffic stops by law enforcement officers. This approach faces many challenges such as safety, traffic disruption, significant personnel cost, and the difficulty of determining cell phone usage or seat belt usage at high speed. Imaging technology and computer vision can provide automated or semiautomated enforcement of these violations.

Security of individuals and property is another factor in the monitoring of transportation networks. Video cameras have been widely used for this purpose due to their low cost, ease of installation and maintenance, and ability to provide rich and direct visual information to operators. The use of video cameras enables centralized operations, making it possible for an operator to "coexist" at multiple locations. It is also possible to go back in time and review events of interest. Many additional benefits can be gained by applying computer vision technologies within a camera network. Consider that, traditionally, the output of security cameras has either been viewed and analyzed in real-time by human operators, or archived for later use if certain events have occurred. The former is error prone and costly, while the latter has lost some critical capabilities such as prediction and prevention. In a medium-sized city with several thousand roadway cameras, computer vision and video analytics allow a community to fully reap the benefits of analyzing this massive amount of information and highlighting critical events in real-time or in later forensic analysis.

In certain security and public safety applications, very rapid analysis of large video databases can aid a critical life or death situation. An Amber Alert or a Child Abduction Emergency is an emergency alert system to promptly inform the public when a child has been abducted. It has been successfully implemented in several countries throughout the world. When sufficient information is available about the incident (e.g., description of captor's vehicle, plate number, and color), a search can be conducted across large databases of video that have been acquired from highway, local road, traffic light, and stop sign monitoring, to track and find the child. Similar to Amber Alert and much more common is Silver Alert, which is a notification issued by local authorities when a senior citizen or mentally impaired person is missing. Statistics indicates that it is highly desirable that an Amber-/ Silver Alert-related search is conducted in a very fast and efficient manner, as 75% of the abducted are murdered within the first 3 h. Consider a statement from the US West Virginia code on Amber Alert 15-3A-7:

> The use of traffic video recording and monitoring devices for the purpose of surveillance of a suspect vehicle adds yet another set of eyes to assist law enforcement and aid in the safe recovery of the child.

Human analysis of video from thousands of camera could take many days, while computer vision methods have the potential to rapidly extract critical information. The speed can be scaled by the available computational power, which is rapidly advancing due to high-speed servers and cloud computing.

Whether it is safety of individuals or security of property, recognition of a vehicle is a key component of a roadway security system. Vehicles traveling on the public roadways in most countries are required by law to carry a clearly visible placard with a unique identifier that is registered with the local government. This placard (license plate) can contain various symbols—letters, numbers, logos, etc.—based on local government regulations and the vehicle class. Given the common requirement for its presence and ease of visibility, the license plate has become the default means for identifying a vehicle and/or its registered operator. Automated license plate recognition (ALPR), also referred to as automated number plate recognition (ANPR), leverages computer vision algorithms to extract license plate information from videos or still images of vehicles. ALPR has become a core technology within modern intelligent transportation systems. Surveillance and police enforcement applications leverage ALPR systems to provide real-time data gathering to support law enforcement efforts. For example, a 2012 study [5] conducted on the usage of ALPR technology by police agencies found that 80% of larger agencies (those with 1000+ sworn officers) were leveraging ALPR in some way. Results from this same study also indicated that police agencies using ALPR reported that the use of technology had increased the recovery of stolen vehicles by 68% and overall arrests by 55%.

A trend that favors improved law enforcement and security in transportation settings is the increasing intelligence across broad camera networks. Cities such as London and Tokyo are said to have over 500,000 government cameras, while Chicago has linked private, school, roadway, and police cameras into a massive interconnected network. Several factors are driving this trend. There is the fight against large-scale terrorism and crime with limited human resources. Camera and network technology are continually becoming more capable and less costly. Cloud computing is also enabling a flexible, scalable architecture for big data analysis. Computer vision becomes the intelligent connecting element that enables the camera network and computing resources to address the societal need.

1.2 Efficiency

The efficiency of a roadway network impacts expended time of individuals, fuel usage, and pollution all of which are key factors contributing to the quality of life of a community. Roadway imaging with computer vision is being increasingly applied to optimize efficiency. Example applications include traffic flow analysis, video-based parking management, open road tolling, and high-occupancy lane management.

Traffic is the movement of people and goods in a public space, where the movement may involve a car, public transport vehicle, bicycle, or foot travel. Data derived from traffic volume studies can help local governments estimate road usage, volume trends, and critical flow time periods. These estimates can be used to optimize maintenance schedules, minimize conflicts, and optimize the timing of traffic enforcement. Real-time traffic flow data can also enable efficient incident management, which consists of incident detection, verification, and response. Traffic variables of interest are flow, speed, and concentration with respect to road capacity. Each of these variables involves detection and recognition and tracking of a particular traveling entity (pedestrian, bicycle, car, etc.) Manual methods involve monitoring a region onsite or via a video feed, both of which are very labor intensive and tend to provide a limited snapshot of information. The past decade has seen a trend toward increased automation, leveraging the ubiquity of roadway cameras and advances in computer vision toward vehicle detection, vehicle classification, and pedestrian detection. This trend is bringing fine-grain and persistent analysis to traffic engineers and, in turn, increasing the efficiency of our roadways.

Urban parking management is receiving significant attention due to its potential to reduce traffic congestion, fuel consumption, and emissions [6, 7]. Real-time parking occupancy detection is a critical component of parking management systems, where occupancy information is relayed to drivers in real time via smartphone apps, radio, the Internet, on-road signs, or GPS auxiliary signals. This can significantly reduce traffic congestion created by vehicles searching for an available parking space, thus reducing fuel consumption. Sensors are required to gather such real-time data from parking venues. Video-based sensing for monitoring on-street parking areas offers several advantages over sensors such as inductive loops, ultrasonic sensors, and magnetic in-ground sensors. One advantage is that one video camera can typically monitor and track several parking spots (see Figure 1.1), whereas multiple magnetic sensors may be needed to reliably monitor a single parking space. Another advantage is that device installation and maintenance is less disruptive in the case of video cameras when compared to in-ground sensors. Video cameras can also support other tasks such as traffic law enforcement and surveillance since they capture a wider range of useful visual

Figure 1.1 Examples of a camera view sensing multiple parking stalls.

information including vehicle color, license plate, vehicle type, speed, etc. A video-based parking occupancy detection system can, therefore, provide a convenient, cost-effective solution to the sensing task and also provide additional functionality for traffic law enforcement and surveillance.

A third problem is highway congestion. Government officials and members of the transportation industry are seeking new strategies for addressing the problems associated with high traffic volumes. One such mechanism to reduce the congestion on busy highway corridors is the introduction of managed lanes such as high-occupancy vehicle (HOV) lanes that require a minimum number of vehicle occupants and high-occupancy tolling (HOT) lanes that set a tolling price depending upon the number of occupants. Due to imposed limitations and fees, HOV/HOT lanes are often much less congested than other commuter lanes. However, the rules of the HOV/HOT lane need to be enforced to realize the congestion reducing benefits. Typical violation rates can exceed 50–80%. Current enforcement practices dispatch law enforcement officers to the roadside to visually examine passing vehicles. Manual enforcement can be a tedious, labor-intensive practice, and, ultimately, ineffective with enforcement rates of typically less than 10% [8]. Besides enduring environmental conditions of snow, darkness, sunlight reflections, and rain, law enforcement officers also have to deal with vehicles traveling at high speeds that may have darkened/tinted glass, reclining passengers, and/or child seats with or without children. As a result, there is a desire to have an automated method to augment or replace the manual process. Practical imaging-based systems have been demonstrated using near-infrared (NIR) illumination and multiple cameras triggered by induction loops or laser break-beam devices (Figure 1.2).

1.3 Driver Safety and Comfort

Many vehicles today employ a so-called *Advanced Driver Assistance System* (ADAS) that uses cameras and various sensors to make inferences about both the driver and the environment surrounding a vehicle. An important element of ADAS is lane departure warning. More than 40% of all fatal roadway accidents in 2001 involved a lane or road departure [9], resulting primarily from driver distraction, inattention, or drowsiness [10]. *Lane Departure Warning* (LDW) systems [11] track roadway markings using a video camera mounted near the rearview mirror or on the dash board of a vehicle so the area in front of the vehicle may be viewed. A warning signal is given to the driver if a vehicle unintentionally approaches a lane marking (i.e., without activating a turn signal). The prevalence of LDW systems is expected to rapidly increase, with various tax incentives being proposed in the United States and Europe for vehicles with LDW systems. LDW algorithms face the daunting task of operating in real-time and under multifarious weather conditions, in order to detect and decipher within this limited field of view a wide assortment of lane markings.

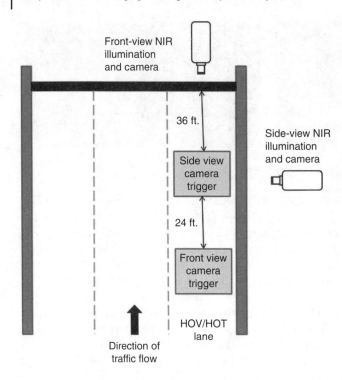

Figure 1.2 Illustration of a gantry-mounted front-view and road-side side-view acquisition setup for monitoring a high-occupancy lane.

Pedestrian detection is another important element of ADAS. While this problem has been extensively studied from the viewpoint of fixed video surveillance cameras, many new challenges arise when a camera is mounted on a common moving vehicle. The detection must comprehend a wide range of lighting conditions, a continuously varying background, changes in pose, occlusion, and variation in scale due to the changing distance. Technologies available today for pedestrian detection use some combination of appearance- and motion-based techniques. A generalization of this problem is collision avoidance with other vehicles, bicyclists, and animals. Feedback can include direct control on vehicle motion, with popular examples being adaptive cruise control and automatic braking.

A third form of ADAS directly monitors the driver's attention with the use of a driver-facing camera and possibly other sensors placed in contact with the driver's body. A significant challenge is making accurate inferences on the state of the driver by monitoring video of facial expression, eye movement and gaze in the presence of vehicle motion, varying illumination, and large variation in affective expression across humans. A promising direction in this domain is the use of adaptive and online learning techniques whose inferences are personalized to a specific driver and vehicle.

A fourth class of ADAS monitors relevant aspects of the static environment, such as road conditions and traffic signs. The European ASSET program has been actively pursuing within-vehicle camera-based methods to detect the friction (conversely, slipperiness) of roads, a significant factor in fatalities worldwide. Imaging systems using polarization filters and IR imaging are being explored to address this problem. Traffic signs provide crucial information about road conditions, and can often be missed by the driver due to weather, occlusion, damage, and inattention. A computer vision system that can quickly and accurately interpret traffic signs can be an invaluable aid to the driver. Challenges pertain to poor image quality due to inclement weather, vehicle motion, the tremendous variety in the scene and objects being captured, and the need for real-time inferences.

ADAS can be classified by the degree of autonomy that is enabled during active driving. In the United States, the NHTSA has defined vehicle automation as having five levels [12] as follows:

No-Automation (Level 0): The driver is in complete and sole control of the primary vehicle controls—brake, steering, throttle, and motive power—at all times.

Function-Specific Automation (Level 1): Automation at this level involves one or more specific control functions. Examples include electronic stability control or precharged brakes, where the vehicle automatically assists with braking to enable the driver to regain control of the vehicle or stop faster than possible by acting alone.

Combined Function Automation (Level 2): This level involves automation of at least two primary control functions designed to work in unison to relieve the driver of control of those functions. An example of combined functions enabling a Level 2 system is adaptive cruise control in combination with lane centering.

Limited Self-Driving Automation (Level 3): Vehicles at this level of automation enable the driver to cede full control of all safety-critical functions under certain traffic or environmental conditions and in those conditions to rely heavily on the vehicle to monitor for changes in those conditions requiring transition back to driver control. The driver is expected to be available for occasional control, but with sufficiently comfortable transition time.

Full Self-Driving Automation (Level 4): The vehicle is designed to perform all safety-critical driving functions and monitor roadway conditions for an entire trip. Such a design anticipates that the driver will provide destination or navigation input, but is not expected to be available for control at any time during the trip. This includes both occupied and unoccupied vehicles.

Full Self-Driving Automation is embodied in what is often referred to as autonomous vehicles, a much publicized example being the Google driverless car. As of September 2015, Google's fleet of autonomous vehicles have logged 1,200,000 driverless road miles. A key technology used in Google's implementation is a roof-mounted Velodyne 64-beam laser, which creates a three-dimensional (3D) map of the surrounding environment in the immediate area of about 50 ft. The 3D image is combined with high-resolution maps that have been programmed into the vehicle's control system. The laser system can differentiate among a large variety of objects, including cars, pedestrians, cyclists, and small and large stationary objects. Four radars (one for front, back, left, and right) sense any fast-moving objects at distances further than the laser detection range and are used to give the car far-sighted vision for handling high speeds on freeways. A front-mounted camera and computer vision algorithms analyze the road ahead of the car, observing road signs and stop lights for information that a human driver typically uses. Other sensors include a GPS, an inertial measurement unit, and wheel encoder. While autonomous vehicle technology poses many challenges to current roadway legislation, it does offer great potential to mobilize citizens with impairments and could make driving safer due to comprehensive sensing and rapid decision making. For further details, the reader is referred to the recent IEEE Spectrum Online article [13].

1.4   A Computer Vision Framework for Transportation Applications

Computer vision can be broadly described as the task of interpreting and making sense of the world around us from images and video. There are many levels and types of interpretation, beginning at the basic level with detecting and recognizing individual objects and their motion, and evolving onto higher levels of inference on interactions and relationships between multiple objects, human–object interactions, and semantic reasoning of human behavior. As such, computer vision falls within the

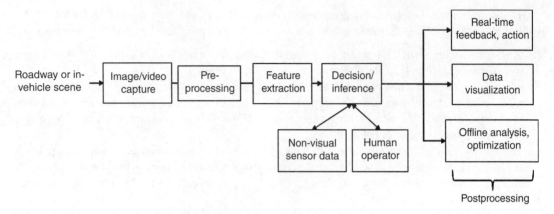

Figure 1.3 Computer vision pipeline for transportation applications.

broader domain of artificial intelligence (AI), and can be thought of as the branch of visual sensing, perception, and reasoning within the field of AI. In the context of transportation systems, computer vision enables the interpretation of the environment outside of and within automotive vehicles, whether it is for the purpose of enhancing safety, efficiency, or law enforcement. In many instances, the computer vision system augments, or if reliable enough, even replaces human interpretation so as to reduce errors, cost, effort, and time. A basic computer vision pipeline is shown in Figure 1.3. Each of the blocks is described next.

1.4.1   Image and Video Capture

The pipeline begins with an imaging system comprising one or more cameras that capture digital images or video of a relevant scene. Infrastructural cameras may be installed on bridges, gantries, poles, traffic lights, etc. In-vehicle cameras are positioned either outside the vehicle, providing front and rearviews of the environment surrounding the vehicle, or within the vehicle usually for the purpose of monitoring the driver's alertness and attention. Factors that are considered in designing the imaging system are field of view, coverage of spectral bands, spatial and temporal resolution, form factor, ruggedness, and cost. A stereo- or multicamera network often requires establishing the relationship among the camera views, and between camera coordinates and 3D world coordinates. Applications utilizing stereo vision techniques in the book include moving violation detection, pedestrian detection, and parking management. Also to be considered are the bandwidth and cost of the data transmission technology used to send video data from the camera to the subsequent computational engines which may be logically and physically co-located with or separated from the imaging system. Another practical factor is the electrical power source. In urban areas, power is typically supplied via lines that support traffic controls and lighting, while more remote regions sometimes harvest energy locally using photovoltaics. The reader will be exposed to different types of imaging systems throughout the course of the book.

1.4.2   Data Preprocessing

The next step in the pipeline is data preprocessing. Well-known image processing operations such as brightness and contrast normalization, distortion compensation, noise filtering, spatial resizing, cropping, temporal frame rate conversion, and motion stabilization may be carried out to prepare the images or video for subsequent analysis. The operations and their parameters are carefully tuned using knowledge of the characteristics of the imaging system, as well as the requirements of the computer vision task at hand. For example, images of vehicle license plates taken with infrastructure cameras require normalization for accurate performance in the presence of widely varying contrast due to ambient lighting and weather conditions. On the other hand, driver-facing video captured within a moving vehicle will likely have to undergo motion deblurring and stabilization due to both vehicle motion and the motion of the driver's head relative to the camera. Another common type of preprocessing identifies one or more regions of interest (ROIs) for subsequent interpretation. An example is human face detection, which is used for driver monitoring and in-vehicle passenger compartment violation detection. A second example is license plate localization, a precursor to plate recognition. Interestingly, such ROI detection can itself be viewed as a computer vision subproblem, involving its own feature extraction and decision operations. Edge detection is a very common preprocessing step for computer vision, and will be encountered in several applications, including lane detection and tracking (Chapter 11). Each chapter discusses data preprocessing unique to a specific problem domain. However, readers who are entirely unfamiliar with the field of image processing are encouraged to review the text by Gonzalez and Woods [14] for an excellent introductory overview of the subject.

Another type of preprocessing involves calibrating for camera characteristics. Two common forms are colorimetric and geometric calibration. Colorimetric calibration transforms the (typically RGB) camera color representation of the input image into a standardized color space such as sRGB, XYZ, or CIELAB. Parameters required to perform such a calibration include knowledge of the spectral sensitivities of the RGB filters, spectral distribution of the incident illumination, and optoelectronic transfer functions relating digital camera values to luminance at each pixel. Colorimetric calibration is important for tasks wherein color provides a critical cue in the inference. An example is locating features within the driver's face, wherein human skin color can serve as a cue. Geometric camera calibration involves relating camera pixel coordinates to real 3D world spatial coordinates. This problem is treated in detail in Chapter 5 describing moving violations, wherein relative motion within the coordinates of the video signal must be translated to absolute motion and speed in world coordinates.

1.4.3 Feature Extraction

The third and fourth steps in the pipeline, namely feature extraction and decision inference, jointly define the computer vision component underpinning the overall system. Feature extraction is the process of identifying from the raw pixel data a set of quantitative descriptors that enable accurate interpretations to be made on the image. Essentially, feature extraction is a mapping from the original set of (possibly preprocessed) image pixels I to an alternate representation x in a suitably defined feature space. One benefit of feature extraction is dimensionality reduction; that is, the dimensionality of x is typically much smaller than that of I, and as such retains only the information that is germane to the inference task at hand. Numerous feature descriptors have been developed in the literature and used successfully in practical applications; the optimal choice is made based on knowledge of the domain, the problem, and its constraints. Feature descriptors fall into two broad categories: global and local features. Global features holistically describe the entire image or video, while local features represent a spatially or spatiotemporally localized portion of the image or video. A simple example of a global feature is a color histogram of an image. Such a feature may be employed for example in a vehicle identification or parking space occupancy detection problem, where color is an important cue. On the other hand, local features represent some spatiotemporally localized attribute such as gradient, curvature, or texture in the image. A common example is the histogram of oriented gradients (HoGs) which will be encountered in a variety of transportation problems, including vehicle classification (Chapter 3), traffic sign recognition (Chapter 14), and pedestrian detection (Chapter 10). The scale-invariant feature transform (SIFT) and Haar descriptors are also commonly used for vehicle classification and other tasks. Note that global features can be modified to operate locally; for example, one can compute color histograms of $N \times N$ image subblocks and concatenate them to form a composite feature x that captures a spatially varying color distribution within the image. Local descriptors may also be aggregated into a pooled feature describing the entire image, using techniques such as Bag-of-Words and Fisher Vector encoding, as will be encountered in one approach to passenger compartment violation detection in Chapter 4.

1.4.4 Inference Engine

The inference stage takes as input the feature descriptors, and emits a hypothesis or decision. The inference operation can be thought of as a mapping $y = h(x)$ that predicts an outcome y from input features x under a hypothesis $h(\cdot)$. Parameters of $h(\cdot)$ are learned offline in a training phase using a set of data samples that are representative of what is to be encountered in the final application. As an example, if the inference task is to infer from an image of a vehicle whether a passenger compartment is occupied or vacant, then $y \in \{y_0, y_1\}$ is a binary class label corresponding to either the "occupied" or "vacant" outcome, and $h(\cdot)$ is a binary decision function whose parameters are learned to minimize the classification error on a set of training images of occupied and vacant compartments. If the task is to interpret a license plate, then features from individual character segments are processed through a character recognition algorithm that outputs one of 36 alphanumerical class labels, $y \in \{y_1, ..., y_{36}\}$. Hypothesis $h(\cdot)$ is similarly learned from training samples of labeled license plate character images. If the task is to predict the speed of a vehicle, then y is a continuous valued variable, and may be predicted using, for example, a linear regression technique.

There are two broad categories of inference techniques using generative and discriminative models, respectively. Generative models attempt to describe the joint statistical distribution $p(x, y)$ of input features and prediction outcomes. Since the joint distribution is often a complex, high-dimensional function, simplifying assumptions are made to arrive at computationally tractable solutions. Commonly, a Bayesian framework is employed, whereby the joint statistics are factored into a conditional likelihood function $p(x|y)$, posterior distribution $p(y|x)$, and a prior distribution $p(y)$ for the output variables, each of which are modeled by tractable parametric forms. At the heart of a generative model is the search for the most likely outcome y given the observed features x. That is, the optimum y maximizes the posterior distribution:

$$y = h(x) = \arg\max p(y \mid x) = \arg\max p(x|y) p(y) \tag{1.1}$$

The last expression in Equation 1.1 is arrived at with Bayes rule. An example of a generative model that is frequently encountered in the book is the Gaussian mixture model (GMM), used in traffic flow analysis, video anomaly detection, driver monitoring, parking space occupancy detection, and other applications. For an excellent introduction to Bayesian generative inference techniques, the reader is referred to Ref. [15].

Discriminative models, on the other hand, attempt to learn $p(y|x)$ or the inference mapping $h(x)$ directly from training data without explicitly characterizing the joint probability distribution. Examples of such models include linear regression, logistic regression, support vector machines (SVMs), and boosting techniques. Various discriminative approaches are encountered in the book in the chapters on parking management, traffic sign recognition, and driver monitoring.

Recently, deep learning methods have gained popularity particularly in various image classification tasks. The basic concept behind deep learning is that the features themselves are learned from the image data to optimally perform a specific classification tasks. Thus, rather than parsing the computer vision task in two phases of feature extraction and inference, deep methods perform the task in one stage, learning both the features and the patterns needed to make inferences. For this reason, these methods are also referred to as representation learning techniques, and have been shown to significantly outperform traditional techniques in certain large-scale image classification tasks. A practical advantage of deep learning over traditional techniques is not having to manually or heuristically design features for a given task and domain. Chapter 10 provides an excellent tutorial and application of deep learning to the pedestrian detection problem.

The chapters in the book introduce a broad variety of machine learning and inference models addressing the various problem domains. The reader will appreciate that the choice of inference technique depends on many factors, including the required accuracy, computational cost, storage and memory requirements, and the cost, labor, time, and effort required to train the system. Those seeking a more basic background on machine learning are referred to Ref. [15] for a fundamental treatment of the subject. Finally, while the focus of the text is on camera-based methods, practical systems often involve input from other non-visual sensors. One example is in-vehicle safety systems that utilize technologies such as radar, LiDAR, and motion sensors to monitor driver and vehicle behavior. Another example is parking management, discussed in Chapter 8, where magnetic, radio frequency identification (RFID), laser, and other sensors are commonly utilized to determine parking space occupancy. Where applicable, discussions are presented on how video-based systems compete with or are complementary to these nonvisual modes.

1.4.5 Data Presentation and Feedback

This is the final step in the computer vision pipeline. The inference engine outputs a prediction of an outcome in the form of a class label probability, or measure of some desired quantity. This outcome must then be acted upon and communicated to the relevant party, be it the driver, law enforcement agent, or urban planner. As would be expected, the method and format of presentation and feedback is strongly dependent on the task and application. In problems relating to driver safety, where the majority of computation takes place in the vehicle, inferences must be communicated in real time to the driver in order to allow immediate action and intervention. Built-in driver assistance systems employ various feedback mechanisms, including visual display on the dashboard, audio warnings, and tactile feedback on the steering wheel. An important consideration here is the design of user interfaces that provide timely feedback without distracting the driver, and that integrate seamlessly with other familiar functions such as entertainment and navigation. Systems that are more proactive can directly alter vehicle motion without driver involvement. Examples are adaptive cruise control, automatic braking in collision avoidance systems and ultimately the driverless vehicle.

Applications relating to transportation efficiency warrant a different type of data visualization. In a video tolling application, the output of a license plate recognition system may be sent to a tolling agency to initiate billing. Additionally, for those plates where automatic recognition has low confidence, the localized plate images may be sent for human interpretation. For an urban planning application, traffic flow metrics generated by the computer vision system may be presented on a dashboard to local agencies, who can then interpret and integrate the analysis into future road planning and optimization efforts. In a parking management system, the vehicle license plate and duration of parking are the relevant metrics that must be made available to the parking administrator for billing purposes. Parking availability at a given location is a metric that must be continuously measured and made available to parking authorities and drivers.

In the domain of law enforcement, image capture is performed within a transportation infrastructure such as a gantry or police vehicle. The computer vision pipeline is often executed on a centralized server, and various data feedback and presentation mechanisms are invoked. For example, a law enforcement officer may want to see a relevant metric such as vehicle speed or occupancy, along with a picture or video clip of the vehicle that can be used for adjudicating a traffic violation, as well as for evidentiary purposes.

One trend that must be recognized and leveraged in the step of data presentation is the ubiquity of mobile technologies such as smartphones, tablets, smartwatches, and other wearable devices that are intimately connected to all human agents in a transportation ecosystem. For example, today it is customary for a smartphone to be connected via Bluetooth to the audio system in the vehicle, or for drivers to use the GPS on their mobile device for driving directions. It is expected that the automobile of the future will seamlessly integrate multimedia information from mobile devices with visual interfaces on the dashboard to present a personalized and context-sensitive driving experience.

We conclude the chapter by observing that while each module in Figure 1.3 can be individually designed and tuned, it is highly beneficial to co-optimize the entire framework at a system level, so as to account for the interplay and interaction among the modules. For example, the choice of features and the design of the inference algorithm are closely coupled, and ideally must be co-optimized for a given computer vision task. Similarly, the choice of imaging system can strongly affect the optimum choice of features and algorithms, as seen in numerous examples in the text. In Chapter 13 on driver monitoring, the choice of data preprocessing and feature extraction method depend on whether active or passive illumination is used within the vehicle. In Chapter 4 on passenger compartment violation detection, cameras capturing single versus multiple spectral bands are compared in terms of detection accuracy. Similarly, Chapter 10 illustrates how thermal infrared cameras can simplify the pedestrian detection task. Also, as mentioned earlier, road condition monitoring benefits greatly from the use of thermal IR imaging and optical polarization. Finally, the means and requirements for information presentation and feedback must be comprehended while designing the preceding steps in the framework.

References

1 Traffic Safety Facts, 2012 Data, Speeding. National Highway Traffic Safety Administration, NHTSA's National Center for Statistical Analysis, Washington, DC, DOT HS 812 021, 2014.

2 The Economic and Societal Impact Of Motor Vehicle Crashes, 2010 (Revised), NHTSA's National Highway Traffic Safety Administration, National Center for Statistical Analysis, Washington, DC, DOT HS 812 013, 2010.

3 L. Thomas, R. Srinivasan, L. Decina, L. Staplin, Safety Effects of Automated Speed Enforcement Programs: Critical Review of International Literature, *Transportation Research Record: Journal of the Transportation Research Board*, 2078, 117–126, 2008.

4 Distracted Driving 2009, Report, U.S. DOT National Highway Traffic Safety Administration, DOT HS 811 379, September, 2010. http://www-nrd.nhtsa.dot.gov/Pubs/811379.pdf (accessed September 9, 2016).

5 D. Roberts, M. Casanova, *Automated License Plate Recognition (ALPR) Systems: Policy and Operational Guidance for Law Enforcement*, Washington, DC: U.S. Department of Justice, National Institute of Justice, 2012.

6 D. C. Shoup, *The High Cost of Free Parking*, Chicago: Planners Press, American Planning Association, 2005, 7.

7 D. C. Shoup, Cruising for Parking, *Transport Policy*, 13(6), 479–486, 2006.

8 S. Schijns, P. Mathews, A Breakthrough in Automated Vehicle Occupancy Monitoring Systems for HOV/HOT Facilities, 12th International HOV Systems Conference, Houston, TX, April 20, 2005.

9 D. P. Jenkins, N. A. Stanton, G. H. Walker, M. S. Young, A New approach to Designing Lateral Collision Warning Systems, *International Journal of Vehicle Design*, 45(3), 379–396, 2007.

10 I-Car Advantage, Lane Departure Warning Systems, September 6, 2005.

11 C. Visvikis, T. L. Smith, M. Pitcher, R. Smith, Study on Lane Departure Warning and Lane Change Assistant Systems, Transport Research Laboratory Project Rpt PPR 374. http://www.trl.co.uk/umbraco/custom/report_files/PPR374.pdf and http://www.trl.co.uk/reports-publications/trl-reports/report/?reportid=6789 (accessed October 13, 2016).

12 U.S. Department of Transportation Releases Policy on Automated Vehicle Development. National Highway Traffic Safety Administration. 30 May 2013. Retrieved December 18, 2013.

13 http://spectrum.ieee.org/automaton/robotics/artificial-intelligence/how-google-self-driving-car-works (accessed September 9, 2016).

14 R. C. Gonzalez, R. E. Woods, *Digital Image Processing*, 3rd Ed., Upper Saddle River, NJ: Prentice Hall, 2008.

15 C. M. Bishop, *Pattern Recognition and Machine Learning*, New York: Springer, 2007.

Part I

Imaging from the Roadway Infrastructure

2

Automated License Plate Recognition

Aaron Burry and Vladimir Kozitsky

Conduent Labs, Webster, NY, USA

2.1 Introduction

Vehicles traveling on the public roadways in the United States, and many other countries, are required by law to carry a clearly visible placard with a unique identifier that is registered with the local government. This placard, most commonly called a "license plate," can contain various symbols—letters, numbers, logos, etc.—based on local government regulations and the class of the vehicle.

Given the common requirement for its presence and ease of visibility, the license plate has become the default means for identifying a vehicle and/or its registered operator across a broad range of applications. For electronic tolling and automated parking solutions, the license plate is used as the unique identifier that links to the account holder for processing of payment transactions. Enforcement applications, such as red light, leverage the license plate information to identify the registered operator as part of processing violations. Surveillance and security systems make use of license plates to identify particular vehicles of interest and to track the movement of individual vehicles—entry and exit from protected areas, patterns of movement across large urban areas, etc.

For some applications license plate information can be extracted manually from images or videos of a passing vehicle. In these cases, human annotators view the footage and produce a hand-coded result for the license plate (both the plate string and often the issuing jurisdiction). Given the costs and turnaround times associated with this type of manual review, industry trends are toward ever-increasing automation.

Automated license plate recognition (ALPR), sometimes also referred to as automated number plate recognition (ANPR), is a technology that leverages computer vision algorithms to extract license plate information from videos or still images of vehicles. ALPR has become a core technology within modern intelligent transportation systems. Surveillance and police enforcement applications leverage ALPR systems to provide real-time data gathering to support law enforcement efforts. In fact, a 2012 study [1] conducted on the usage of ALPR technology by police agencies found that 80% of larger agencies (those with 1000+ sworn officers) were already leveraging ALPR in some way. Results from this same study also indicated that police agencies who were using ALPR reported that the use of technology had increased recovery of stolen vehicles by 68% and overall arrests by 55%. In the tolling industry, over 90% of the tolled roadways in the United States leverage some form of electronic tolling [2]. Many of these electronic tolling systems rely primarily on radio frequency

Computer Vision and Imaging in Intelligent Transportation Systems, First Edition.
Edited by Robert P. Loce, Raja Bala and Mohan Trivedi.
© 2017 John Wiley & Sons Ltd. Published 2017 by John Wiley & Sons Ltd.
Companion website: www.wiley.com/go/loce/ComputerVisionandImaginginITS

identification (RFID) tags to identify account holders, with ALPR systems used as backup in cases where tags are not properly read for a passing vehicle. However, there have been several recent successful deployments of video-only tolling solutions in New Zealand [3] and Sweden [4]. In these video-only systems, all transactions are processed via captured imagery—leveraging some combination of ALPR and manual review of the images—thereby eliminating the need for the deployment and maintenance costs associated with RFID tags. Automated billing and access for off-street parking and enforcement of on-street parking are other areas that have seen recent growth in adoption and deployment of ALPR-based solutions.

An ALPR solution must find and recognize the license plate information within an image of a larger scene containing the vehicle of interest. At a first glance, the ALPR problem appears similar to that of text recognition from scanned documents—both applications require finding and recognizing text within images. However, there are several key differences between the challenges posed in these two domains. One of the primary differences is how much more controlled the image capture environment is for scanning documents. In contrast, images captured from real-world transportation infrastructure can contain a wide array of noises. These include heavy shadows, poor contrast, non-uniform illumination (from one vehicle to the next, daytime vs. nighttime, etc.), challenging optical geometries (tilt, shear, or projective distortions of the license plate and its characters), plate frames and/or stickers partially touching characters, partial occlusion of characters from things like trailer hitches, and dirt or other contamination of the license plate surface.

For ALPR systems deployed in the United States in particular, the inherent variation of license plate formats further adds to the difficulties for automated recognition systems. Such variations can include background color, background uniformity (i.e., flat background vs. a pictorial of some sort), character color, character font, character width, character spacing, and the presence of special logos. The level of variation in license plate formats is demonstrated by the fact that some US states, such as Florida, have well over one hundred different types of license plate designs in circulation [5].

When recognizing words in images of scanned documents, various techniques have been developed for leveraging known dictionaries for both detecting and correcting errors in automated text recognition systems (see [6–9], e.g., approaches). Unfortunately, the information contained on license plates is much more weakly constrained, with far more possible combinations of letters and numbers. Thus, to achieve the same levels of overall accuracy, the fundamental recognition technologies used for ALPR must be more accurate.

2.2   Core ALPR Technologies

The block diagram shown in Figure 2.1 illustrates the set of processing operations comprising a typical ALPR system. Cameras along a roadway or other observation area capture images (or videos) of passing vehicles, which are then submitted to the ALPR system for evaluation. The first step in the ALPR process is the localization of the subregion of the image containing the license plate. Candidate

Figure 2.1 Block diagram of a typical ALPR processing flow.

license plate regions are then sent to a character segmentation module to extract and crop subimages containing valid characters, logos, and other license plate symbols. The individual segmented images are then sent to a recognition module (often generically referred to as optical character recognition or OCR) to determine the license plate code. Finally, a state identification module leverages the resultant license plate code and other information extracted from the license plate image—the presence or absence of logos, color information, character spacing and font, etc.—to determine the source jurisdiction that issued the plate.

The following sections will describe in greater detail some of the more common approaches for implementing each of these fundamental capabilities within an ALPR solution. The goal is not an exhaustive review of the academic research published in this field—for that see the survey papers of [10] and [11]. Instead, the following will describe several of the most common methods behind each subsystem in the ALPR framework. The emphasis here is on giving the reader a basic understanding of the key requirements, challenges, and the basic methodologies.

2.2.1 License Plate Localization

Depending on the specific application and the image capture geometry, the source images from the capture device may contain varying amounts of the vehicle and surrounding background scene in addition to the license plate region of interest. The first step in the ALPR processing pipeline is therefore an identification of the coarse subregions of the overall image that are likely to contain a license plate.

Extracting these region-of-interest (ROI) subimages and limiting subsequent processing to only these areas provides several major benefits. First, constraining the segmentation and recognition operations to a more limited region of the image can in fact result in higher accuracy results for these operations. In other words, the localization operation can crop off regions of the image that might, in some sense, serve as distractions for segmentation and/or character recognition. A second benefit of starting the ALPR process flow with plate localization is a substantial reduction in the total computational overhead of subsequent processing steps. To realize this gain, however, the localization step itself must obviously be computationally efficient.

Beyond computational efficiency, another critical performance metric for the license plate localization method is the missed detection rate (i.e., failure to identify a visible license plate within an image). Since the plate localization is the first in the processing sequence, a missed detection here will eliminate any possibility for successful ALPR for that image. Depending on the application, this could mean lost tolling revenue, a missed red light violator, etc. Conversely, false alarms at this stage of the process (i.e., candidate subregions of the image that do not actually contain visible license plates) still provide the opportunity for subsequent processing operations in the pipeline to identify the correct ALPR result. Note that these additional ROI candidates do however increase the overall computational burden and can lead to false recognitions—for example, recognizing text on a bumper sticker rather than from the actual license plate. Thus, there is a design trade-off to be made in terms of increasing the detection rate for valid license plate regions as much as possible while not incurring undue negative impacts (false recognitions and processing overhead) to the overall ALPR system performance.

Various methods have been demonstrated in the literature for implementing license plate localization. These approaches can be organized into several major categories: color-based methods, edge- or texture-based methods, and methods that utilize image-based classifiers. A more detailed description of the types of algorithms underlying each of these general approaches is given in the following sections.

2.2.1.1 Color-Based Methods

Some countries or local jurisdictions make use of specific color combinations in the design of their license plates. For instance, the license plate most common for passenger cars in the state of New York is an orange background with dark blue letters. One approach to localization is then to leverage this *a priori* knowledge to search for subregions of an image that conform to these criteria. The idea here is that the color combination of a plate and characters is unique, and this combination occurs almost only in a plate region [12]. A prototypical example of this type of approach is outlined in Ref. [13], wherein a neural network is first used to classify input pixels into one of four color groups (based on localized pixel information). Color projection histograms in both horizontal and vertical dimensions are then used to identify likely plate regions.

Although color-based approaches are intuitive, there are several major drawbacks to using color information as the primary discriminator for license plate subregions of an image. First, some image capture systems for transportation applications make use of narrower band, near-infrared (NIR) illumination sources and, therefore, output only grayscale images. Obviously color information is not a viable localization solution for images obtained with these systems.

Color-based localization methods can also be inherently sensitive to variations in illumination. Unfortunately, these types of noises are quite common for images obtained for real-world transportation applications. Nonuniform illumination can be caused by changes in background illumination due to time of day, location-to-location variability in image capture setup and native levels of ambient lighting, within-image variations due to shadows, etc. Some efforts have been made to formulate the separation of background regions from foreground license plate regions in alternative color spaces. For instance, in Ref. [14] an input RGB image is converted to the hue, saturation, and intensity (HSI) color space and a set of pixel level constraints on each color channel are used to identify pixels as either foreground or background based on learned statistics for several common license plate designs. Another approach aimed at obtaining robust use of color information is illustrated in Ref. [15]. Here, the authors compute edge information only for certain color pairs known to be prominent in the license plate designs they considered: black–white, red–white, and green–white.

Perhaps the biggest limitation of using color as the primary feature for localization is that this type of an approach is highly tuned to a specific design of the license plate. Unfortunately, there is a high degree of variability in the format of license plates between states in the United States and from country to country in other parts of the world. For example, the dominant license plate format for passenger cars is a yellow background with black letters in New Jersey, a white background with dark blue letters in Pennsylvania, and yellow letters on a dark blue background for the state of Delaware. Since travel across state boundaries is quite common in the United States, these examples illustrate a key challenge with relying on color as the primary cue for plate localization. More specifically, any color-based localization approach must comprehend a sufficiently broad set of license plate designs to handle the mixture of jurisdictions that will be presented to it.

2.2.1.2 Edge-Based Methods

Regions of images that contain text tend to have locally steep gradients due to the sharp transition from characters to background. Since license plate content is mainly text, edge information can be used as a good feature to distinguish license plate regions from other regions in a source image, even for reasonably complex scenes. Edge information tends to be more robust to changes in illumination than other features like color, and the computation and analysis required for the class of edge-based approaches tends to be relatively low overhead. Because edge-based methods are extremely popular, the remainder of this section describes the basic operations and key challenges of several of these approaches in more detail.

Figure 2.2 Original and resulting edge image.

Figure 2.3 Edge projection-based detection for license plate localization.

The computation of the vertical and horizontal edge information, often using the Sobel operators, is the first step for most edge-based methods. A threshold is then applied to eliminate spurious edge information, leaving only "strong" edges. An example of a partially redacted real-world vehicle image and the corresponding edge image after the Sobel filtering and a threshold have been applied are shown in Figure 2.2. Here, the contrast between the strong local edge content within the license plate region, as compared to the surrounding image regions, can be clearly seen.

Having computed an edge image, a number of methods can then be used to identify candidate license plate subregions of the image. Vertical and horizontal projections of the edge information can be computed (see Figure 2.3), with license plate candidates then taken as those having strong vertical edge content within a localized region. Similar approaches based on vertical edge density [16] and local edge statistics [17] are common.

Another strategy for plate localization based on edge information leverages mathematical morphology and connected-component analysis (CCA) to identify the candidate subregions. There are many variations on this basic theme, but a typical sequence of operations follows the outline in Ref. [18]. Morphological filtering operations are used to connect the strong edges in the license plate into a contiguous foreground "blob," while at the same time eliminating spurious edge information in other areas of the image.

The most common morphological operations on binary images are erosion, dilation, opening, and closing. The erosion of a binary image A by the binary structuring element B is given as follows:

$$A \ominus B = \{x : B_x \subset A\} \tag{2.1}$$

Here, B_x is the translation of binary set B by a point x,

$$B_x = \{b + x : b \in B\} \tag{2.2}$$

Likewise, the dilation of set A by structuring element B can be represented as follows:

$$A \oplus B = \bigcup_{b \in B} A_b \qquad (2.3)$$

The opening operation is defined as an erosion followed by a dilation,

$$A \circ B = (A \ominus B) \oplus B \qquad (2.4)$$

while the closing operation is defined as a dilation followed by an erosion,

$$A \cdot B = (A \oplus B) \ominus B \qquad (2.5)$$

The results of applying a morphological *close* operation to the edge image of Figure 2.2 are illustrated in Figure 2.4. Here, the license plate region is seen to be well connected into a foreground blob. However, there are a number of spurious connections to other foreground blobs as well. Further morphological filtering, such as an *open* operation, can then be applied to help further isolate the license plate from other spurious edge information in the image. The results for this image are illustrated in Figure 2.5.

After morphological filtering, CCA can be applied to screen candidate blob regions based on known general characteristics of license plates. For instance, it can be expected that license plates will be rectangular in shape and substantially horizontal in orientation. In addition, the aspect ratio

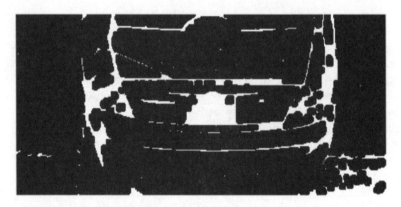

Figure 2.4 Results of morphological close operation on edge image.

Figure 2.5 Results of additional morphological filtering operations.

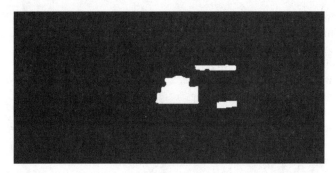

Figure 2.6 Results of applying CCA to reduce candidate blob regions.

Figure 2.7 Extracted sub-image regions.

of license plates is roughly $2:1$ (the width is twice the height). Candidate blobs that are too large or too small to be of interest can be removed by leveraging some prior knowledge of the image capture setup and expected license plate size in the image. The results after applying this type of CCA can be seen in Figure 2.6.

These remaining foreground blobs are then used as selection masks to extract the ROI subimages for the candidate license plate regions, as illustrated in Figure 2.7. Note that in this example several false regions are identified along with the correct license plate subimage. As mentioned previously, there is a design trade-off between ensuring that the plate containing region is identified and minimizing the number of such false alarms.

2.2.1.3 Machine Learning–Based Approaches

In addition to the image processing-based methods described before, learning-based approaches for object localization can also be applied to the problem of identifying license plate subregions. Here, there are two main types of approaches. The first category of approaches involves training a classifier that is used to directly scan the image for the license plate containing subregions. This methodology can be thought of as analogous to that of scanning an image looking for faces, as in the commonly referenced Viola–Jones detector [19]. Following an approach similar to Viola–Jones, the authors of Ref. [20] trained an Adaboost classifier to detect license plates by sweeping the detection window across the image. To obtain some robustness to the size of the license plate in the image, it is common practice to sweep the classifier at multiple scales (different magnifications of the image relative to the classifier's template size). Beyond the simple Haar-like features used in Ref. [20], a multitude of other features can be considered for detecting license plates. In Ref. [21], the authors build a descriptor for an image patch that is composed of the covariance between simple image features (pixel coordinates, pixel intensity, and first- and second-order derivatives in both x and y dimensions) at different pixel coordinates within the patch. The bag-of-visual-words method [22, 23] is another common technique for image-based classification. Extending this concept, the authors in Ref. [24] attempt to abstract the most distinctive scale-invariant feature transform (SIFT) [25] features. Here, a set of labeled training images of segmented license plate characters is analyzed. The most distinctive SIFT features, called principal visual words (PVWs), for each symbol in the dictionary are then identified. At test time, SIFT features are identified for the full scene image and the previously learned dictionary of PVWs are used for screening—those candidate character locations that do not closely match a PVW in the dictionary are discarded. Regions of the image that contain a high density of candidate characters with similar geometric orientations are then considered likely license plate regions.

The second category of learning-based approaches for plate localization involves first generating candidate license plate regions ("proposals"), for instance, using the types of methods outlined in Section 2.2.1.2, and then applying a feature-based classification on these subimage regions to either rank order them (most likely to least likely) or to screen unlikely candidates from the pool. This type of an approach is typically taken when the performance of the sweeping classifier type of method alone is unacceptable. Poor performance here could mean many things—too many missed plates, too many false alarms on nonplate regions, or too computationally expensive to calculate features and sweep the classifier template across the entire image.

As a prototypical example of this type of an approach, Cano and Pérez-Cortés [26] use Sobel filtering followed by application of a threshold to generate edge-based candidates for license plate regions (once again very similar to approaches outlined in Section 2.2.1.2). These candidate proposals are then evaluated with a trained k-nearest neighbor classification approach, using the gray values within a defined window as the input feature vectors.

In principle, any of the techniques that could be applied for directly detecting the license plate subregion could be applied to this secondary task of evaluating candidate region of interest proposals.

2.2.2 Character Segmentation

Given a roughly cropped license plate ROI image, such as that shown in Figure 2.8, the character segmentation module must extract individually cropped sub-images for each of the characters in the license plate. This can be a difficult task given that the input image often has poor contrast, perspective distortion, and license plate frames or other artifacts that can touch or even partially occlude some of the characters. Although it may be attacked directly, the overall character segmentation problem for license plate images is often broken into two separate steps, as outlined in the following sections.

Figure 2.8 Cartoon of a coarsely cropped license plate ROI image output from the localization module.

Image processing techniques for extracting text from images is a well-studied field. Good survey papers highlighting the breadth of methods in this space can be found in Refs. [27–29]. The intent of the description in the following sections is to highlight the particular challenges of text segmentation for license plate images, as well as some of the most common methods for addressing these challenges.

2.2.2.1 Preprocessing for Rotation, Crop, and Shear

The first step is to obtain a much tighter crop of just the subregion of the image containing the characters. This tightly cropped image will also be tilt and shear corrected as much as possible in an effort to achieve consistent, vertically oriented characters as shown in Figure 2.9. This preprocessing normalizes the characters in an effort to provide a more consistent, more easily segmented image to the subsequent step of identifying the actual "cuts" between the individual license plate characters.

For the more general problem of detecting text within an image scene, the line structure of words can be exploited to assist localization [30]. Applied to license plates, this line structure constraint can be leveraged by first locating strong candidates for license plate characters within the image, and then using the location and size information of these character candidates to perform cropping, de-rotation, and de-shearing of the image. One common approach to locating candidate characters in the loosely cropped ROI image is to binarize the image and then leverage a set of screening techniques—typically a sequence of morphological filtering and CCA, much like the edge-based methods for localization. Examples of this type of approach can be found in Refs. [31] and [32]. Because of their common usage, the basic sequence of operations for this class of methods will be described in more detail in order to highlight the underlying techniques and some of the key challenges.

The candidate license plate region is first converted to a binary image to differentiate between foreground and background pixels. Common methods for binarization of the plate region of interest image are the statistical methods of Otsu [33], Savola [34], and Niblack [35], or a gradient-based approach such as that outlined in Ref. [36]. The images in Figure 2.10 are for a partially redacted original source and the resulting binary image for a plate region. Similar to the edge-based methods for plate localization, the process flow outlined in Refs. [31] and [32] then leverages a combination of morphological filtering and CCA applied to the binary image to eliminate unlikely candidates from the character pool. Here, the characters are known to be substantially vertical, roughly the same size,

Figure 2.9 Extracting a tightly cropped ROI image around the characters.

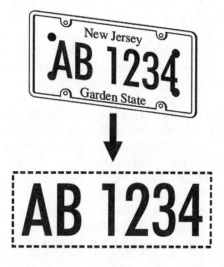

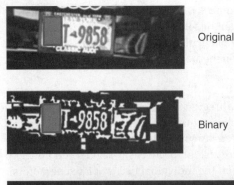

Original

Figure 2.10 Original and binary images for a coarsely cropped ROI around a license plate.

Binary

Figure 2.11 Remaining foreground blobs after morphological filtering and CCA.

Figure 2.12 Remaining foreground blobs after pruning based on centerline fit method of candidate character blobs.

Figure 2.13 Vertically cropped ROI image for a partially redacted license plate.

and have about a 2:1 aspect ratio. The image of the resulting foreground blobs after applying these operations is given in Figure 2.11.

To assist with elimination of erroneous foreground blobs, Burry et al. [32] use the residual error contribution to the centerline fit from each candidate character to estimate the likelihood of a foreground blob being a valid license plate character. Methods such as Random sample consensus (RANSAC) [37] could also be used to remove outliers from the centerline fit as a method for pruning the candidates. The result of pruning based on centerline fit for the example case of Figure 2.11 is given in Figure 2.12. As illustrated in this result, it isn't necessary to obtain accurate location information for each character within the license plate at this stage—just a subset is sufficient to identify the line along which the license plate text lies within the image. Once the set of candidate characters has been pruned, the centerline fit to the remaining foreground blobs is used to determine the required de-rotation to obtain a horizontally oriented license plate in the image. Based on the size of the reduced set of candidate character blobs, a tight cropping of the image in the vertical dimension can then be performed (see the example result in Figure 2.13).

In order to obtain the final horizontal crop of the license plate, one common method is to leverage gradient information to identify the strong contrast that typically exists between the plate and the

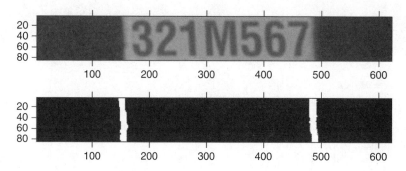

Figure 2.14 A simple gradient example.

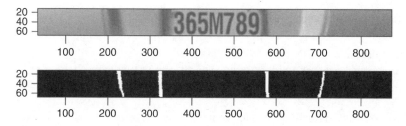

Figure 2.15 An example of a more complex vehicle background.

surrounding vehicle body. An example of a dominant license plate on a simple background for which a gradient-based approach is intuitive is shown in Figure 2.14. Here, both the original image and a gradient map, after imposing a minimum threshold, are provided. Clearly for an example such as this, the gradient information alone is sufficient to accurately locate the license plate boundaries. A more complex scenario is provided in Figure 2.15. Note the occurrence of dominant edge information well outside the boundaries of the license plate text region in this example. Accurate determination of the license plate region is clearly more difficult for this type of image. Thus, when relying on image gradient information, the degree of contrast between the plate region and the surrounding vehicle, as well as the potential occurrence of extraneous hard edges outside the plate region, can complicate the detection of the horizontal boundaries of the license plate.

An image-based classification approach can also be used to determine the horizontal boundaries of the license plate text in the image. For instance, one method is to train a detector that recognizes the difference between valid license plate characters and noncharacter content. By sweeping this detector horizontally across the image, a profile of detection scores will be generated (see Figure 2.16). The license plate subregion can then be determined as the region with the highest contiguous detection scores. One of the biggest challenges for this type of an approach is the strong similarity between the text symbol "1" used in many license plate fonts and other hard edges in the image around the plate. Due to this confusion, a classifier-based approach can suffer from over- or under-cropping of the license plate region.

To avoid the pitfalls of either approach, hybrid approaches that leverage information from both the raw gradient and an image-based classification can be used. An example of such a hybrid method is described in Ref. [38]. Here, the image-based classifier is first used to provide a coarse boundary selection, with the gradient information then allowing for a fine-tuning of the identified borders.

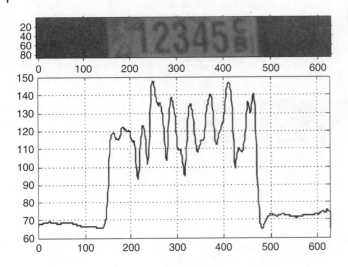

Figure 2.16 Original image and detection scores for valid character detector.

2.2.2.2 Character-Level Segmentation

The second step in the segmentation process is identifying the cut points between characters to achieve properly separated sub-images. The desired results of this operation can be seen in the sketch of Figure 2.17.

Various methods have been proposed in the literature for achieving this character-level segmentation of license plate images. One of the simplest and most commonly used approaches is based on the vertical histogram of the image [10, 11]. This histogram profile, effectively the mean profile of the image computed in the vertical direction, can be calculated either for the original source gray values or for a binary version of the tightly cropped license plate image. Cropping points are then identified in the zero-valued regions of the profile between characters. An illustration of the approach is provided on a partially redacted license plate image in Figure 2.18.

Unfortunately, the vertical histogram approach is sensitive to localized artifacts such as license plate frames and heavy shadows. An example of the issues that can occur is illustrated in Figure 2.19. Here, the adjacent background content near the plate characters is sufficient to lead to a significant, nonzero value for the projection profile of the binary image between the characters "3" and "5." To combat this, it is possible to make use of a nonzero or a locally adaptive threshold value. Other advancements on this basic concept employ additional techniques for identifying over- or under-segmentations of characters. For instance, in Refs. [39] and [40] the uniformity of the character spacing on the plate is used to identify outliers in the candidate segmentation cuts. More specifically, over-segmented characters will manifest as too narrow a spacing between consecutive character centers while under-segmented regions can be identified by too large a spacing. The candidates that have been identified as erroneous in this fashion are then either merged with neighbors or split as required.

2.2.3 Character Recognition

Given an image of a cropped license plate character, the character recognition subsystem must identify the correct label for this image. At a high level, this problem is synonymous to the OCR problem associated with labeling/recognizing characters from images of scanned documents. However, for a number of reasons the similarities end at this broad level. For instance, in the ALPR scenario there is little control over lighting, especially in open toad tolling (ORT) applications where cameras are

Figure 2.17 Segmentation cut points as cropping boundaries between characters.

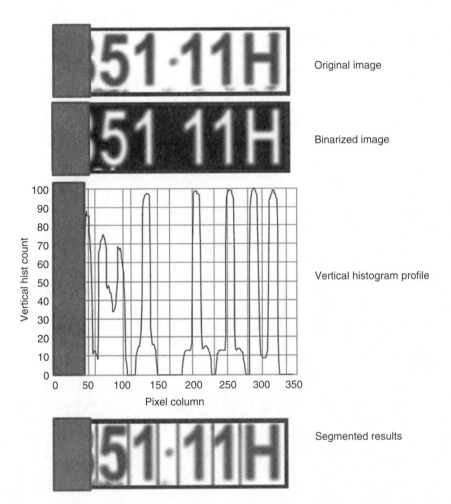

Original image

Binarized image

Vertical histogram profile

Segmented results

Figure 2.18 Example of vertical projection approach for character segmentation.

mounted on gantries above a roadway of free-flowing traffic. Another complication seen in the ALPR application is that license plate text backgrounds can vary substantially across plate types and juris-dictions. In addition, noises such as shadows, partial occlusions, and perspective distortions contrib-ute to wide variations in the appearance of characters. Further, standard document OCR engines heavily leverage a natural language dictionary of common words to ensure optimal performance [6–9], whereas no such dictionary is available for ALPR.

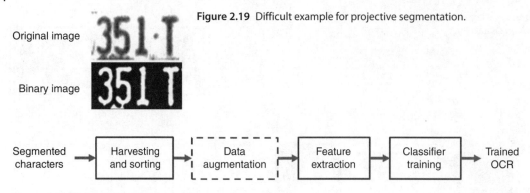

Figure 2.19 Difficult example for projective segmentation.

Figure 2.20 Typical OCR training workflow.

As illustrated in Figure 2.20, the standard workflow in developing an OCR engine comprises the following three major steps: character harvesting and sorting, feature extraction, and classifier training. The optional data augmentation step leverages collected examples to artificially increase the images available for training.

2.2.3.1 Character Harvesting and Sorting

Given a set of license plates, a functional segmentation algorithm will provide segmented characters that can be used for training an OCR engine. Segmentation algorithms vary and each produces characters with unique characteristics or bias. This bias comes in several forms such as scaling, cropping, inconsistent centering, shear, and rotation. Typically, these biases are label dependent and cannot be automatically corrected. Given this, a dependency exists between the segmentation algorithm and the OCR engine. Thus, it is important to use the same segmentation algorithm for character harvesting as well as run-time evaluation to ensure that representative noises are included.

Before starting the harvesting process, it is useful to have some ground truth information available for each license plate. The minimum information required is license plate jurisdiction at a country and province/state level since character fonts vary across provinces and states. Thus, with license plates organized by jurisdiction (i.e., font), character images can be extracted via a segmentation algorithm. A human observer will then typically need to sort this set of segmented characters into smaller subsets organized by labels such as "0," "1," and "A". This is an extremely tedious process that can be made easier with the application of an external OCR engine (i.e., one that was trained for a different jurisdiction/font) to automatically sort the images. At this stage, this alternate OCR engine will likely have a substantial error rate, given that it was not tuned specifically for the jurisdictions being processed. For this reason, the initial output from this external OCR engine will typically still need to be manually validated.

To simplify the sorting process further, it is helpful to have a second level of ground truth that includes the license plate code (number). The ground truth code can be used to guide the harvesting and sorting process in conjunction with an external OCR engine through string alignment methods. Given a set of segmented images from a single license plate, the external OCR will produce a result. An edit distance algorithm with backtrace such as Levenshtein distance can be applied to the OCR result and ground truth code to determine the number of insertions, deletions, and substitutions required for the two to match. Insertions occur when the segmentation algorithm returns noncharacter imagery such as logos, symbols, and off-plate content. Deletions arise when real character images are not returned by segmentation—this typically occurs at plate boundaries. Substitutions are introduced by OCR errors or in combination with deletions when segmentation erroneously returns a single image containing two individual characters.

A threshold applied to the edit distance value can be used to guide automatic sorting. Typical license plates have seven characters and setting a threshold at less than or equal to 3 ensures that many characters can be automatically sorted. License plates that don't meet the edit distance threshold can have characters extracted to a catch-all bin that a human may review and sort to gather sufficient examples. Setting the threshold too high runs the risk of accepting non license plate content such as stickers, banners, and other extraneous text content. Non-license plate regions will at times be generated by the localization algorithm and they may be segmented effectively but would pollute a training data set for OCR. It is only with plate code-level ground truth that these cases can be automatically rejected.

If the automatic sorting process produces enough examples, there is no need to review the unsorted bin. How many examples are sufficient? This depends on various factors such as the number of labels, the type of features/classifier used for OCR, the effectiveness of segmentation in normalizing characters, and the required level of performance. In general, more data will lead to better performance with decreasing accuracy improvement returns past 1500 samples per dictionary symbol.

2.2.3.2 Data Augmentation

Collecting large real-world data sets is expensive and ground truth generation is error prone—often requiring double-blind review of images to ensure error rates less than 1%. Data augmentation methods can be applied to supplement the real world data, thereby artificially enlarging the training sets. In general, training set variability typically rises with sample size. However, improvement via augmentation for already large sets will be minimal or worse performance may actually result as the addition of synthetic examples may in fact bias the classifiers to unrealistic imagery—effectively swamping the nominal range of noises in the initial "real" image sample set. A large amount of work has been done in modeling degradation of characters for document based OCR [41, 42] where the population of fonts is known and limited. Parameterized models can be used to generate a large volume of character images drawing from a statistical distribution of likely noises [43, 44].

Multiple augmentation options exist—the first is illustrated in Figure 2.21. Existing real world data is distorted through application of vertical and horizontal spatial offsets, scaling, blur (through down-sampling followed by up-sampling), injection of general image noise, and rotation. This can be extended to include other residual noise sources that are not completely removed by the segmentation algorithm, such as perspective distortions or cropping biases. In Ref. [45], the authors computed mean images for each class and then applied weighted amounts of seven distortions to the mean images to generate augmented examples. For convolutional neural network (CNN)-based classifiers, spatial offsets combined with horizontal reflections have been found to work well [44].

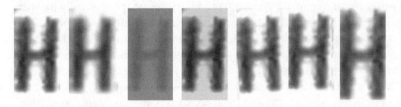

Figure 2.21 Distorted versions of real characters (original is first).

Figure 2.22 Synthetically generated license plates.

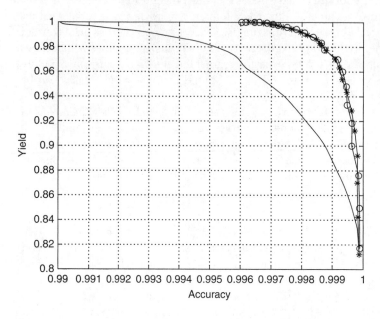

Figure 2.23 OCR performance for three sets of training data: (i) 700 real SPC (cross-marks), 2000 synthetic SPC (dashed curve), and (ii) a mix of 2000 synthetic and 100 real SPC (circle marks).

The second type of augmentation requires knowledge of character font. The font can be derived from matching to a font library, or it can be designed from collected examples. Once the font is known, the distribution of noise sources can be learned from collected character images and then used to generate a much larger set. In Ref. [46], the authors carry out the synthesis on a plate level starting with blank templates and then rendering the plate code text onto the template. The synthesis includes a color to infrared (IR) transform as most cameras for ALPR systems operate in the IR spectrum, while real-world license plates contain a variety of color content. The images in Figure 2.22 show a few synthetically generated examples and Figure 2.23 illustrates how the addition of 2000 synthetic samples per character (SPC) is equivalent to having an additional 600 real SPC for training.

2.2.3.3 Feature Extraction

Once characters have been harvested and sorted, they need to be resized to a fixed template size before the feature extraction process can begin. Well-segmented character images typically have an aspect ratio of 2:1 (height:width), and this aspect ratio should be maintained for the template. The size of the template is driven by the median character size such that a minimum number of characters will need to be significantly up-sampled or down-sampled. If the segmentation algorithm is effective at generating well-centered characters, a good normalization strategy is to resize each image to template height allowing the width to float, then to either crop or pad the sides preserving the center.

Choice of template size is driven by several factors. The main factor is native character resolution. If the native resolution is low, the best option is to avoid down-sampling as important information

Figure 2.24 Example of SMQT robustness to gain and offset distortions.

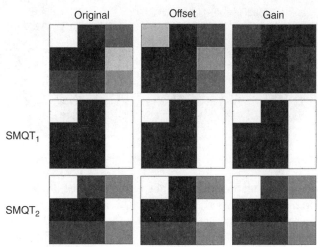

will be lost. Provided that input resolution is sufficiently high, down-sampling will be effective at reducing the number of pixels without compromising recognition performance. The optimal size can be derived experimentally as there is an interaction between template size and feature type. For typical license plates that include 36 labels (0–9 A–Z) performance will begin to suffer notably when templates drop below 24×12 pixels. The greatest impact in performance will be observed in decreased accuracy for close number/letter pairs such as 0/O, 0/D, 1/I, 2/Z, 5/S, and 8/B. For number plates (0–9), it has been observed that the OCR template size can be reduced down to 16×8 provided that contrast is high and blur is minimal.

In general, feature extraction is a dimensionality reduction operation where redundancy on a pixel level is removed and replaced with a higher lever descriptor. A goal of feature extraction is to reduce intra-class variability while increasing it across classes to produce data that will enhance classification performance. A number of feature extraction methods have been applied to the OCR problem. These include template matching, geometric and Zernike moments, contour and spline profiling, zoning, and projection histograms [47]. Most of the documented features rely on expert insight and in turn are hand-crafted.

Edge and gradient-based features have been found to work well for ALPR as these are robust to illumination variation that is typical in unconstrained imaging scenarios. Given variability in character segmentation, features that are robust to spatial variation are preferred as well. The successive mean quantization transform (SMQT) was originally proposed for face detection by Nilsson [48] and is used to compute local gradients for each pixel by removing local gain and bias to reveal the underlying object structure. Nilsson compared this transform to local histogram equalization (HE), local binary patterns (LBPs), and the modified census transform (MCT) finding that the SMQT provided a desirable trade-off between number of quantization levels and computational load.

In Ref. [49], the SMQT was applied to the recognition of license plate characters with good results. The authors found that a single-level (SMQT1) 3×3 pixel local neighborhood was sufficient to handle illumination variability while ensuring fast processing time. An example illustrating the consistency of the SMQT output across both gain and offset distortions is given in Figure 2.24. The SMQT-based algorithm was both four times faster than a nearest-neighbor image distortion-based model in Ref. [50] as well as more accurate given the same number of training examples.

Original image   HOG features   HOG inverse

Figure 2.25 Extracted HOG features for license plate characters.

Another feature that has been used successfully for real-world text recognition is histogram of oriented gradients (HOGs) [51]. The HOG feature, like SMQT, extracts edge-based information (here using actual gradient values), thereby providing strong robustness to illumination variation. The input image is represented by blocks of cells. Illumination and contrast is normalized on a block level. Each block comprises multiple cells where for each cell a histogram of gradients is computed. Cells are made up of pixels with cell sizes varying as a function of resolution. By pooling the gradients on a cell level, the HOG feature provides robustness to small spatial variation. The HOG feature was originally shown to work well for object detection, namely pedestrians, but has recently been applied to OCR [51, 52]. An example illustrating the resultant HOG features for a portion of a license plate image is given in Figure 2.25.

Another popular feature used in object detection and recognition is the SIFT [25]. The basis of SIFT is a keypoint descriptor where each is represented as a 128-element vector. Information captured for each keypoint includes spatial position, scale, and orientation. Once a keypoint is identified, a descriptor is generated comprising of a 4×4 spatial histogram of local gradients about the keypoint. The advantage of SIFT is robustness to scale, orientation, affine distortion, and limited illumination. However, this robustness creates problems for detecting shape-based objects like characters where relative spatial position of features is more important than mere presence of these in the image. This problem was demonstrated in the work done by de Campos [53] for cropped character classification where SIFT and other appearance-based features fell short.

In Ref. [54], the authors compare recognition performance for hand-crafted features like SMQT, and HOG against methods where the features are learned directly from data in an unsupervised manner. The two learning methods that were compared were sparse auto-encoders and k-means feature learning [55]. Once the features were learned, they were extracted from training images and classifiers were trained using the typical supervised approach. The authors found that for the data set of house numbers captured by Google Streetview, the learned features outperformed hand-crafted variants. This finding is consistent with recent work showing superiority of learned features particularly those generated by CNNs for object recognition tasks [44, 56].

2.2.3.4 Classifiers and Training

Once character images have been normalized and features extracted, a number of classification algorithms and architectures are available. Classifier algorithms include nearest-neighbor (kNN), template matching, artificial neural networks (ANNs) such as multilevel perceptrons (MLPs), support vector machines (SVMs), and CNNs, among others. Classifier architectures include one-vs-one (OvO), one-vs-all (OvA), and cascades.

In kNN-based classification, all training examples are stored as feature vectors. At test time, a distance is computed between the test vector and all training vectors. The nearest "k" training examples are selected and the most frequent label amongst "k" is the classification result. The distance metric has a significant impact on kNN classification. Common distance metrics include Euclidian and Hamming. Learned metrics such as the large margin nearest neighbor (LMNN) where a Mahalanobis distance metric is learned by semidefinite programming improve classification performance [57]. One drawback of kNN is that training set features must be available at test time and performance is positively correlated with training set size. This can lead to large memory utilization and run-time expense as the training set grows. Dimensionality reduction such as PCA and LDA is typically applied, but the issue persists. Keysers [50] proposed a two-stage kNN classifier for character recognition where the first stage used Euclidian distance of Sobel edge maps to find 500 neighbors and a computationally expensive image distortion model (IDM) distance was used for the second stage to find three neighbors. They showed competitive results and fast execution.

ANNs are statistical learning methods inspired by neuron structure of animal brains. In ANNs, neurons are modeled by perceptrons where each neuron takes one or more input values from other neurons, or the initial feature values if in the first layer, applies an activation function, and then outputs the result to one or more connected downstream neurons. Typically, the logistic function,

$$f(x) = \frac{1}{1 + e^{-x}} \tag{2.6}$$

is used for activation because it is differentiable, thereby enabling the computation of cost and gradient for each neuron. Backpropagation can then be used to propagate errors at the output classification layer back through the network and adjust the weights of the hidden layer neurons. Typical structure of ANNs involves at least three layers with the first "input" layer accepting feature values, the second "hidden" layer, and the final "output" layer. The hidden layer can comprise one or more layers with multiple hidden layers typically termed "MLPs." Consistently high accuracies have been achieved leveraging MLPs for ALPR in Refs. [58] and [59].

The sparse network of winnows (SNoWs) classifier was used successfully with SMQT features in Refs. [49, 60] for ALPR character classification. The SNoW classifier is particularly useful for classification tasks involving a very large number of features. The feature space for a single 40×20 pixel character using a single-level SMQT 3×3 pixel representation is $40 \times 20 \times 2^9 = 410k$. While most of the feature values are zero for any given test image, the SNoW classifier effectively learns linear classification boundaries on this sparse network. As a simplification, the SNoW classifier is similar to a single-layer neural network where the update rule is multiplicative (winnow) instead of additive (as in the perceptron).

SVM-based classifiers learn a discriminating hyperplane between two classes and ensure that the hyperplane has the largest minimum distance to the training examples. The margin is defined as twice the minimum distance, and often the SVM is referred to as a maximal margin classifier. To improve linear separability and obtain a higher margin hyperplane, kernel functions are at times used to remap input features to higher dimensionality feature spaces. The standard formulation for calculating the raw output of an SVM classifier is

$$f(x) = \sum_i y_i \alpha_i k(x_i, x) \tag{2.7}$$

where x represents the input feature vector (sample) to be classified, k is a kernel function, the x_i are the support vectors (feature vectors for a subset of the training samples), the y_i are the binary class membership values (+1 or −1 for binary classification tasks) for the x_i support vectors, and the α_i are the weights learned during training that provide maximal separation of the two classes. For binary classification problems, a threshold is then applied to this raw output to determine the SVM classification result. Typical kernel functions are polynomial, sigmoid, and Gaussian radial basis function (RBF). Parameters for kernel functions also need to be tuned on a held-out validation set, and in general the use of kernel functions can lead to significant increase in training and test time. Typically, performance is evaluated using the linear kernel, and if inadequate, kernel functions are introduced with RBF being the most popular. In Ref. [61], the authors split each character template into top and bottom halves and trained 4 OvA SVM classifiers for each half with separate classifiers for numbers and letters achieving 97.2% character recognition accuracy on Korean license plates.

Recently, CNN-based classifiers have started to dominate the object detection and recognition problem sets. In Ref. [44], Krizhevsky introduced a CNN algorithm to classify images that provided a large gain in accuracy over baselines at that time (2012). In general, CNNs rely on various techniques to overcome issues that arise when applying standard ANNs. One such technique is weight-tying where the network uses multiple copies of the same neuron to express large models while keeping the number of parameters low. A pooling layer where the max or average value of a feature for an image region is returned allowing for robustness to small translations. The rectified linear unit (ReLU) activation function,

$$f(x) = \max(0, x) \tag{2.8}$$

in place of the logistic function has also been used to help speed up network computation. Finally, a dropout method where in training, nodes are randomly dropped or not updated to minimize overfitting. Dropout also has a side benefit of speeding up training. Training speed is key as the network in Ref. [44] is eight layers deep with 60 million parameters.

One significant advantage of CNNs is completely unsupervised feature learning. It has been shown that features learned by a CNN increase in complexity from first layer to last [62]. The early layers learn simple edge-based features, middle layers learn combinations of edges, and the final layers learn complex object structure such as faces and characters. Typically, the last layer of a CNN is a softmax classification layer, representing the following posterior probability calculation:

$$P(y = j|x) = \frac{e^{\theta_j^T x}}{\sum_{l=1}^{K} e^{\theta_l^T x}} \tag{2.9}$$

Here, there are k candidate class labels in the multiclass problem, the x values (the observations) are the outputs of the previous layer, and the θ values are the learned weights within the softmax layer itself. As stated previously, CNN methods using this softmax output classification layer have been shown to have state-of-the-art performance on data sets such as the MNIST handwritten digit classification challenge.

The features before this layer can also be extracted and used with an alternate classifier such as SVM to carry out classification for a completely different problem. Here, the benefit of the unsupervised learning of features in the prior layers is leveraged on a new classification task. Indeed, CNNs like LENET-5 [63] trained for MNIST character recognition can be used to generate effective features for use in training ALPR OCR. Although the performance has been observed to improve if LENET-5 is retrained specifically for ALPR, the overall improvement is modest.

Many classifiers are binary in nature—learning to separate one class from another. The OCR problem is, by definition, a multiclass problem of identifying one class label from a dictionary of possible symbols. Various methods exist for leveraging binary classifiers for multiclass classification. Two options are shown in Figure 2.26. For the OvA case, 36 classifiers are trained, one for each class, where the positive side contains examples from the class of interest and the negative side contains all other examples. At test time, all 36 classifiers are applied to the unknown example and the classifier with the maximum score is selected as the answer. For OvO, $N(N-1)/2$ binary classifiers are

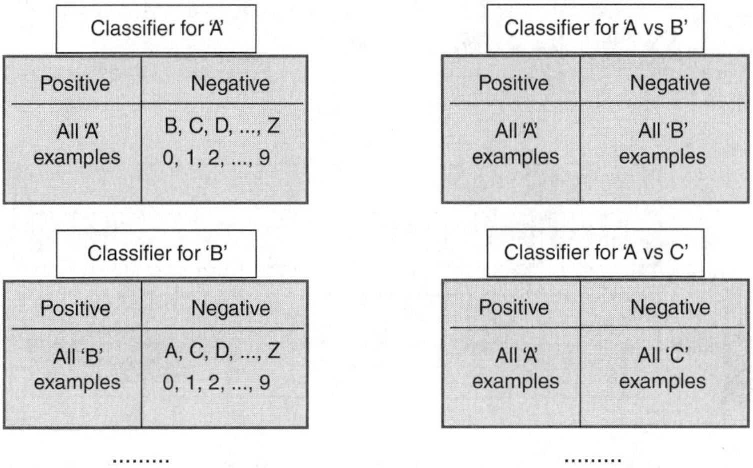

Classifier for 'A'	
Positive	Negative
All 'A' examples	B, C, D, ..., Z 0, 1, 2, ..., 9

Classifier for 'A vs B'	
Positive	Negative
All 'A' examples	All 'B' examples

Classifier for 'B'	
Positive	Negative
All 'B' examples	A, C, D, ..., Z 0, 1, 2, ..., 9

Classifier for 'A vs C'	
Positive	Negative
All 'A' examples	All 'C' examples

....

Figure 2.26 One-vs-all (left) and one-vs-one (right) classifier architecture.

trained—630 for a symbol dictionary of size 36 (10 digits and 26 letters). Each classifier learns to discriminate between only two classes. At test time, all classifiers are evaluated and the votes for each class recorded. The class with the most votes is selected as the answer.

The concept of score or confidence is an important output of a classifier. This is especially true for the ALPR problem space, wherein the confidence score associated with a result is often used to determine its expected accuracy or validity. In fact, for applications such as tolling complex business rules are often used to determine under what conditions the ALPR's automated result should be accepted and when an image should still be forwarded for manual review.

For kNN classification approaches, the average distance amongt the k neighbors is a good measure of classification confidence. For ANN and MLP, the output score is typically acceptable as a confidence measure. For SVMs, particularly kernel-based SVMs, the classier score or distance from margin boundary isn't a directly useful measure of confidence. One way to overcome this limitation is to transform the raw SVM score from Equation 2.7 to a posterior probability. This can be achieved, for instance, using Platt's method [64] as follows:

$$P(y=1|f) = \frac{1}{1+\exp(Af+B)} \tag{2.10}$$

Here, A and B are the parameters that are learned on a held-out training set. The resulting posterior probability value can be used as a comparative metric to determine the confidence associated with each resulting classification.

2.2.3.5 Classifier Evaluation

At test time, segmented characters are presented to the OCR module, features are extracted, and classifiers are evaluated to determine the predicted character label. A typical workflow is captured in Figure 2.27.

Many ALPR installations involve traffic from multiple jurisdictions. Since fonts can vary substantially across jurisdictions (see example in Figure 2.32), multiple OCR engines typically need to be trained to ensure optimal performance. For the most part, a general OCR trained across all fonts will not perform as well as one trained for a specific font.

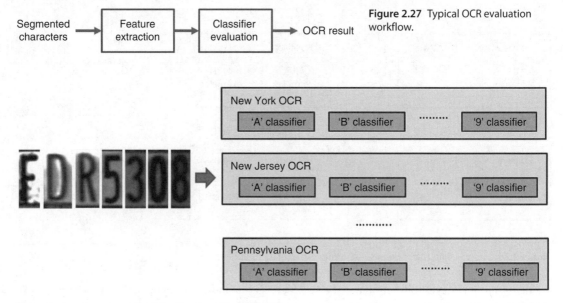

Figure 2.27 Typical OCR evaluation workflow.

Figure 2.28 OCR with multiple trained fonts.

Table 2.1 Selection of optimal result given multiple fonts.

OCR	Result	Conf 1	Conf 2	Conf 3	Conf 4	Conf 5	Conf 6	Conf 7	Mean Conf
New York	FDR5308	0.8	1	1	1	1	0.6	0.8	0.886
New Jersey	EDR5308	0.2	0.9	1	0.7	0.88	0.75	0.5	0.704
Pennsylvania	EDR5308	0.4	0.85	1	0.8	0.7	0.65	0.6	0.714

Figure 2.28 illustrates the scenario where multiple OCR classifiers have been trained, one for each font. The same test characters are presented to each OCR engine, and results from the OCR engines along with character confidences are recorded as shown in Table 2.1. The average of character confidence for each OCR is calculated, and the result from the OCR with the highest average character confidence is selected. Another good measure for font selection is the maximum of minimum character confidences as the OCR result is only as good as the worst character. It may be tempting to select the OCR result for each position by selecting results across OCR engines. The danger with this approach is that a font-specific OCR engine may sometimes return a high confidence, incorrect score for a single character from another font.

2.2.4 State Identification

The combination of letters and numbers on the license plate is typically unique within a jurisdiction, but could be repeated in other jurisdictions. For instance, license plates issued in New York and Pennsylvania, two neighboring states, can have the exact same plate codes. As a result, for many applications the accurate decoding of the license plate text is insufficient without also recognizing the state or jurisdiction that issued the plate.

Figure 2.29 Partially redacted plate image showing a clearly visible state name.

Figure 2.30 Partially redacted plate image illustrating the challenges of low contrast and blur in the state name.

Figure 2.31 Partially redacted plate images with occlusion of the state name by license plate frames and car body features.

License plates from different jurisdictions can be visually differentiated based on a number of features—logos, pictorial backgrounds on the plate, state mottos, and the name of the jurisdiction printed across either the top or bottom of the plate. Given that the ALPR process has already properly located, segmented, and recognized the license plate characters themselves, one approach would be to apply these same techniques to attack the problem of identifying the jurisdiction by directly decoding the printed state name. There are a number of challenges associated with this type of an approach. First, the characters that make up the state name (or the state motto) are typically printed in a significantly smaller font (see the partially redacted image in Figure 2.29). Thus, the number of pixels per stroke width of the font would be insufficient for standard segmentation and OCR methods. This problem is further complicated by the fact that license plate images can often suffer from poor contrast and blur. These noise sources make the state name even harder to decode, as can be seen in Figure 2.30. The prevalence of license plate frames today is another complication for directly recognizing the state name or motto. As shown in the example on the left of Figure 2.31, these frames often cover the upper and lower portions of the license plate where the state name is printed. The image on the right of Figure 2.31 illustrates how features of the car body itself can also occlude the state name or motto on the license plate.

Table 2.2 License plate formats for passenger cars across a sample of US states.

US state	Passenger car plate format
Colorado	NNN LLL
Connecticut	NLL LLN
Florida	LLL NNN
Georgia	*LLL NNNN*
Illinois	LNN NNNN
New York	*LLL NNNN*
Pennsylvania	*LLL NNNN*
Louisiana	LLL NNN
Texas	*LLL NNNN*

Given the many challenges associated with direct decoding of the printed jurisdiction information on the plate, it is typical to leverage other features on the plate to assist with state identification. These features can include the color of the characters and the plate background (which are often unique from one jurisdiction to another), the presence (or absence) of a specific state logo, and character sequence information. Although all of these features can provide clues as to the issuing jurisdiction, most can also suffer from the same challenges of low resolution, poor contrast, and partial occlusion. One key feature that is known to be available for state identification is the character sequence itself. After all, if the license plate code itself cannot be accurately recognized—due to occlusion, poor contrast, etc.—the success or failure of the state identification stage is a moot point for most ALPR applications.

In the United States, many states have their own requirements for the information on the license plate. In particular, there are restrictions on whether letters or numbers may occur at different positions of the plate code. For instance, some states restrict standard license plates for passenger cars to be formatted as three letters followed by four numbers. However, these layouts can vary substantial from state to state. To illustrate this, Table 2.2 shows the standard passenger car license plate formats for a number of US states, where "L" and "N" refer to whether letters or numbers may occur at that position in the plate sequence. Note that across many states, there is clear separation based solely on this format information. Unfortunately, as highlighted in the table for the "LLL NNNN" layout, significant overlap in formats across multiple states can also occur.

Within this kind of basic format template, most jurisdictions issue license plates serially—starting with lower values and progressing numerically and alphabetically in sequence (e.g., AAA-1000, AAA-1001, AAA-1002, …, AAB-1000, AAB-1001, AAB-1002, …, and BAA-1000, BAA-1001, BAA-1002, …). This means that there are typically only a limited subset of all possible combinations of a particular layout in circulation for a given jurisdiction. In fact, some license plate enthusiast groups actually track the "high" and "low" values for different plate designs that have been observed by group members on the roadway. For instance, according to Ref. [65], the highest issued passenger car license plate for the state of New York (as of August 2015) contains "HAV" as the first three letters—with all plate sequences below that having been issued at some point and those above not yet (observed) in circulation.

Due to a number of factors, the valid license plate sequences that are actually in circulation for different states do not always overlap, even for the same plate formats. As an example although New

York and Virginia share the same basic passenger car template (LLL-NNNN), the plates in circulation in each state occupy a different range within the overall set of possible plate sequences. More specifically, New York passenger car plates don't extend beyond those having a first letter of "H" (e.g., "HAV-1000"), whereas Virginia's plates for this same format occupy the range starting with a first letter of "V" and above (e.g., "VAA-1000").

Thus, sequence information of this form can be used to help distinguish source jurisdictions, even for the same basic format. Leveraging such information as evidence, one could attempt to manually construct rules that separate states based on the patterns in the input character sequences. However, a more efficient and effective approach to the state identification challenge is to construct a classifier using machine learning techniques. This classifier will accept the OCR output character labels as input and will provide a predicted source jurisdiction as output. The training data for the classifier consists of license plate codes (inputs) and source jurisdictions (outputs)—typically extracted from a database of license plates that have been observed on the roadways. There are various underlying methodologies that can be used to construct the classifier: simple decision trees, random forests, hidden Markov models (HMMs), etc.

One important consideration in the training of this classifier is that the overall distribution of jurisdictions in the training data should be relatively consistent with the expected distributions at the installation/application site. For instance, if the classifier is trained with data that has been observed on a roadway with mostly California license plates, the performance of the derived classifier is not likely to be very good if it is then applied for an ALPR system in operation in Florida (with mostly Florida plates). Thus, these type of approaches based on character sequence information typically require an investment in retraining the state identification module for each local installation to achieve optimal performance. In addition, as states continue to issue new plates, the valid sequences across states will slowly "drift." This requires that the state identification module be retrained periodically to keep up with these shifting distributions of valid sequences. The sensitivity to this type of prior information is a general limitation of most sequence-based approaches for state identification.

Another limitation is the fundamental overlap that exists between plate formats and sequences across states. For applications such as tolling or parking wherein the traffic is expected to be dominated by a local jurisdiction, along with a limited set of surrounding states, sequence information can prove sufficient for achieving accuracy targets. However, between certain states there would simply be no separation based solely on the sequence of plate characters. In addition, even where separation does exist today, as states continue to issue more and more license plates they can in a sense "use up" the separation that once existed between their valid plate sequences.

To combat this sensitivity to circulation distributions and the fundamental lack of separation between valid plate sequences across certain jurisdictions, other sources of information beyond the plate code can be incorporated into the state identification decision. As illustrated in Figure 2.32, font information can be a very distinctive feature between states. One method for incorporating font information into the state identification decision is to leverage the selection made in the OCR

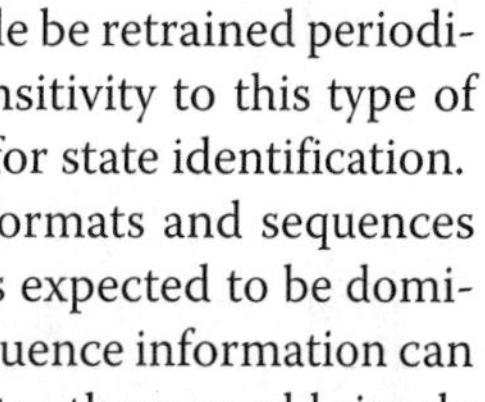

Figure 2.32 Comparison of fonts for New York and California license plate characters: "A" and "4."

A 4   New York

A 4   California

module. Referring back to Figure 2.28 and Table 2.1, a common approach for OCR in the ALPR space is to train a highly tuned OCR engine for each jurisdiction of interest. At test time, the individual segmented character images are then provided to each trained OCR engine and the "winner" is selected based on the overall confidence scores across all of the plate characters. To assist with state identification, the state whose font was selected during the OCR process can be considered. In some cases, this font information alone can be used to conclude the source jurisdiction for the plate, without regard to the underlying character sequence. In other approaches, the OCR/font selection can be combined with the sequence information—for instance, using a Bayesian merging of these two pieces of "evidence" (see for instance, Ref. [66]).

References

1 Roberts, D. and Casanova, M. *Automated License Plate Recognition (ALPR) Systems: Policy and Operational Guidance for Law Enforcement.* Washington, DC: U.S. Department of Justice, National Institute of Justice, 2012.

2 Toll Facilities in the United States, US Department of Transportation, Federal Highway Admin, Publication no. FHWA-PL-13-037, September 2013.

3 Samuel, P. (2009). New Zealand gets first modern pike—video tolled. *TOLLROADSnews.*

4 Samuel, P. (2007). Central toll in Stockholm restarts with IR cameras only, 700k transponders dumped. *TOLLROADSnews.*

5 2014 Specialty License Plate Rankings, Florida Department of Highway Safety and Motor Vehicles. Available: http://www.flhsmv.gov/specialtytags/tagsales.pdf (accessed September 10, 2016).

6 Schulz, K. U. and Mihov, S. (2002). Fast string correction with Levenshtein automata. *International Journal on Document Analysis and Recognition*, 5(1), 67–85.

7 Kukich, K. (1992). Techniques for automatically correcting words in text. *ACM Computing Surveys (CSUR)*, 24(4), 377–439.

8 Zobel, J. and Dart, P. (1995). Finding approximate matches in large lexicons. *Software: Practice and Experience*, 25(3), 331–345.

9 Takahashi, H., Itoh, N., Amano, T., and Yamashita, A. (1990). A spelling correction method and its application to an OCR system. *Pattern Recognition*, 23(3), 363–377.

10 Anagnostopoulos, C.-N. E., Anagnostopoulos, I. E., Psoroulas, I. D., Loumos, V., and Kayafas, E. (2008). License plate recognition from still images and video sequences: A survey. *IEEE Transactions on Intelligent Transportation Systems*, 9(3), 377–391.

11 Du, S., Ibrahim, M., Shehata, M., and Badawy, W. (2013). Automatic license plate recognition (ALPR): A state-of-the-art review. *IEEE Transactions on Circuits and Systems for Video Technology*, 23(2), 311–325.

12 Shi, X., Zhao, W., and Shen, Y. Automatic license plate recognition system based on color image processing. *Proceedings of the International Conference on Computational Science and Its Applications (ICCSA 2005)*, Singapore, May 9–12, 2005. Springer, Berlin/Heidelberg, pp. 1159–1168.

13 Lee, E. R., Kim, P. K., and Kim, H. J. Automatic recognition of a car license plate using color image processing. *Presented at the IEEE International Conference on Image Processing, 1994 (ICIP-94)*, Austin, TX, November 13–16, 1994, vol. 2, pp. 301–305. doi: 10.1109/ICIP.1994.413580.

14 Deb, K. and Jo, K.-H. HSI color based vehicle license plate detection. *Presented at the International Conference on Control, Automation and Systems, 2008. ICCAS 2008*, ICCAS 2008, Seoul, South Korea, October 14–17, 2008. IEEE, pp. 687–691. doi: 10.1109/ICCAS.2008.4694589.

15 Chang, S.-L., Chen, L.-S., Chung, Y.-C., and Chen, S.-W. (2004). Automatic license plate recognition. *IEEE Transactions on Intelligent Transportation Systems*, 5(1), 42–53.

16 Zheng, D., Zhao, Y., and Wang, J. (2005). An efficient method of license plate location. *Pattern Recognition Letters*, 26(15), 2431–2438.

17 Bai, H. and Liu, C. A hybrid license plate extraction method based on edge statistics and morphology. *Presented at the 17th International Conference on Pattern Recognition*, Cambridge, August 23–26, 2004, pp. 831–834. doi: 10.1109/ICPR.2004.1334387.

18 Suryanarayana, P. V., Mitra, S. K., Banerjee, A., and Roy, A. K. A morphology based approach for car license plate extraction. *2005 Annual IEEE INDICON Conference*, Chennai, India, December 11–13, 2005, pp. 24–27. doi: 10.1109/INDCON.2005.1590116.

19 Viola, P. and Joes, M. Rapid object detection using a boosted cascade of simple features. *Presented at the 2001 IEEE Computer Society Conference on Computer Vision and Pattern Recognition, 2001 (CVPR 2001)*, Jauai, Hawaii, December 8–14, 2001. Vol. 1. IEEE. doi: 10.1109/CVPR.2001.990517.

20 Zhang, H., Jia, W., He, X., and Wu, Q. Learning-based license plate detection using global and local features. *Presented at the 18th International Conference on Pattern Recognition, 2006 (ICPR 2006)*, Hong Kong, 2006. Vol. 2. IEEE, pp. 1102–1105. doi: 10.1109/ICPR.2006.758.

21 Porikli, F. and Kocak, T. Robust License Plate Detection Using Covariance Descriptor in a Neural Network Framework. *Presented at the IEEE International Conference on Video and Signal Based Surveillance, 2006 (AVSS'06)*, Sydney, November 22–24, 2006. IEEE. doi: 10.1109/AVSS.2006.100.

22 Csurka, G., Dance, C., Fan, L., Willamowski, J., and Bray, C. Visual Categorization with Bags of Keypoints. Workshop on Statistical Learning in Computer Vision, ECCV. Vol. 1. No. 1–22. 2004.

23 O'Hara, S. and Draper, B. A. (2011). Introduction to the bag of features paradigm for image classification and retrieval. arXiv preprint arXiv:1101.3354.

24 Zhou, W., Li, H., Lu, Y., and Tian, Q. (2012). Principal visual word discovery for automatic license plate detection. *IEEE Transactions on Image Processing*, 21(9), 4269–4279.

25 Lowe, D. G. Object recognition from local scale invariant features. *Presented at the 7th International Conference on Computer Vision*, Corfu, Greece, 1999. doi: 10.1109/ICCV.1999.790410.

26 Cano, J. and Pérez-Cortés, J.-C. Vehicle license plate segmentation in natural images. In Perales, F. J., Campilho, A., Pérez, N., and Sanfeliu, A. (eds.), *IbPRIA 2003*. LNCS, vol. 2652. Berlin/Heidelberg: Springer, 2003, pp. 142–149.

27 Jung, K., Kim, K. I., and Jain, A. K. (2004). Text information extraction in images and video: A survey. *Pattern Recognition*, 37(5), 977–997.

28 Casey, R. G. and Lecolinet, E. (1996). A survey of methods and strategies in character segmentation. *IEEE Transactions on Pattern Analysis and Machine Intelligence*, 18(7), 690–706.

29 Sumathi, C. P., Santhanam, T., and Gayathri, G. (2012). A survey on various approaches of text extraction in images. *International Journal of Computer Science and Engineering Survey*, 3(4), 27–42.

30 Epshtein, B., Ofek, E., and Wexler, Y., Detecting text in natural scenes with stroke width transform. *Presented at the IEEE Conference on Computer Vision and Pattern Recognition (CVPR)*, New York, June 13–18, 2010, pp. 2963–2970. doi: 10.1109/CVPR.2010.5540041.

31 Yoon, Y., Ban, K. D., Yoon, H., and Kim, J., Blob extraction based character segmentation method for automatic license plate recognition system, *Proceedings of the IEEE International Conference on Systems, Man, and Cybernetics*, Anchorage, AK, 2011, pp. 2192–2196. doi: 10.1109/ICSMC.2011.6084002.

32 Burry, A., Fillion, C., Kozitsky, V., and Fan, Z. Method and System for Robust Tilt Adjustment and Cropping of License Plate Images. US Patent App. No. 13/453144, filed April 23, 2012.

33 Otsu, N. (1979). A threshold selection method from gray-level histograms. *IEEE Transactions on Systems, Man, and Cybernetics*, 9(1), 62–66.

34 Sauvola, J. and Pietikäinen, M. (2000). Adaptive document image binarization. *Pattern Recognition*, 33(2), 225–236.

35 Niblack, W. *An Introduction to Image Processing*. Englewood Cliffs, NJ: Prentice-Hall, 1986, pp. 115–116.

36 Gonzalez, R. C. and Woods, R. E. *Digital Image Processing*. Reading, MA: Addison-Wesley, 1993, pp. 454–455.

37 Fisher, M. and Bolles, R. (1981). Random sample consensus: A paradigm for model fitting with applications to image analysis and automated cartography. *Communications of the ACM*, 24(6), 381–395.

38 Burry, A., Fillion, C., Kozitsky, V., Bala, R., and Fan, Z. Robust Cropping of License Plate Images. US Patent App. No 13/448976, filed April 17, 2012.

39 Jia, X., Wang, X., Li, W., and Wang, H. A novel algorithm for character segmentation of degraded license plate based on prior knowledge. *Presented at the 2007 IEEE International Conference on Automation and Logistics*, 2007. IEEE, pp. 249–253. doi: 10.1109/ICAL.2007.4338565.

40 Burry, A., Fillion, C., and Kozitsky, V. Robust Character Segmentation for License Plate Images. US Patent 8,934,676, issued January 13, 2015.

41 Baird, H. S. The state of the art of document image degradation modelling. In *Digital Document Processing*. London: Springer, 2007, pp. 261–279.

42 Kanungo, T., Haralick, R., and Phillips, I. (1994). Nonlinear global and local document degradation models. *International Journal of Imaging Systems and Technology*, 5(3), 220–230.

43 Ho, T., and Baird, H. Evaluation of OCR accuracy using synthetic data. *Proceedings of the 4th Symposium on Document Analysis and Information Retrieval*, UNLV, Las Vegas, Nevada, April 24–26, 1995, pp. 413–422.

44 Krizhevsky, A., Sutskever, I., and Hinton, G. E. ImageNet classification with deep convolutional neural networks. *Proceedings of the 26th Annual Conference on Neural Information Processing Systems (NIPS)*, Lake Tahoe, December 3–6, 2012, Curran Associates, Red Hook, New York, pp. 1106–1114.

45 Mecocci, A. and Tommaso, C. Generative models for license plate recognition by using a limited number of training samples. *Proceedings of the IEEE International Conference on Image Processing*, Atlanta, GA, October 8–11, 2006, pp. 2769–2772.

46 Bala, R., Zhao, Y., Burry, A., Kozitsky, V., Fillion, C., Saunders, C., and Rodriguez-Serrano, J. Image simulation for automatic license plate recognition. *Proceedings of the SPIE 8305, Visual Information Processing and Communication III*, Burlingame, CA, USA, January 22, 2012, 83050Z.

47 Trier, O., Jain, A., and Taxt, T. (1996). Feature extraction methods for character recognition: A survey. *Pattern Recognition*, 29(4), 641–662.

48 Nilsson, M., Nordberg, J., and Claesson, I. Face detection using local SMQT features and split up snow classifier. *Presented at the IEEE International Conference on Acoustics, Speech, and Signal Processing, (ICASSP 2007)*, Honolulu, HI, April 15–20, 2007, vol.2, pp. II-589–II-592. doi: 10.1109/ICASSP.2007.366304.

49 Paul, P., Burry, A., Wang, Y., and Kozitsky, V. Application of the SNoW machine learning paradigm to a set of transportation imaging problems. *Proceedings of SPIE, Visual Information Processing and Communication III*, International Society for Optics and Photonics, 2012, February 15, 2012, 830512. doi: 10.1117/12.912110.

50 Keysers, D., Deselaers, T., Gollan, C., and Ney, H. (2007). Deformation models for image recognition. *IEEE Transactions on Pattern Analysis and Machine Intelligence*, 29(8), 1422–1435.

51 Dalal, N. and Triggs, B. Histograms of oriented gradients for human detection. *Presented at IEEE Computer Society Conference on Computer Vision and Pattern Recognition, CVPR 2005*. doi: 10.1109/CVPR.2005.177.

52 Bissacco, A., Cummins, M., Netzer, Y., and Neven, H. PhotoOCR: Reading text in uncontrolled conditions. *Presented at the IEEE International Conference on Computer Vision, (ICCV 2013),* Sydney, December 2013. doi: 10.1109/ICCV.2013.102.

53 de Campos, T., Babu, B., and Varma, M., Character recognition in natural images. *Presented at the International Conference on Computer Vision Theory and Applications (VISAPP),* Lisbon, Portugal, February 2009. doi: 10.5220/0001770102730280.

54 Netzer, Y., Wang, T., Coates, A., Bissacco, A., Wu, B., and Ng, A. Y. Reading Digits in Natural Images with Unsupervised Feature Learning. NIPS Workshop on Deep Learning and Unsupervised Feature Learning, 2011, Curran Associates, Red Hook, New York.

55 Coates, A., Lee, H., and Ng, A. Y. An analysis of single-layer networks in unsupervised feature learning. *Proceedings of the Fourteenth International Conference on Artificial Intelligence and Statistics,* Gordon, G., Dunson, D., and Dudík, M. eds. JMLR W&CP, vol. 15, Fort Lauderdale, April 11–13, 2011.

56 Jaderberg, M., Simonyan, K., Vedaldi, A., and Zisserman, A. Reading text in the wild with convolutional neural networks. arXiv:1412.1842v2, 2014.

57 Weinberger, K. Q. and Saul, L. K. (2009). Distance metric learning for large margin nearest neighbor classification. *Journal of Machine Learning Research (JMLR),* 10, 207–244.

58 Wang, M.-L., Liu, Y.-H., Liao, B.-Y., Lin, Y.-S., and Horng, M.-F. A vehicle license plate recognition system based on spatial/frequency domain filtering and neural networks. In *International Conference on Computational Collective Intelligence.* Berlin, Heidelberg: Springer, 2010, pp. 63–70.

59 Kocer, H. E. and Cevik, K. K. (2011). Artificial neural networks based vehicle license plate recognition. *Procedia Computer Science,* 3, 1033–1037.

60 Burry, A., Kozitsky, V., and Paul, P. License Plate Optical Character Recognition Method and System. US Patent 8,644,561 issued February 4, 2014.

61 Kim, K. K., Kim, K. I., Kim, J. B., and Kim, H. J. Learning-based approach, for license plate recognition. *Neural Networks for Signal Processing X, 2000. Proceedings of the 2000 IEEE Signal Processing Society Workshop,* Sydney, December 11–13, 2000, vol. 2, pp. 614–623. doi: 10.1109/NNSP.2000.890140.

62 Zeiler, M. D. and Fergus, R. Visualizing and understanding convolutional networks. CoRR, abs/1311.2901, 2013.

63 LeCun, Y., Boser, B., Denker, J. S., Henderson, D., Howard, R. E., Hubbard, W., and Jackel, L. D. (1989). Backpropagation applied to handwritten zip code recognition. *Neural Computation,* 1(4), 541–551.

64 Platt, J. Probabilistic outputs for support vector machines and comparisons to regularized likelihood methods. In Smola, A. J. (ed.), *Advances in Large Margin Classifiers.* Cambridge, MA: The MIT Press, 2000, pp. 61–74.

65 License Plate News. (n.d.). From: http://www.licenseplates.cc/ (accessed August 6, 2015).

66 Burry, A. and Kozitsky, V. Method and System for Automatically Determining the Issuing State of a License Plate. US Patent 9,082,037, issued July 14, 2015.

3

Vehicle Classification

Shashank Deshpande, Wiktor Muron and Yang Cai

Carnegie Mellon University, Pittsburgh, PA, USA

3.1   Introduction

Automatic vehicle classification (AVC) is a method for automatically categorizing types of motorvehicles based on the predominant characteristics of their features such as length, height, axle count, existence of a trailer, and specific contours. AVC is an important part of intelligent transportation system (ITS). Automatic toll collection at toll plazas is one of the earliest functions that sparked the development of AVC. Toll collection systems require AVC when tolls vary by vehicle class. A freeway section with a heavy truck flow may require special rules to manage truck traffic to avoid vehicle congestion, etc. Pavement management processes pay special attention to trucks because they can damage the road surface. The growing traffic density of highways rapidly fueled the need to develop an automated management system for discriminating various types of motor vehicles to enable more efficient toll computation. Early AVC systems involved mounting sensors such as radar, infrared detectors, weight sensors, magnetic sensors, loops, and light curtains into the roadbed at toll plazas. These systems usually classified vehicles on the basis of weight in motion (WiM), number of wheels, vehicle contour, or the distance between axles. However, these systems have drawbacks including high installation costs, ongoing maintenance of detectors, and damage to the road surface. They also fail to provide additional information on the vehicles which can be further used for identification or tracking.

As the applications of AVC grew from automatic toll collection toward surveillance, abnormality detection, tracking, and autonomous driving, the demand for more sophisticated algorithms and cost-efficient techniques increased rapidly. Today, there are numerous algorithms available for automatic detection and classification of vehicles which are real-time and utilize cost-efficient sensors such as cameras for video-based classification, LiDAR for depth analysis, and sensor fusion. Although the criteria for classifying different types of vehicles are still under debate, the recent results of AVC show promising accuracy.

In this chapter, we introduce a few recent developments in visible-light cameras and thermal and LiDAR-based AVC techniques, including related algorithms, and the differences among them.

Computer Vision and Imaging in Intelligent Transportation Systems, First Edition.
Edited by Robert P. Loce, Raja Bala and Mohan Trivedi.
© 2017 John Wiley & Sons Ltd. Published 2017 by John Wiley & Sons Ltd.
Companion website: www.wiley.com/go/loce/ComputerVisionandImaginginITS

3.2 Overview of the Algorithms

There are a significant number of algorithms developed for AVC. They can be classified into types of sensors used for identifying key differences among vehicle types. Figure 3.1 shows an overview of the methods developed for AVC. In this chapter, we focus on movable sensors, including RGB video cameras, thermal imaging sensors, and LiDAR imaging sensors. The different categories of algorithms include heuristics and statistics-based, shape-based, and feature-based methods.

3.3 Existing AVC Methods

Many existing AVC systems use special devices such as laser beam curtains (see Figure 3.2), radar, WiM, stereovision cameras, and so on. The TDS AVC system uses a laser light curtain, a Doppler radar, and a fiber treadle-based axle detection system [1]. Each radar message reports the distance

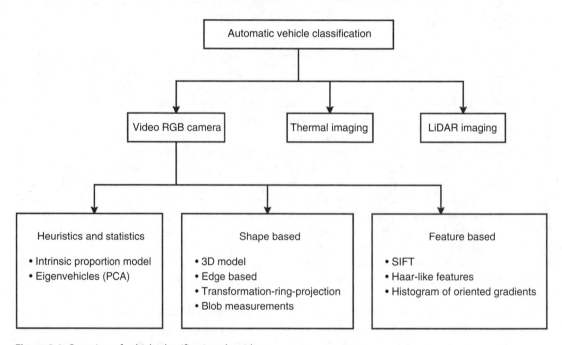

Figure 3.1 Overview of vehicle classification algorithms.

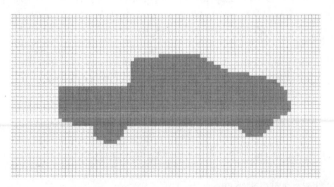

Figure 3.2 Visualization of a vehicle profile generated by a system using data from radar/light curtains. *Source*: Adapted from [1].

and velocity that the radar is currently sensing in its beam. Each light curtain message provides a report of the beam status of each of the 120 beams. The treadle inputs are sampled each time a light curtain message is received. The vehicle detection process begins when the light curtain reports sufficient penetration concurrent with a radar report of an object moving in the vicinity of the lane in the path of the light curtain beams. This approach does not work well in multilane cases. Migma uses audio, thermal imaging, and visibility to classify vehicles [2]. Their vehicle detection and classification system are based on stereoscopic video cameras in which two calibrated cameras and specialized software are needed for a single installation. The Central Federal Lands Highway Division (CFLHD) AVC systems WiM, closed-circuit television (CCTV), and roadside detection (RS-D) systems are used to classify vehicles [3]. Again, RS-D and WiM do not scale well in multilane situations. Trafficon AVC systems require thermal imaging cameras to distinguish between bicycles and motorcycles [4]. Laser-based vehicle detection and classification (LVDC) also requires infrared laser scanners and special processing software.

3.4 LiDAR Imaging-Based

Although camera-based AVC techniques are robust and yield considerable accuracy in classification, they are far from being 100% accurate due to inclement weather, illumination changes, shadows, occlusion, etc. Dealing with these limitations often results in more complex algorithms, which drastically effect processing speed. LiDAR sensors attracted growing attention recently because of their applications in autonomous driving vehicles. LiDAR can be applied to AVC as well. The point cloud from the LiDAR sensor provides depth information for object recognition and classification. LiDAR produces accurate, high-density spatial data. Recent research has moved toward fusing LiDAR and vision-based approaches. Here, we will divide our LiDAR-based AVC techniques into two parts: algorithms that use solely LiDAR sensors and algorithms that combine LiDAR and vision sensors.

3.4.1 LiDAR Sensors

LiDAR sensor-based AVC techniques follow a general pipeline as shown in the Figure 3.3. The algorithms differ in the techniques used in all three stages of the pipeline. The LiDAR sensor collecting the data can either be fixed (i.e., mounted on pavement, a surveillance structure, a traffic light pole, or a parked vehicle on the roadside), or it can be mounted on a moving vehicle. If the sensor is mounted on a moving vehicle, the AVC algorithms take into account the speed and, by extension, the displacement of the host vehicle in the pre-processing stage. The interpretation of LiDAR data generally involves dimensionality transformation. The transformed LiDAR data can either directly yield useful information, or definitive features can be extracted from the data later. This information is then used to categorize different vehicle types.

Data acquisition depends on the quantity and type of LiDAR sensors used. There are two major types of LiDAR sensors: airborne LiDARs and terrestrial LiDARs (see Figure 3.4). In terrestrial LiDARs, algorithms can use multiple stationary LiDAR sensors positioned at a distance in the same

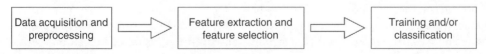

Figure 3.3 Typical steps seen in AVC systems using LiDAR sensors.

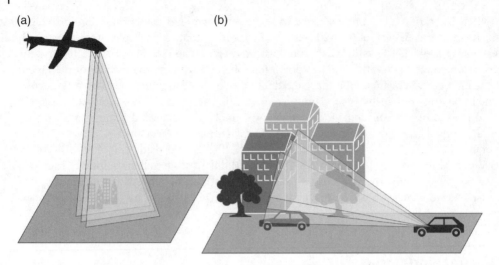

Figure 3.4 (a) With airborne LiDAR, the infrared laser light is emitted toward the ground and is reflected back to the moving airborne LiDAR sensor. (b) Terrestrial LiDAR collects very dense and highly accurate points, facilitating more precise identification of objects. These dense point clouds can be used to model the profile of a vehicle.

horizontal plane, or a single revolving LiDAR sensor, which scans a horizontal plane across the scene. A few algorithms use multilayered stationary or revolving LiDAR sensors, which provide scans of multiple horizontal planes across the scene. The major preprocessing involves noise removal, synchronization of data received from multiple sensors, LiDAR image stitching when using revolving LiDAR sensors, and displacement adjustment when using moving LiDAR sensors.

Once the data collected by the LiDAR is cleaned, the depth data is transformed into more useful key descriptors in the feature extraction and selection step. These descriptors magnify subtle differences which stand out for different vehicle classes. Since vehicle categories are more often defined by the size of the vehicle, most of the AVC algorithms that only use LiDAR data tend to use shape, height, length, or width parameters. These characteristics can easily be extracted from LiDAR scans as predominant features.

The training and/or classification model utilizes the extracted features and segments vehicle types using various techniques. The model may use a complex technique, such as a hyperplane, for division in multidimensional feature space, or a simple technique such as thresholding depending upon the quality of the extracted feature. Table 3.1 shows extracted key features and the classification method employed to categorize vehicle types from significant recent work on LiDAR sensor-based AVC algorithms.

3.4.2 Fusion of LiDAR and Vision Sensors

Vision-based and LiDAR-based AVC algorithms have matured separately over the years. In recent years, the limitations of both techniques have surfaced as AVC applications have demanded more accurate classification methods. However, the separate strengths of these two methods are capable of overshadowing their shortcomings. For example, LiDAR-based classification techniques are invariant to lighting and weather conditions, and vision-based techniques provide sensory information such as color and texture. Therefore, in recent years, an effort to utilize the complementary properties of these techniques has emerged. AVC algorithms based on the fusion of LiDAR and vision

Table 3.1 Comparison of papers according to key features and classification methods for LiDAR AVC systems.

Key features	Classification method	Comments
Size and shape measurements include [5]: • Vehicle length (VL) • Vehicle height (VH) • Detection of middle drop (DMD) • Vehicle height at middle drop (VHMD) • Front vehicle height (FVH) • Front vehicle length (FVL) • Rear vehicle height (RVH) • Rear vehicle length (RVL)	Thresholding of size and shape measurements based on a defined classification tree	LiDAR sensors mounted in a side-fire configuration. Key features of size and shape were extracted for each vehicle and these features were used to classify the vehicle into one of six categories defined in the paper. The elaborate use of size- and shape-based features strengthens the classification model.
Geometric parameters of vehicles to derive shape [6]: • Height • Length/Width	PCA clustering and shape-based classification	Reference [6] discusses the limitations of vehicle classification using only geometric parameters. It argues that using geometric parameters to derive a shape profile which are then used for profile-based classification can provide better classification.
Size parameters [7]: • Vehicle length • Vehicle height: subdivided into Group 1 (front bumper and roughly the center of the vehicle) And Group 2 (roughly the center of the vehicle and the rear bumper)	Thresholding of vehicle length and height using predefined classification rules	Based on the measured speed of the vehicle, the length is extracted. A thresholding-based classification approach is then employed to classify vehicle types.

sensors loosely follow a general pipeline as shown in the Figure 3.5. Algorithms differ greatly, both in the type of features extracted in the vision and LiDAR modules, and in the decision-making/classification module.

The key in AVC fusing vision and LiDAR sensors is the synchronization of the input data from both sensors. Even a slight delay in frames can cause an imbalance in the mapping, which may result in the two modules processing different time stamps. This leads to substantial classification errors. Generally, less time is needed for camera data acquisition than for LiDAR (by a factor of 2). Since we are dealing with dynamic environments, both the sensors and the target/scene drastically change with respect to time, which generates problems in data fusion. Thus, the synchronization of data from both sensors is absolutely necessary.

Data acquired from both sensors after synchronization generally undergoes a preprocessing step, such as contrast enhancement, noise removal, geometric transformations, or sharpness enhancement. This step intrinsically adds to feature enhancement for the extraction step. For example, in shape-based classification methods using visual features, sharpness enhancement results in cleaner edges that can be more easily detected. The preprocessed data at this stage have been used in various ways. For example, LiDAR data can be used as a distance filter to form a rigid region of interest (ROI) or field of view before vision-based features are extracted. Vision-based features can also be used for preliminary vehicle classification and LiDAR features can be used to strengthen the hypothesis. Finally, vision-based features and LiDAR-based features can be merged to form a highly dimensional feature vector for classification purposes.

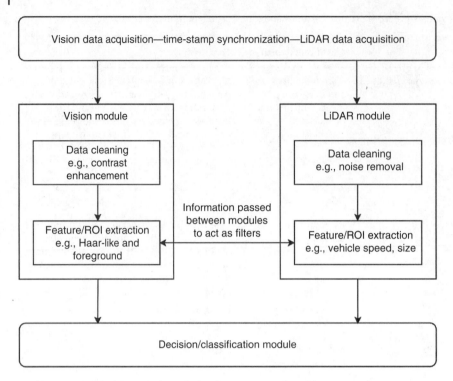

Figure 3.5 Simplified diagram for AVC that fuses vision and LiDAR sensors.

Table 3.2 Comparison of papers according to vision- and LiDAR-based features and classification techniques for vision + LiDAR fusion AVC systems.

Vision features	LiDAR features	Classification techniques
Haar-like features + AdaBoost [8]	• Shape	Support vector machines (SVM)
Haar-like features representing the following characteristic properties [9]: • Edge • Line • Symmetry	• Contours • Size • Location • Orientation	AdaBoost
Haar-like features [10]	• Segment centroid • Radius of a circle • Median	• AdaBoost for vision-based features • GMM classifier of LiDAR-based features • Bayesian decision-making

The overlay of these two modules varies greatly with the algorithm used and application of AVC. Similarly, the classification or decision-making module also varies with respect to the weight assigned to each module. Table 3.2 surveys key features extracted from both modules, and the classification methods employed to categorize vehicle types based on significant recent works on the fusion of vision and LiDAR sensor-based AVC algorithms.

The fusion of LiDAR and vision-based AVC algorithms has just started to unfold. The symbiotic nature of this fusion is already working wonders with AVC by closing in on the perfect classification accuracy. As autonomous vehicle research and development increases, the need for such fusion is undeniable. Since the margin of error in such applications is close to zero, the ability for LiDAR and vision-based AVC algorithms to nullify the limitations of each other seems to be the best way forward.

3.5 Thermal Imaging-Based

A thermal camera can be used to overcome some of the drawbacks of visible-light image sensors. Vehicles tend to cast shadows, so in the case of visible-light sensors, sophisticated algorithms can be used to eliminate shadows from the image before further processing. Furthermore, visible-light cameras cannot be used at night due to insufficient lighting and additional occlusion effects caused by the vehicle headlights. Other low-visibility conditions, such as thick fog or precipitation, may also decrease the level of detection accuracy. However, it is possible to limit the effects of these obstacles and improve detection robustness through the usage of a thermal camera. A thermal-based approach enables the system to work in scenarios when the vehicle is occluded by any nonthermal insulating obstacle such as metals (as opposed to using visible light, where even partial occlusion of the vehicle may deteriorate detection accuracy).

We can divide systems using thermal imagery into two main groups: those using a stationary camera with a nonchanging background and those with a camera mounted on a vehicle or an unmanned aerial vehicle (UAV). In the latter system, the background changes constantly. Unintended camera shaking, changes in global intensity contrast, or undetermined obstacles make video recording and processing all the more difficult. It leads to a different approach in designing vehicle detection algorithms, sometimes involving hybrid systems that employ both thermal and visible-light cameras. On the other hand, using a stationary camera with a static background enables us to simplify further processing (detection, tracking, and optional classification) by employing one of the background subtraction methods to obtain motion segmentation images. These images depict only the masked moving vehicles, and can be used to select proper ROIs. Figure 3.6 shows multiple views from a UAV thermal imaging sensor.

3.5.1 Thermal Signatures

A thermal imaging camera registers the radiation intensity of the infrared portion of electromagnetic spectrum. In order to use thermal photographs for further image processing, they are converted to a

(a) (b) (c)

Figure 3.6 Different viewpoints of the thermal camera: (a) side view, (b) overpass view [11], and (c) UAV view [12].

Figure 3.7 Example frames from a thermal camera for different vehicles: (a) small car, (b) SUV, (c) pickup truck, and (d) bus.

grayscale mode, in which every pixel value corresponds to a respectively scaled radiation value. In the examples shown in Figure 3.7, areas of higher temperature are represented by the brightest pixels.

The heat signature characteristic of vehicles visible in thermal images is dependent on the position of the camera. As a result, the algorithm used to perform further detection/classification processing must be adapted to the available thermal image. For instance, we would analyze image features differently when using camera footage taken from an overpass, as opposed to from the side of the road. However, in order to develop a successful traffic surveillance algorithm, we tend to use overpass/bridge-positioned cameras as video sources, since their image consists of separable road lanes. Therefore, we can limit the ruinous effect of mutual occlusion between vehicles in heavy multilane traffic.

Depending on the method, vehicle detection is based on finding thermal imagery frame features. Each element of the vehicle, such as the windshield, tires, and engine, has distinct heat signatures, forming the vehicle's specific thermal identity [13]. Since different groups of vehicles were created for various purposes, all with varying sizes, shapes, and positions of windshields; differing engine powers; and cooling systems, we are often able to determine what category a vehicle belongs to by analyzing its thermal identity, or, at worst, to simply detect the presence and approximate location of a vehicle. For instance, some publications [13, 14] analyze such features (engine heat area and windshield area; see Figure 3.8) and their locations to classify the vehicles into groups. This classification

Figure 3.8 Vehicle temperature regions in thermal imagery.

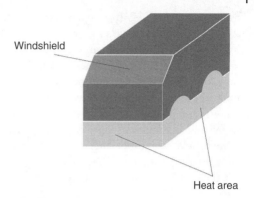

Windshield

Heat area

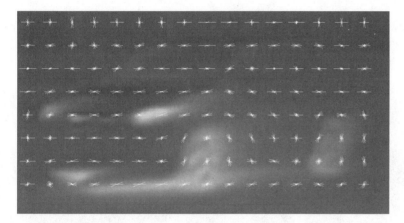

Figure 3.9 HOG features extracted from a single region containing a car.

may rely on relative localization, shape, and temperature of heat areas, windshield shape, and the space between. It is also possible to detect the location of windshields and their surroundings using a Viola–Jones detector, applying such features as a recognition target.

Another issue that must be addressed is the fact that temperature relationships between vehicles and the outside environment may vary depending on the season. Namely, windshield temperature is usually lower than that of the road surface (frequently, even lower than the outside air temperature) in warmer seasons, while on cold winter days, this relationship is reversed [14]. To resolve this variation, we can use a simple complement operation on intensity images to reverse this relationship into a default state, thus enabling us to perform uniform algorithms during all seasons. It can be obtained by transformation,

$$I_T(i,j) = I_{max} - I(i,j)$$

where I_{max} is the maximum value of a pixel in the intensity image (by default $I_{max} = 255$ in 8-bit grayscale), $I(i,j)$ is a pixel value of the original image at (i,j) position.

One approach that can be used to detect the thermal signature of a car was proposed in Ref. [15], where HOG (see Figure 3.9) is calculated at multiple scales and can be used to distinguish cars from other objects (i.e., animals). The algorithm is based on analyzing the regions containing moving objects. The analysis is performed by separating every moving object into a set of regions, and extracting the HOGs from each region. This approach, involving using the same moving object at different resolution levels for gradient calculation, allows us to describe and process vehicles at varying distances from the camera. To differentiate between classes (vehicles/nonvehicles), we use a learned set of vehicle signatures stored in a database. We can use, for example, multinomial pattern matching (MPM) [16] to compare the current signature with the patterns in the database. In order to make the algorithm more robust, every moving object is tracked in sequential frames, and the matching is based on a sequential probability ratio test (SPRT) [17]. This combines matching decisions for individual instances of an object in frames into the final classification decision. To overcome the issue of relative temperature between vehicle and environment, we can modify the HOG algorithm so that we do not differentiate between positive and negative gradient orientation, thereby immunizing the method to inconsistencies.

3.5.2 Intensity Shape-Based

Since thermal cameras represent higher temperature regions as brighter pixels, we can use the characteristic shape of these regions to detect and classify various types of vehicles [13]. Considering an algorithm without the use of background subtraction (i.e., if the camera is stationary), we can take the following steps. First, we threshold the intensity image with a desired threshold level (corresponding to proper temperature) so that an initial heat signature blob is obtained (see Figure 3.10). A threshold level can be fixed or adapted according to maximum pixel intensity in the image, analysis

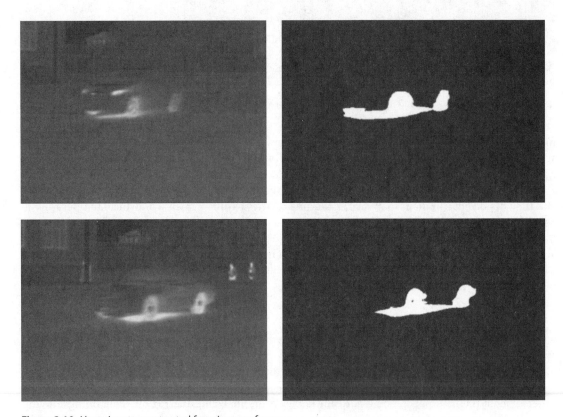

Figure 3.10 Heat signature extracted from image of a car.

of histogram, etc. Second, we use morphological opening to reduce noisy singular pixels and improve the quality of detected blobs. If the result is not satisfactory, we can utilize one or a few points from detected blobs as seed points in a mophological watershed algorithm to obtain a more accurate mask, and then merge both versions. Third, to analyze the mask and separate the blobs, we use connected component labeling. Afterward, we can filter out the blobs that do not fit certain criteria (i.e., too small or too big).

Fifth, raw blobs by themselves do not convey enough useful information for further classification. We can describe these blobs using shape feature descriptors, such as the bounding box, minimal area bounding box, fitted ellipse, blob area, filled blob area, blob perimeter, convex-hull area and perimeter, centroid position and central moments, compactness, eccentricity, etc. These features and their explanations are covered more extensively in the following section on shape-based vehicle classification.

Sixth, we can choose to perform feature dimension reduction. This can be useful when the number of features is high, increasing the computational cost, and/or when we are not able to manually specify which features are discriminative between classes or which ones are simply overdescribing vehicles, since some extracted features may be correlated or even redundant. To reduce feature vector dimensions, we can use linear discriminant analysis (LDA) [18] or principal component analysis (PCA) [19]. The main difference between PCA and LDA is that PCA uses data sets (our training group of manually classified vehicle data) to find a linear combination of features. This maximizes the total variance in data, which may sometimes lead to loss of discriminative data between classes. On the other hand, LDA explicitly models the differences between classes by finding combinations of features and separating them in the best way possible. In order words, LDA tries to maximize the separation between the means of each class in a new subspace and, simultaneously, minimizes the variance within each class separately. Finally, we perform learning and further classification with calculated features. Popular classifiers used in publications include the following: support vector machine (SVM), k-nearest-neighbor algorithm (k-NN), random forest classifier, [20] etc.

If the camera is stationary, we can alter some parts of the algorithm by using a background subtraction method. Many systems use an algorithm (recursive or nonrecursive) that in some way models the background, and then subtracts it from the current frame. As shown in Ref. [13], we can also subtract free road intensity instead, provided that we model (i.e., using Gaussian modeling) its temperature gray level so that it is robust enough to distinguish the road properly, yet limited enough not to cover heated parts of the vehicle. Once we have access to the actual vehicle shape, or at least its bounding box, we can determine which part of a vehicle has a strong heat signature and where it is positioned within a vehicle shape. For example, buses have much stronger heat signatures in the back, while big trucks are warmer in front of the vehicle, corresponding directly to the position of their engines. Values such as relative displacement between the vehicle shape centroid and its heat signature, and the ratio between heat signature area and vehicle shape area can be derived from this information, improving vehicle classification robustness.

Regardless, any background segmentation method used must be prepared to handle unpredictable changes in lighting and weather conditions, while at the same time being able to produce satisfactory results in real time. The main goal is to separate the moving parts of the image, presumably vehicles in motion (foreground), from static roads and sidewalks (background). In the case of thermal imagery, this approach is not ideal, since pixel gray level corresponds to temperature, so often a part of the vehicle and the road have similar intensities. Because of that, foreground masking can be highly corrupted and certain parts of the vehicle will be misinterpreted as background elements). In order to improve the level of foreground segmentation accuracy, we can use slightly extended ROI covering the initial approximation of vehicle shape, and perform edge detection within the extended ROI. The

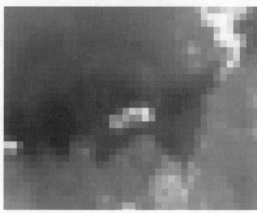

Figure 3.11 Thermal image [12] with a pixelated car showing Haar features.

edge detector parameters should be set to correspond to the given image quality and noise, so that it extracts only desired contours. Then, we can use logical and morphological operations on masks from ROI to merge them, fill in regions, remove noise, and improve the accuracy of the detected shape.

For aerial thermal images gathered from UAV for instance, the advantageous thermal signature from viewing the lower portion of the vehicle is not available. Furthermore, the heat signatures of the tops of vehicles are not as unique as the ones visible from the side or front of the vehicle [21]. This signature may also vary substantially depending on weather conditions. One of the approaches to resolving this issue is to use a hybrid system involving both thermal and visible-light cameras [21]. To detect the silhouettes of vehicles (see Figure 3.11), starting with a visible-light camera, we can use four Haar classifiers trained on sample vehicle images manually captured and separated in four angular positions: 0°, 45°, 90°, and 135°. Using trained classifiers, we can detect the positions of the vehicles in the image with a certain level of confidence. As mentioned before, an approach using Haar features in a Viola–Jones classifier has also been used in the detection of the windshield and its surroundings when the thermal camera views the vehicle from above. Therefore, this method is not limited to a UAV camera only.

When the thermal signature is strong enough to be detected due to sunlight or engine heat transferred to the top of the vehicle, we can confirm previously detected positions, thereby increasing the accuracy of detections. To get such confirmation, we can perform steps one to three as described previously with a different set of parameters: thresholding temperature, the use of structuring elements in morphological operations, and the size range of detected blobs. In addition, if the shapes of detected blobs are corrupted, it is possible to use structural elements as seed points in the watershed algorithm to capture more accurate silhouettes.

3.6   Shape- and Profile-Based

Many algorithms of shape-based vehicle classification employ very different approaches to solving the problem at hand (see Table 3.3). Often, these methods depend on the position of the camera, which implies further algorithm foundations. For instance, using the width/height ratio of a bounding box around the detected vehicle may yield two very different results, depending on whether the camera is positioned above the vehicle rather than on the side. Also, the steps performed in each

Table 3.3 Comparison of selected papers related to shape-based vehicle classification.

View	Methods	Classification	Applications
Static camera, distant, side-top view [22]	Edge detection, SIFT descriptor, constellation model	Sedan, minivan, taxi	Video surveillance
Static camera, distant side view [23]	TRP (transformation-ring-projection), DWT, MFC	Eight classes, including hatchback, motorcycle, and bus	Video surveillance
Moving camera, distant aerial view [21]	Oriented Haar classifiers, hybrid visible-light/thermal cameras used to confirm detections	Vehicle/person	UAV surveillance
Moving camera, rearview [24]	Haar features, AdaBoost	Vehicle/non-vehicle	Vehicle mounted camera
Moving camera, distant aerial view [25]	HOG features, AdaBoost, SVM linear classifier	Vehicles/other objects	Low-attitude UAV surveillance
Moving camera, side view [26]	HOG features, SVM classifier	Sedan, minivan, pickup, none	Vehicle mounted camera
Moving camera, various viewpoints [27]	HOG features, SVM classifier	Vehicle/person	Vehicle mounted camera
Static camera, distant, side-top view [28]	PCA + SVM/eigenvehicles	Car, van, truck	Video surveillance
Static camera, distant, upper view [29]	Segmentation by GMM, combination of silhouette shape and HoG features trained SVM	Car, van, bus, motorcycle	Video surveillance
Static camera, distant, upper view [20]	Multiple silhouette shape features to train SVM	Car, van, bus, motorcycle	Video surveillance
Static camera, distant, upper view [30]	Background subtraction, simple dimension features	Cars, trucks	Video surveillance
Static camera, distant, upper view [31]	17 blob shape features, LDA and k-NN	Seven classes, including sedan, truck, SUV etc.	Video surveillance
Static camera, distant, upper view [32]	Shape features, two-step classification (coarse and fine)	Seven classes, including sedan, van, pickup etc.	Video surveillance
Rearview [33]	Features such as taillights, license plate positions and dimensions and dynamic Bayesian networks	Sedan, pickup, SUV, and others	Video surveillance
Moving camera, distant aerial view [34]	3D models used for matching, HOG-based features classifier	Nine classes, including compact sedan, station wagon, and truck	UAV surveillance
Static camera, side view [35]	Edge-based, FLD	–	Video surveillance
Static camera, upper view [36]	11 blob shapes features, LDA, k-NN	People/vehicles, estimation of color	Video surveillance
Static, omnidirectional camera [37]	Hybrid: Silhouette + k-NN and HOG + SVM	Car, van, motorcycle	Video surveillance

particular method differ significantly based on general assumptions of the algorithm (i.e., whether it is silhouette-based, using three-dimensional (3D) model matching, edge analysis, etc.).

However, we can specify a simplified procedure describing general steps, which are incorporated in such an algorithm. First, we want to select the ROIs containing each vehicle or object visible in a scene. If the camera is stationary, we can use any background subtraction method to mask the moving objects, separating vehicles from the outside scene (road, signs, etc.). Then, we perform the necessary preprocessing on the mask to make it cleaner and easier to decipher. Preprocessing may include various noise removal techniques, morphological filtering, and filling holes. For example, we can use connected component labeling to separate the blobs and filter out ones which are not likely to be vehicle candidates based on certain criteria. Depending on the approach, we can extract multiple blobs with a bounding box around each blob, as ROIs to be used in the next steps. Often, we implement tracking of the blobs in consecutive frames for velocity measurement and other purposes, such as occlusion avoidance. Next, we use one of the approaches specified in following sections to extract features from each ROI. After further preprocessing of the set of features, we occasionally perform dimension reduction using an LDA or PCA method. Finally, we train the specified classifier with a manually labeled data set. Popular classifiers include SVM and k-NN.

3.6.1 Silhouette Measurements

One of the basic approaches that has been used in the field of vehicle classification research is to focus on features extractable from an ROI, including a binary blob of separated object/vehicle. There are various ways to extract blobs containing objects. One method is using an algorithm for background subtraction [38], which, combined with proper binary filtering, gives the mask of approximated blobs containing the vehicles. These methods are covered in greater detail in previous sections of this book. As mentioned in Ref. [37], if the vehicle blob in a particular frame is noisy and does not present enough repeatability in separate frames, we can calculate an average silhouette of a vehicle. To do so, we track a blob of the same vehicle in consecutive frames, add them to each other so that the centroids are overlapping, and divide them by the number of frames used to obtain an average blob. Next, we threshold the intensity of the image to filter out the parts of the blob that did not occur frequently enough. In Ref. [37], such an approach was used to process omnidirectional camera frames, but it can also be used in standard cameras to improve the quality of vehicle blob segmentation.

Occasionally, we try to track blobs in consecutive frames and estimate the velocity of the vehicles during the first phase after obtaining the blobs. A method for doing this uses the constant velocity Kalman filter on the centroids of the blobs [29]. This may be helpful in cases where separate vehicle blobs are very close to the point of merging. In such instances, the predicted centroid position of neighboring merged vehicles in the next frame differs significantly from the ones modeled by the Kalman filter. Thus, the merged blob can be rejected from further analysis. Another approach to detecting this type of occlusion as described in Ref. [32] is based on analysis of the motion field. This can be obtained by performing block matching to search the motion vector for each pixel. To improve both computational efficiency and the motion field approximation accuracy, we can perform it according to a fixed direction because the vehicle's velocity direction is not expected to change. Due to the perspective effect, the motion vectors may differ even for the same object. Objects that are further from the camera seem to be moving more slowly. To reduce this effect, we can use a transformation, multiplying a motion vector by a specified scale factor according to its position on the image so that the perspective effect is compensated. In other words, we can make the motion field of one blob homogenous. A simple analysis of motion fields of all objects can show if two separate objects

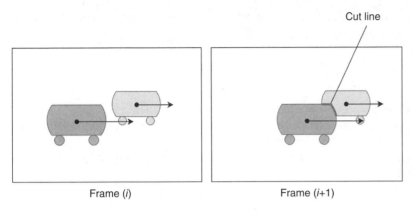

Cut line

Frame (*i*)   Frame (*i+1*)

Figure 3.12 Motion field calculated for two vehicles with different velocities in two consecutive frames, indicating the cut line used later to divide the merged blob.

Figure 3.13 Bounding box, minimal area bounding box (rotated), and fitted ellipse of a blob.

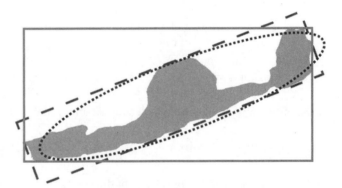

have merged because one blob will have an increased variance of motion vector magnitudes. Once we detect such occlusion, we can divide two distinct objects from one blob according to the different motion fields (see Figure 3.12).

In previous research [20, 30–32, 36, 37], various sets of binary silhouette features have been used as shape descriptors [39], including the following:

- Bounding box: Smallest rectangle with vertical and horizontal sides that can contain the entire blob, from which we can calculate features such as width, height, aspect ratio of the box, and/or its area (see Figure 3.13).
- Minimal area bounding box: Similar to the ordinary bounding box, but rotated to more accurately fit the shape of the blob (can be a decent approximation of object orientation) [40].
- Fitted ellipse: Ellipse that best fits the shape (or its contour) in the least-square sense (see Ref. [41]).
- Simple blob shape features: Including blob area (sometimes also filled blob area), blob perimeter or convex hull area [40], and perimeter.
- Blob centroid position: Center of mass of the blob, which can be used to represent the position of mass in the ROI of the shape,

$$x_{\text{centroid}} = \frac{1}{N}\sum_{i=1}^{N} x_i, y_{\text{centroid}} = \frac{1}{N}\sum_{i=1}^{N} y_i.$$

- Central moments: Image moments calculated about the centroid of such image. Such moments are only translation invariant. Defined by Sonka et al. [40]

$$\mu_{p,q} = \sum_{i=1}^{N}\left(x_i - x_{\text{centroid}}\right)^p \left(y_i - y_{\text{centroid}}\right)^q$$

where N number of pixels in a blob, $(p+q)$ specify order of central moment.
- Hu-invariant moments: Set of moments described in Ref. [42], which are translation, rotation, and scale invariant.
- Compactness: Indicates similarity of a shape to a circle. It is independent of linear transformations [40] and can be calculated using

$$\text{Compactness} = \frac{L^2}{A}, \text{ where } A \text{ is area, } L \text{ is the perimeter of a blob.}$$

- Eccentricity, elongation: Measures of the aspect ratio of a shape that can be calculated by the principal axes method [39] or more simple minimal bounding box approach (smallest rectangle containing every point of a blob, and can be rotated). Eccentricity is calculated as the length ratio of major and minor axes of the shape.

$$\text{Elongation} = 1 - \frac{W}{L}, \text{ where } W \text{ is width and } L \text{ is length of mentioned box.}$$

- Solidity: Ratio of pixels of a blob and its convex hull [39] (see Figure 3.14), defined by

$$\text{Solidity} = \frac{A_S}{A_{\text{CH}}}, \text{ where } A_S \text{ is the area of a blob and } A_{\text{CH}} \text{ is the area of convex hull of this blob}$$

- Convexity: Ratio of the perimeter of the convex hull of a blob to the perimeter of a blob [39]:

$$\text{Convexity} = \frac{L_{\text{CH}}}{L_S}, \text{ where } L_S \text{ is the perimeter of a blob and } L_{\text{CH}} \text{ is the perimeter of convex}$$
hull of this blob.

The problem with using such a high number of features is that upon initial inspection, it is difficult to determine which are truly useful in discriminating between various classes of vehicles. Some of the features might be correlated or even redundant [31]. In addition, it is beneficial to keep the

(a)   (b)   **Figure 3.14** A blob (a) and its convex hull (b).

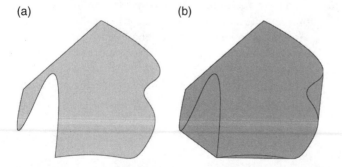

number of features at a reasonable level so that the computational cost is not too high. We can use methods such as LDA to reduce the feature space to a lower dimension, which actually designs a new feature space so that the distance between distinct classes is maximized, emphasizing the discriminative features. The next step of this approach is to teach the classifier with the manually labeled data set. Popular classifiers used in previous works [29, 31, 43] are SVM and k-NN.

As mentioned in Refs. [32] and [26], sometimes we can implement a so-called hierarchical classification method and select obviously discriminative features to perform coarse classification first, prior to the final fine classification (see Figure 3.15). With a certain level of confidence, we can say that features such as blob areas or lengths (where the camera is positioned toward the side of the vehicle) are discriminative enough to separate blobs into large and small vehicle classes. The next step of classification can be focused on finer classification between members of these high-level classes separately, thereby more accurately distinguishing between more similar subclasses (i.e., between a sedan and a small van in a small vehicle group, or between a bus and a truck in a large vehicle group).

Vehicle classification can be further refined by using additional features beyond those based on silhouettes. For example, Ref. [29] extracted intensity-based HOG features for each blob-containing vehicle, and combined them with silhouette-based features to train the SVM. Research proved such combination to be beneficial to overall classification accuracy.

As mentioned in Ref. [33], in cases where the camera is facing the rear of the vehicle, the diversity of blob shapes is much lower than in cases where the camera is aimed toward the side or even top side of the vehicle. This is quite obvious since we can only perceive the almost universal, roughly rectangular shape of the back of a vehicle. To overcome this issue, we can focus on purely simple, low-level features extractable from the rearview of a vehicle. These include approximate bounding box of the vehicle (representing car rear width and height), taillight positions and dimensions, license plate position, and relative distances between them all (see Figure 3.16). Such an approach reduces the need to have high-resolution, close-up images of the rear of the vehicle. Even relatively low-resolution images can suffice to extract the desired features.

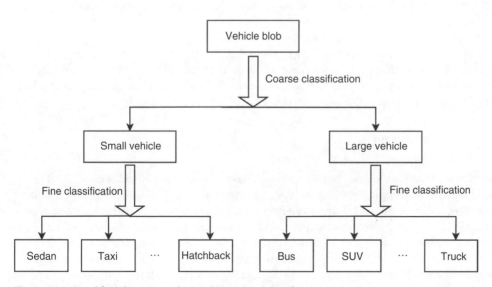

Figure 3.15 Simplified diagram explaining hierarchical classification.

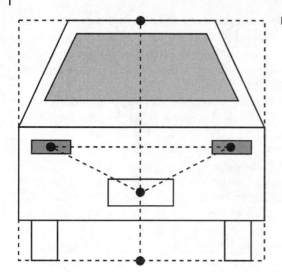

Figure 3.16 Visualization of rear car features.

There are multiple ways to detect a license plate described in previous works [33, 44, 45]. For example, we can detect edges and use them to generate binary blobs, which we can filter using our criteria (i.e., similarity to rectangular shape and size similar to that of a license plate), or use so-called sliding concentric windows to segment plate shapes [44]. To increase the confidence level of our detection, we can perform double-checking with another algorithm and combine both results. Detecting rear lights can be performed by locating regions with a high concentration of red pixels in the vehicle silhouette (prior to grayscale conversion [33]). Unfortunately, such an approach would fail when applied to a red vehicle, so a histogram can be computed to determine if imaging is in the infrared. If so, we can perform extra steps to segment color regions and filter out regions that are too big or too small. Only if these regions meet the criteria, we can then effectively isolate the taillights. After we obtain all of the values describing the rear of a vehicle, we can set it as a feature vector that can be used in further classification, for example, using a hybrid dynamic Bayesian network [33].

Another approach, presented in Ref. [23], employs a method called transformation-ring-projection (TRP) [46] to describe the shape of a blob. This method is both rotation and scale invariant and reduces a 2D pattern to 1D. After performing a normalization of image size (to a minimum bounding square given that its center is a blob centroid), we can use the following formula to compute the TRP value for a specified radius (again, with blob centroid as a center of polar coordinates):

$$f(r) = \int_{0}^{2\pi} P(r\cos\theta, r\sin\theta)d\theta$$

Here, r is the radius for which we calculate integral sum, and P is the normalized image.

The formula is used to compute $f(r)$ for $r \in [0, R_{max}]$, where R_{max} is half of a square-normalized image side. It gives a distribution of shape presence along concentric circles, which can be represented as a simple 1D sequence. In order to improve the robustness and reduce the level of detail we lose during transformation, a variant of TRP called transformation-semi-ring-projection (TSRP; see Figure 3.17) can be computed. TSRP is obtained by modification of the TRP formula so that we compute the integral for semi-rings (range equal to $[0, \pi]$ instead of $[0, 2\pi]$), and the starting angles for it are in eight evenly distributed angular locations (every $\pi/4$). We can decompose eight computed 1D

(a)   (b)

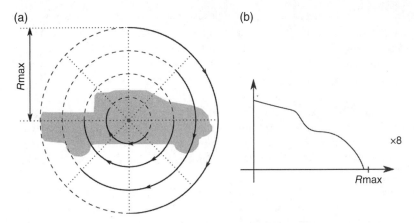

Figure 3.17 Transform-semi-ring-projection used to describe one 2D blob image (a) into eight 1D vectors (b).

Figure 3.18 Border edge of vehicle shape used for scanning to create a sequence of segments.

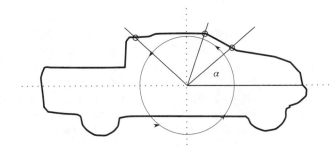

sequences using discrete wavelet transform (DWT) coefficients, which smoothes and decreases the level of noise present in the eight sequences. Next, a step of coarse and fine classification (by fuzzy c-means clustering [47]) is performed to separate the classes.

3.6.2 Edge-Based Classification

Another approach to vehicle classification is based on edge detection. This method relies on the assumption that detected edges themselves may be more immune than other techniques to variables such as lighting and color changes [22]. However, approaches developed in previous works vary significantly in their interpretation of edges and further implementation. Most approaches based on detecting edges use a camera with the field of view facing the sides or top sides of the vehicle. From these angles, the shape of a vehicle varies more between different classes, thereby facilitating the extraction of more significant edge features.

A simple algorithm was described in Ref. [33], which is concerned with describing the external outline of a vehicle. First, we prepare the blobs of moving objects by foreground segmentation and shadow removal. Second, we condition extracted blobs so that their silhouette is not interrupted by extensive noise. Third, we obtain the boundary edges of a blob using morphological or other edge detection. To describe the outer edge of a shape, we can create a pseudo time series by scanning the edge from the center of a shape, which assigns a radius distance to the external border of a blob for each angle (see Figure 3.18).

It gives us a 1D time series, which is the signature of a vehicle shape. To compare the series, we can use a dynamic time warping (DTW) algorithm [48] and determine the distance between them. DTW is a method of measuring the similarity between two temporal sequences, which may vary in speed or timing. It is often used in automatic speech recognition applications [48]. Once we have a way to calculate the similarity between such "shapes," we can use a k-NN classifier to perform actual classification.

However, this method relies heavily on the assumption that the shape and class of a vehicle can be effectively determined solely by the outline edge of a vehicle blob, which may itself be corrupted. Also, none of the vehicle's internal edges (i.e., windshields and side windows) is contributing to the 1D signature of the car. As a result, this method may not be sufficiently robust for certain applications.

To overcome this issue, we can analyze all the edges present on a ROI belonging to a single vehicle (see Figure 3.19). There are methods based on Hausdorff distance [49] or Chamfer matching [50], which are examples of global edge map matching methods. However, as stated in Refs. [22] and [35], because of variations between individual edge pixels within the same classes, such an approach is not robust enough to be used to accurately compare vehicles.

In research conducted by Ma et al. [22], we assume that even though vehicles of the same class have very strong similarities of edge masks (some of the points are repeatable), there are still certain variations in such edge structures which, as stated before, makes it inefficient to use global edge maps to distinguish them. The solution is to group similar edge points together. To determine the similarity between edge points, a scale-independent feature transform (SIFT) method with a few modifications can be used. Namely, edge points can be used as anchor points to create SIFT descriptors in a corresponding grayscale image, but in order to make the method most applicable for vehicle classification, we do not differentiate between orientations which have a difference equal to 180°. In other words, instead of the default eight bins for a histogram of orientations, we only use four, which are

Figure 3.19 Vehicles and detected edges.

unpolarized. This is meant to increase detection efficacy against unexpected changes in lighting and contrast. Another modification is to use χ^2 distance instead of Euclidean distance, because when comparing two values for the same bin, the former distance makes the difference relative to the values of this bin, thereby giving us a better way to compare two distributions. Edge points which are close to each other and also have similar SIFT descriptors can be grouped to create segments. Such segments are advantageous, since their positions are more repeatable within classes than just the edge points considered separately. The next step is to combine all of the segments' information (edge point coordinates, their descriptors, the average descriptor as a feature vector) into a group of features describing one vehicle. Alternatively, other features such as difference-of-Gaussian (DoG) or saliency [22, 51, 52] features can be computed instead of SIFT. Then, we can use a constellation model [22] to compile the classes for further classification. Such an approach is a model that consists of a collection of vehicle parts in which appearance and positions can be flexible.

Additional research by Shan et al. [35] attempts to solve the issues of edge map comparison in a different way. First, we try to perform a transformation on both edge maps to project them onto a coordinate system with minimized shape differences. During initial calculations, we can use Chamfer distance to approximate distance. We want to align the edge maps through analysis of edge map centroid translation and dominant gradient orientation to determine rotation parameters. Given the approximate transformation, we perform a coarse search to fine-tune the parameters of it. Next, a step involving an iterative closest point (ICP)-based refinement [35] is carried out. Afterward, a 6D measurement vector is obtained for every pair of edge maps based on distance and angular and magnitude difference. To obtain a more discriminative classifier, we employ Fisher linear discriminant and Gibbs sampling.

3.6.3 Histogram of Oriented Gradients

Generally, HOG [53] lets us extract features that describe local intensity gradients and/or edge directions, which represent the object appearance in the image. As stated in Ref. [26], the HOG approach is not exactly the best method to classify cars between various classes, but rather to determine the presence of a vehicle (or other object) in an ROI. However, it may be applied to classify objects in groups such as vehicles and people. Normally in HOG-based detection of cars, we could use sliding windows moving and of gradually increasing size for the whole frame to calculate the features [37]. However, it is not usually computationally feasible in real-time applications such as traffic surveillance (without additional features and sliding window method improvements [25]). Therefore, another method must be employed. As an example, we can use a bounding box of a shape blob easily detectable by background subtraction. A key element is the chosen approximation of a bounding box of a vehicle. This is crucial to make further image-based measurements effective and accurate. Generally, searching on the image for potential regions with objects we want to classify (performed to decrease the computational cost when compared with sliding window approach for all image) is called focus of attention's mechanism [27]. There are various methods to generate focus of attention, and these may fit into one of the following categories: knowledge-based, motion-based, and stereo-based. To the first group belong methods generally deriving from our knowledge of the wanted object (color, shape, vertical symmetry, etc.), while the second group often uses, for example, optical flow. A stereo-based category can employ methods such as inverse perspective mapping [54, 55], which is used to approximate locations of interesting locations such as vehicles and obstacles.

Once we have the approximate bounding box of an object (potentially vehicle), we can use the HOG method to extract features describing the object inside it (see Figure 3.20). First, we can use the scale-space approach described in Ref. [15] to obtain a gradient of the image. Scale space represents a set of

Figure 3.20 HOG features extracted from a gray-scale image of a vehicle.

pictures at various resolutions and processes them with Lindeberg's [56] method to create a gradient map, which is then used to handle structures (i.e., vehicles) visible in the image at different scales. To calculate the features from the gradient image, we divide the bounding box into $M \times N$ subregions, and for each subregion we quantize the directions of gradients in K bins corresponding to $[0°, 180°]$ range. After further postprocessing of histogram arrays, we can concentrate them to create one feature vector of the object inside of the bounding box. As a classifier, we can use SVM, but, as stated earlier, the overall accuracy of the classification between various vehicle classes is poorer than in using methods such as edge or shape features. Thus, the HOG-based approach can be used to coarsely classify between completely different categories such as vehicle, person, or other object, rather than between objects in similar classes, such as car and small van. Research conducted in Refs. [29, 43] attempted to overcome this issue by employing a method combining two approaches: the first approach using vehicle silhouette features (as described in previous sections) and k-NN classifier, and the second method using HOG with SVM, which proved to be superior to separate ones in their experiments.

When we cannot clearly state what view of a vehicle we are expecting to encounter, (i.e., when the camera is not facing the side of a road, but rather is mounted on a vehicle), we can try to improve the accuracy of our HOG-based vehicle classifier by creating a few classifiers instead of a general one [27]. Each of these classifiers corresponds with a different viewpoint of a vehicle (i.e., front, rear, front-right side, and front-left side) and is taught separately with the selected training images from a proper view category (but with the same negative training examples). Another modification by Han et al. [27] is to extend HOG description by assigning spatial localization inside subregions during its calculation, thereby better representing structured objects such as vehicles. To achieve that, in addition to accumulating orientations inside orientation bins, for each angle bin we also bin distance of the pixel to the center containing its subregion. Other work by Cao et al. [25] proposed a different improvement of the HOG-based method, which was achieved by using discrete AdaBoost algorithm to decrease the dimensionality of feature vectors, and to speed up detection.

3.6.4 Haar Features

Haar-like features (see Figure 3.21) are used in cascaded Haar classifiers, which were first described in Ref. [58]. There were many papers introducing ways of using such classifiers to perform facial detection, which was later generalized to detect different objects such as vehicles [21, 24, 57, 59].

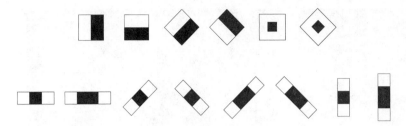

Figure 3.21 Haar-like features used in vehicle detection: edge features, center features, and line features. *Source*: Bai et al. [57]. Reproduced with permission of IEEE.

Cascaded classifiers use a moving search window for multiple positions and scales to detect possible locations of vehicles in an image. The process is constructed so that the most discriminative features corresponding to individual weak classifiers are calculated first (later, further cascades of features). Then, if the current cascade of these classifiers does not qualify the sample image window as a vehicle, the remaining features do not need to be calculated. Such an approach drastically reduces the computational cost, which would be extensively high if we attempted to verify all of possible combinations of type, location, and scale of the Haar features in the image. Examples of Haar-like features used for vehicle classification [21, 57] are presented in Figure 3.221. Another potential improvement is if the camera is mounted on a vehicle or a static platform, and we could approximate angle of the camera and how the road is presented on the image. Namely, we can narrow down the range of searching window sizes and positions depending on the part of the image where we search for vehicles. An approach called "adaptive sliding window" [59] searches via larger search windows in close proximity to the camera, and smaller ones when the part of the image corresponding to further distances is analyzed.

In Ref. [24], the approach is slightly different because we do not use Haar features to search for candidates in the whole image, but rather to determine whether the given individual image sample contains a vehicle or a nonvehicle. In such an approach, a method for candidate locations generations has to be performed, for instance, by using a symmetry evaluation approach (given that the camera is facing rears of the cars).

As mentioned in Ref. [21], when acquiring the images from the air (see Figure 3.22), we cannot expect the vehicles to be rotated and positioned in one fixed orientation. To improve detection accuracy, we can use an independent set of cascade Haar classifiers and train them separately for different angular positions. If a higher number of classifiers confirm the presence of the vehicle in a particular location in the image, we can be quite confident in such detection.

3.6.5 Principal Component Analysis

PCA is a statistical method that uses an orthogonal transformation to project a set of data (observations described with possibly correlated variables) into a new space so that the axes of new space are the computed principal components (which are linearly uncorrelated) and, therefore, represent more important information about observations. The principal components are defined in such a way that the first one has the greatest variance in the data set and the next ones are orthogonal to the previous ones and sequentially represent the largest variance remaining in the data. Apart from better representation and visualization of data, we can also compress data by rejecting the least-contributing components from the new PCA space. The principal components can be calculated as eigenvectors of covariance matrix computed from given data [19]. We can divide PCA-based vehicle classification

Figure 3.22 Image with pixelated car manifesting distinct Haar features.

approaches into three main categories [28, 60] depending on what kind of purpose the PCA algorithm is used for during the description and classification process:

1) PCA used as a classifier on normalized vehicle images—eigenvehicles approach [28]
2) PCA used as a classifier on features (i.e., Haar wavelet features) extracted from images [60]
3) PCA used as a method to describe vehicle image with a low-dimensional feature vector before further classification (i.e., SVM) [28]

In every approach, we assume that we have a rectangular bounding box around the vehicle. Bounding boxes can be created manually on static frames (for training purposes), or we can use methods of background subtraction to extract moving objects and their bounding boxes. In any case, it is advised to perform preprocessing on all vehicle images. First, if needed, we convert the vehicle image to grayscale. Second, we rescale every image so that they have the same width and height. However, if the camera does not face a particular side of a vehicle, we can employ a method of orientation and scaling so that each vehicle sample is uniform and oriented in the same direction before scaling [28]. By using arbitrarily set lane orientation, analysis of the vehicle blob's elongation, and tracking its motion direction, we can determine the angle to rotate the image. Third, we try to reduce the impact of lighting changes, which can be done by histogram equalization, normalization, or using z-scores [19, 61]. For example, we can use the following transformation for each pixel in every image: $p_z(i,j) = \left[p(i,j) - \mu(i,j) \right] / \left[\sigma(i,j) \right]$, where μ is average image of all vehicles (see Figure 3.23) and $\sigma(i,j)$ is the standard deviation of (i,j) pixel intensity in every image (because we consider it as a variable, or a feature which is describing each vehicle in multidimensional space, in which every pixel is 1D). An additional step that we can perform depends on whether we have access to accurate blob data of the vehicle. We can set the pixel value of rest of the bounding box as 0, so that it does not contribute to PCA calculations. Typically, we can perform hierarchical classification and, as a first step, only simple features such as length and area to roughly divide vehicles into more specific groups. We only employ PCA-based methods to classify vehicles in groups in a finer manner after rough classifications have been made [28].

The first approach derives from a method of face recognition using so-called eigenfaces [28, 62]. Similarly, eigenvehicles are eigenvectors of a set of sample training vehicles, meaning that any of the initial training vehicles can be projected onto a new PCA space (in other words, built with the proper combination of principal components) and later reconstructed using these vectors. Eigenvehicles (see Figure 3.24) represent invariant characteristics of vehicle silhouettes that can be found in various

Figure 3.23 Calculated mean of normalized images.

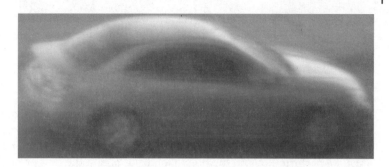

(a)

(b)

Figure 3.24 (a) Examples of vehicle pictures used to calculate eigenvehicles of sedan class, (b) first eight principal components—eigenvehicles of sedan class.

vehicles. Because primary components contribute most to representing variation in all training images, we can pick a few of the first eigenvehicles as a new base of lower dimension space in which we represent samples. However, if a new vehicle sample is to be projected to new space (i.e., built with limited number of principal components), we will surely encounter a certain level of error in case of back projection to primary space, which can be properly interpreted as a sign of belonging or not belonging to a class for which given eigenvehicles were computed.

The algorithm is as follows: for each class, we gather the samples of vehicles. We obtain normalized images (bounding boxes) by the same method we used while obtaining bounding boxes in PCA as a classifier approach. We represent each image of size $m \times n$ as a vector of size $1 \times mn$ (which can be easily restored to an image anytime). Then, we store (for each class separately) all k sample images (as $1 \times mn$ vectors) in a matrix of size $k \times mn$, on which we perform PCA, which gives us eigenvehicles of a particular class. If we want to classify a new sample vehicle, we perform the same normalization steps on its image, projecting it to PCA-based space for each class, and then calculating Euclidean distances between the new sample and the training set (for every class separately in a proper space). For example, if the mean distance is small enough, we can then assume that the new sample belongs to that class. Another way to determine what class a vehicle belongs to is to inspect the truncation error we encounter when we project a new sample to a dimensionally reduced class subspace and then project it back. The class for which the truncation error is the smallest is chosen as a class of new sample [60]. However, in a paper by Wu and Zhang [60], instead of using whole vehicle images as input data for PCA (obtaining eigenvehicles), they used a method employing prior extraction of Haar features. Then, we use such features to compute PCA and obtain eigenvectors rather than eigenvehicles. Apart from that, the rest of approach is very similar.

Another approach using PCA is based on the use of different segmented images of vehicles [28]. First, we interpret the vehicle image (without the background) as a point cloud of 3D space. Each point consists of three values: the x and y coordinates, and the intensity of the corresponding pixel. Then, we perform PCA on the point cloud and obtain the first principal component (first eigenvector), which describes the distribution of points best. We follow this approach for each class and every training image separately, and then we train one class SVM classifier [28] using principal components as feature vectors. It gives us a set of one-class SVM classifiers for every class of vehicles we want to classify, which we use to determine to which one a new sample belongs.

3.7 Intrinsic Proportion Model

The automatic vehicle classification system described here is a pattern recognition model that relies on the proportion of the bounding box of a vehicle. To classify vehicle sizes independently from view angles, we can use an Analogia graph [63] to represent the relationship between vehicles and the road. An Analogia (Greek, "proportion") graph is an abstraction of a proportion-preserving mapping of a shape. Assume a connected nonrigid graph G, has an edge with a length u. The rest of the edges in G can be normalized as $p_i = v_i/u$.

We first create an Analogia graph of objects. In our case, we consider classifying vehicle types such as trucks and cars on the known width of a lane. Given n set of observations of vehicle objects, we have an m-dimensional feature vector for each vehicle: $V_i = (v_{i1}, v_{i2}, ..., v_{im})$, which are normalized by a common scale factor u, for example, the highway lane width. We have the proportional features,

$$V_i' = \frac{V_i}{\mu} = \left(\frac{V_{i,1}}{\mu}, \frac{V_{i,2}}{\mu}, ..., \frac{V_{i,m}}{\mu} \right)^T$$

The vehicle classification is to classify the n observations into k vehicle types. This can be formulated as an unsupervised machine-learning problem, based on the proportional features. The objective function is to minimize the inner-class distances to the mean of the class,

$$\min \sum_{i=1}^{k} \sum_{j=1}^{n} \left\| V_j' - \bar{V_i'} \right\|$$

where V_i' is the average of the features in k classes.

The algorithm starts with randomly assigning k samples to be centroids. For each remaining sample, another sample is assigned to the cluster to which it is the most similar, based on the distance between the object and the cluster mean. Next, the new mean is computed for each cluster. This process iterates until the criterion function converges.

Figure 3.25 depicts screenshots for the classification of trucks (large boxes) and cars (small boxes) of the live videos from three cameras in different angles. Using adaptive proportionality, we

Figure 3.25 Screenshots for the classification of trucks (large boxes) and cars (small boxes) of the live videos from three cameras in different angles.

Table 3.4 Vehicle classification test results.

	Camera 1 (%)	Camera 2 (%)	Camera 3 (%)
Positively identified	95	94	93
False positives	4	3	6
False negatives	2	3	1
Mean-squared errors	8	6	9

can scale the truck to car ratio according to the location and angle of the camera based on the metadata. Therefore, the proportional relationship stays the same even if we move the camera or use it in a different location. Our preliminary results are very encouraging. Tested with 10,000 frames from PennDOT's live videos from three cameras, we have results in Table 3.4. Figure 3.25 shows screenshots for the classification of trucks and cars of the live videos from three cameras in different angles.

3.8 3D Model-Based Classification

There are several similar prior studies on vehicle classification using dimensional features. For example, the method in Ref. [64] projects 3D vehicle shapes into 2D line contours and then makes dimensional measurements in absolute values. The results are dependent on the distance to the vehicles and the angle of the camera. This does not appear to be very flexible when operating in a large urban traffic camera network, where pan, tilt, and zoom cameras are common. The authors' solution does not have such limitations. The 3D modeling approach enables detailed vehicle feature descriptions, but it needs photorealistic 3D modeling and training samples. It also needs very accurate background segmentation, which is computationally expensive. Table 3.4 shows that the vehicle classification results are quite stable despite different camera angles, distances to vehicles, and locations.

3.9 SIFT-Based Classification

SIFT [51] is a DoG gradient-based key-point feature detection algorithm that requires training samples. The advantage of using this method is that it is scale and rotation invariant. However, it relies on the image resolution and the similarities to the training samples. The team [65] trained a SIFT model with 1000 frames and tested it with different sample frames that came from the same camera. With 100 frames for the main lane only, we correctly identified three trucks and 19 cars in the video. However, the algorithm falsely identified one car as a truck on the exit ramp and combined two cars side by side and identified them as a truck. We found that the SIFT feature provided no separation between cars and trucks based on feature count compared to training data. From our results, we found that the Intrinsic Proportion Model is the fastest and most robust method for PTZ cameras. SIFT needs training samples in similar imaging conditions. It can work with still images, however, it is rather slow and less robust.

3.10 Summary

The technologies for automated vehicle classification have been evolving over decades. It began with hardware solutions such as light curtains and embedded loops merely for single lane and fixed location systems. Those systems are relatively laborious in terms of installation and operation. With rapidly growing affordable sensors such as CCTV cameras, LiDAR, and even thermal imaging devices, we are able to detect, track, and categorize vehicles in multiple lanes simultaneously. Meanwhile, we are also able to extract useful information such as license plate characteristics, color, estimated vehicle velocity, and much more.

The vehicle classification solutions introduced in the chapter each have their advantages and disadvantages. Thermal and near-infrared LiDAR imaging works well at night. However, they might encounter noisy signals in the daytime or high failure rate in extreme weather conditions. Conversely, RGB cameras (or visible cameras) work better in broad daylight, but perform poorly at night. LiDAR imaging creates a 3D model of the vehicles; devices can be installed on cars and unmanned aviation vehicles; it by far holds the most potential as a solution for automated vehicle classification. However, LiDAR needs to solve the field of view problem by either rotating its sensory orientation, or merging the views from multiple sensors.

The shape-based vehicle classification methods, such as HOG, Haar-like features require substantial training images and carefully designed preprocessing algorithms, such as deformable model, sample light and size normalization, and sample rotation resampling. The challenge emerges when multiple vehicles are clustered with other in the view of the camera. The silhouette-based methods such as vehicle contours are one of the simpler classification methods. However, they need relatively clean blobs to start with. For stationary CCTV cameras, this kind of approach is feasible using background segmentation methods. Simple methods such as proportion-based modeling were successful methods in our highway case studies. However, the amount and variety of vehicles it can handle is very limited due to lack of proportion specifications for every vehicle type.

In many cases, we can also utilize a hierarchical classification, which allows us to divide the classification process into a coarse and a fine stage. The former involves a rough filtering based on silhouette measurement, while the latter stage is responsible for further distinction between objects in more similar subclasses.

The accuracy of vehicle classification can be improved by combining multiple sensors such as thermal imaging, LiDAR imaging, and RGB visible cameras. When possible, we can even incorporate acoustic sound signals to detect unique vehicle signatures, such as distinctive sounds made by motorcycles and trucks.

References

1 Model 110 Automatic Vehicle Classification System, http://tds-its.com/pdf/avc_model_110.pdf (accessed September 12, 2016).
2 MigmaWalktime™, http://www.migmapd.com/migmawalktime.htm (accessed July 30, 2013).
3 Automatic Vehicle Classification, Tracking and Speed Estimation, https://www.youtube.com/watch?v=MvPY1CBapE4 (accessed July 1, 2013).
4 Kapsch Product Description, https://www.kapsch.net/ktc/Portfolio/products/video-sensor/classification-products (accessed November 17, 2015).
5 H. Lee and B. Coifman. Side-fire LiDAR-based vehicle classification. *Transportation Research Record: Journal of the Transportation Research Board*, 2012, 2308: 173–183.

6 T. Lovas, C.K. Toth, and A. Barsi. Model-based vehicle detection from LiDAR data. Department of Photogrammetry and Geoinformatics, Budapest University of Technology and Economics (2004).

7 R. Yang. *Vehicle Detection and Classification from a LiDAR Equipped Probe Vehicle.* Dissertation, The Ohio State University, 2009.

8 F. Zhang, D. Clarke, and A. Knoll. Vehicle detection based on LiDAR and camera fusion. *Proceedings of the 2014 IEEE 17th International Conference on Intelligent Transportation Systems (ITSC),* Qingdao, October 8–11, 2014. IEEE, New York City, NY.

9 L. Huang and M. Barth. Tightly-coupled LiDAR and computer vision integration for vehicle detection. *Proceedings of the 2009 IEEE Intelligent Vehicles Symposium,* Xi'an, June 3–5, 2009. IEEE, New York City, NY.

10 C. Premebida, et al. A LiDAR and vision-based approach for pedestrian and vehicle detection and tracking. *Proceedings of the IEEE International Conference on Intelligent Transportation Systems, (ITSC 2007),* Seattle, WA, September 30 to October 3, 2007. IEEE, New York City, NY.

11 B. Issa, et al. Vehicle Location by Thermal Image Features http://cs426team11.github.io/vltif/ (accessed December 9, 2015).

12 SDMS, DAAB05A24D, CD-R with thermal UAV images: https://www.sdms.afrl.af.mil

13 A. Sangnoree. Vehicular separation by thermal features' relative angle for nighttime traffic. *Proceedings of the 6th International Conference on New Trends in Information Science and Service Science and Data Mining (ISSDM),* Taipei, Taiwan, October 23–25, 2012. IEEE, New York City, NY.

14 Y. Iwasaki, M. Misumi, and T. Nakamiya. Robust vehicle detection under various environments to realize road traffic flow surveillance using an infrared thermal camera. *The Scientific World Journal,* 2015, 2015: 947272.

15 M.W. Koch and K.T. Malone. A sequential vehicle classifier for infrared video using multinomial pattern matching. *IEEE Conference on Computer Vision and Pattern Recognition Workshop, 2006 (CVPRW'06),* Portland, June 23–28, 2006. IEEE, New York City, NY.

16 K.M. Simonson. Multinomial pattern matching: A robust algorithm for target identification. *Proceedings of Automatic Target Recognizer Working Group (ATRGW),* Huntsville, AL, October 12–14, 1997.

17 A. Wald. *Sequential Analysis.* New York City: Courier Corporation, 1973.

18 R.O. Duda, P.E. Hart, and D.G. Stork. *Pattern Classification.* New York: John Wiley & Sons, Inc., 2012.

19 H. Abdi and L.J. Williams. Principal component analysis. *Wiley Interdisciplinary Reviews: Computational Statistics,* 2010, 2(4): 433–459.

20 Z. Chen, T. Ellis, and S. Velastin. Vehicle type categorization: A comparison of classification schemes. *Proceedings of the 14th International IEEE Conference on Intelligent Transportation Systems (ITSC), 2011,* Washington, DC, October 5–7, 2011. IEEE, New York City, NY.

21 A. Gaszczak, T.P. Breckon, and J. Han. Real-time people and vehicle detection from UAV imagery. *Proceedings of IS&T/SPIE Electronic Imaging,* San Francisco, CA, USA, January 23–27, 2011. International Society for Optics and Photonics.

22 X. Ma, W. Eric, and L. Grimson. Edge-based rich representation for vehicle classification. *Proceedings of the Tenth IEEE International Conference on Computer Vision, 2005 (ICCV 2005),* Beijing, China, October 17–21, 2005. Vol. 2. IEEE, New York City, NY.

23 D. Zhang, S. Qu, and Z. Liu. Robust classification of vehicle based on fusion of TSRP and wavelet fractal signature. *IEEE International Conference on Networking, Sensing and Control, 2008 (ICNSC 2008),* Sanya, April 6–8, 2008. IEEE, New York City, NY.

24 G.Y. Song, K.Y. Lee, and J.W. Lee. Vehicle detection by edge-based candidate generation and appearance-based classification. *IEEE Intelligent Vehicles Symposium,* Eindhoven, the Netherlands, June 3–6, 2008. IEEE, New York City, NY.

25 X. Cao, et al. Linear SVM classification using boosting HOG features for vehicle detection in low-altitude airborne videos. *Proceedings of the 18th IEEE International Conference on Image Processing (ICIP 2011)*, Brussels, Belgium, September 11–14, 2011. IEEE, New York City, NY.

26 T. Gandhi and M.M. Trivedi. Video based surround vehicle detection, classification and logging from moving platforms: Issues and approaches. *IEEE Intelligent Vehicles Symposium, 2007*, Istanbul, Turkey, June 13–15, 2007. IEEE, New York City, NY.

27 F. Han, et al. A two-stage approach to people and vehicle detection with HOG-based SVM. *Performance Metrics for Intelligent Systems 2006 Workshop*, Gaithersburg, MD, USA, August 21–23, 2006.

28 C. Zhang, X. Chen, and W.-B. Chen. A PCA-based vehicle classification framework. *Proceedings of the 22nd International Conference on Data Engineering Workshops*, Atlanta, GA, USA, April 3–8, 2006. IEEE, New York City, NY.

29 Z. Chen, T. Ellis, and S. Velastin. Vehicle detection, tracking and classification in urban traffic. *Proceedings of the 15th International IEEE Conference on Intelligent Transportation Systems (ITSC)*, Alaska, USA, September 16–19, 2012. IEEE, New York City, NY.

30 S. Gupte, et al. Detection and classification of vehicles. *IEEE Transactions on Intelligent Transportation Systems*, 2002, 3(1): 37–47.

31 B. Morris and M. Trivedi. Improved vehicle classification in long traffic video by cooperating tracker and classifier modules. *Proceedings of the IEEE International Conference on Video and Signal Based Surveillance, 2006 (AVSS'06)*, Sydney, Australia, November 22–24, 2006. IEEE, New York City, NY.

32 C.-L. Huang and W.-C. Liao. A vision-based vehicle identification system. *Proceedings of the IEEE 17th International Conference on Pattern Recognition*, Cambridge, UK, August 23–26, 2004. Vol. 4, IEEE, New York City, NY, pp. 364–367.

33 M. Kafai and B. Bhanu. Dynamic Bayesian networks for vehicle classification in video. *IEEE Transactions on Industrial Informatics*, 2012, 8(1): 100–109.

34 S.M. Khan, et al. 3D model based vehicle classification in aerial imagery. *Proceedings of the IEEE Conference on Computer Vision and Pattern Recognition (CVPR)*, June 13–18, 2010. IEEE.

35 Y. Shan, H.S. Sawhney, and R. Kumar. Unsupervised learning of discriminative edge measures for vehicle matching between nonoverlapping cameras. *IEEE Transactions on Pattern Analysis and Machine Intelligence*, 2008, 30(4): 700–711.

36 O. Hasegawa and T. Kanade. Type classification, color estimation, and specific target detection of moving targets on public streets. *Machine Vision and Applications*, 2005, 16(2): 116–121.

37 H.C. Karaimer, I. Cinaroglu, and Y. Bastanlar. Combining shape-based and gradient-based classifiers for vehicle classification. *Proceedings of the 2015 IEEE 18th International Conference on Intelligent Transportation Systems (ITSC)*, Las Palmas, Spain, September 15–18, 2015. IEEE, New York City, NY.

38 S. Sen-Ching, S. Cheung, and C. Kamath. Robust techniques for background subtraction in urban traffic video. *Proceedings of the Video Communications and Image Processing, SPIE Electronic Imaging*, San Jose, CA, January 20–22, 2004. International Society for Optics and Photonics.

39 M. Yang, K. Kpalma, and J. Ronsin. A survey of shape feature extraction techniques. In P.-Y. Yin (ed.) *Pattern Recognition Techniques, Technology and Applications*. Vienna: I-Tech, 2008, 43–90.

40 M. Sonka, V. Hlavac, and R. Boyle. *Image Processing, Analysis, and Machine Vision*. Stanford: Cengage Learning, 2015.

41 A.W. Fitzgibbon and R.B. Fisher. A buyer's guide to conic fitting. *DAI Research paper*, University of Edinburgh, Edinburgh, UK, 1996.

42 M.-K. Hu. Visual pattern recognition by moment invariants. *IRE Transactions on Information Theory*, 1962, 8(2): 179–187.

43 Z. Chen, et al. Road vehicle classification using support vector machines. *Proceedings of the IEEE International Conference on Intelligent Computing and Intelligent Systems, 2009 (ICIS 2009)*, Shanghai, China, November 20–22 2009. Vol. 4. IEEE, New York City, NY.

44 C. Anagnostopoulos, et al. A license plate-recognition algorithm for intelligent transportation system applications. *IEEE Transactions on Intelligent Transportation Systems*, 2006, 7(3): 377–392.

45 H.-J. Lee, S.-Y. Chen, and S.-Z. Wang. Extraction and recognition of license plates of motorcycles and vehicles on highways. *Proceedings of the 17th International Conference on Pattern Recognition, 2004 (ICPR 2004)*, Cambridge, UK, August 23–26, 2004. Vol. 4. IEEE, New York City, NY.

46 Y.-Y. Tang, et al. Ring-projection-wavelet-fractal signatures: A novel approach to feature extraction. *IEEE Transactions on Circuits and Systems II: Analog and Digital Signal Processing*, 1998, 45(8): 1130–1134.

47 D.P. Mukherjee, P. Pal, and J. Das. Sodar image segmentation by fuzzy c-means. *Signal Processing*, 1996, 54(3): 295–301.

48 H. Sakoe and S. Chiba. Dynamic programming algorithm optimization for spoken word recognition. *IEEE Transactions on Acoustics, Speech and Signal Processing*, 1978, 26(1): 43–49.

49 D.P. Huttenlocher, G. Klanderman, and W.J. Rucklidge. Comparing images using the Hausdorff distance. *IEEE Transactions on Pattern Analysis and Machine Intelligence*, 1993, 15(9): 850–863.

50 A. Thayananthan, et al. Shape context and chamfer matching in cluttered scenes. *Proceedings of the 2003 IEEE Computer Society Conference on Computer Vision and Pattern Recognition*, Madison, WI, USA, June 16–22, 2003. Vol. 1. IEEE, New York City, NY.

51 D.G. Lowe. Distinctive image features from scale-invariant keypoints. *International Journal of Computer Vision*, 2004, 60(2): 91–110.

52 R. Fergus, P. Perona, and A. Zisserman. Object class recognition by unsupervised scale-invariant learning. *Proceedings of the 2003 IEEE Computer Society Conference on Computer Vision and Pattern Recognition*, Madison, WI, USA, June 16–22, 2003. Vol. 2. IEEE, New York City, NY.

53 D. Dalal and B. Triggs. Histograms of oriented gradients for human detection. *IEEE Computer Society Conference on Computer Vision and Pattern Recognition, 2005. CVPR 2005*, San Diego, CA, June 20–26, 2005. Vol. 1. IEEE, New York City, NY.

54 S. Tan, et al. Inverse perspective mapping and optic flow: A calibration method and a quantitative analysis. *Image and Vision Computing*, 2006, 24(2): 153–165.

55 H.A. Mallot, et al. Inverse perspective mapping simplifies optical flow computation and obstacle detection. *Biological Cybernetics*, 1991, 64(3): 177–185.

56 T. Lindeberg. Scale-space theory: A framework for handling image structures at multiple scales. *Proceedings of the CERN School of Computing*, Egmond aan Zee, September 8–21, 1996.

57 H.-L. Bai, T. Wu, and C. Liu. Motion and Haar-like features based vehicle detection. *Proceedings of the 12th International Conference on Multi-Media Modelling*, Beijing, China, January 4–6, 2006. IEEE, New York City, NY.

58 P. Viola and M. Jones. Rapid object detection using a boosted cascade of simple features. *Proceedings of the 2001 IEEE Computer Society Conference on Computer Vision and Pattern Recognition, 2001 (CVPR 2001)*, Kauai, HI, December 8–14, 2001. Vol. 1. IEEE, New York City, NY.

59 A. Haselhoff and A. Kummert. A vehicle detection system based on Haar and triangle features. *IEEE Intelligent Vehicles Symposium*, 2009. IEEE.

60 J.-W. Wu and X. Zhang. A PCA classifier and its application in vehicle detection. *Proceedings of the International Joint Conference on Neural Networks (IJCNN'01)*, Washington, DC, USA, July 15–19, 2001. Vol. 1. IEEE, New York City, NY.

61 E. Kreyszig. *Advanced Engineering Mathematics*. New York: John Wiley & Sons, Inc., 1988.

62 M. Turk and A.P. Pentland. Face recognition using eigenfaces. *Proceedings of the IEEE Computer Society Conference on Computer Vision and Pattern Recognition (CVPR'91)*, Maui, Hawaii, June 3–6, 1991. IEEE, New York City, NY.

63 J. Laws, N. Bauernfeind, and Y. Cai. Feature hiding in 3D human body scans. *Information Visualization*, 2006, 5(4): 271–278.

64 A.H.S. Lai, G. Fung, and N. Yung. Vehicle type classification from visual-based dimension estimation. *Proceedings of the 2001 IEEE Intelligent Transportation Systems*, Oakland, CA, August 25–29, 2001. IEEE, New York City, NY.

65 Y. Cai, et al. Adaptive feature annotation for large video sensor networks. *Journal of Electronic Imaging*, 2013, 22(4): 041110.

4

Detection of Passenger Compartment Violations

Orhan Bulan[1], Beilei Xu[2], Robert P. Loce[2] and Peter Paul[2]

[1] *General Motors Technical Center, Warren, MI, USA*
[2] *Conduent Labs, Webster, NY, USA*

4.1 Introduction

A moving violation is any violation of the law committed by the driver of a vehicle while it is in motion. Moving violations can typically be identified by observing the behavior of a vehicle such as its speed or running a stop sign or red traffic light. However, several moving violations require observation into the passenger compartment of a vehicle. Failure to wear seat belts and operating a handheld telecommunications device (i.e., cell phone) while driving are two common passenger compartment violations concerning safety. Another type of passenger compartment violation is related to efficient use of roadways. Managed lanes such as high occupancy vehicle (HOV) and high occupancy tolling (HOT) lanes require a minimum number of occupants within the vehicle or a tolling price that varies depending upon the number of occupants. Imaging technology and computer vision can provide automated or semiautomated enforcement of these violations.

Seat belt use in motor vehicles is the single most effective traffic safety device for preventing death and injury to persons involved in motor vehicle accidents. Unrestrained drivers and passengers of motor vehicles involved in accidents frequently suffer major injury due to their partial or complete ejection from the vehicle as well as general displacement and impact within the vehicle. Generally, the severity of these injuries is easily lessened, if not prevented, by utilizing a seat belt. According to the National Highway Traffic Safety Administration (NHTSA), wearing a seat belt can reduce the risk of crash injuries by 50%. Statistics shows that seat belts save lives [1].

- In total, 14,154 occupants were saved by seat belts in 2002.
- A total of 260 additional lives would have been saved if (front seat daytime) belt use had been one point higher (76%).

Evidence shows that strong enforcement is the major driving force behind the upward trend of compliance of seat belt use. For example, the study in Ref. [2] found that seat belt use continued to be higher in the United States where vehicles can be pulled over solely for occupants' nonuse of seat belts ("primary law states") as compared to states having weaker enforcement laws ("secondary law states") or without seat belt laws. A recent report indicated that seat belt violations are primary enforcement in 33 states in the United States [2]. With more than 10 years of the Click-It-Or-Ticket campaign

Computer Vision and Imaging in Intelligent Transportation Systems, First Edition.
Edited by Robert P. Loce, Raja Bala and Mohan Trivedi.
© 2017 John Wiley & Sons Ltd. Published 2017 by John Wiley & Sons Ltd.
Companion website: www.wiley.com/go/loce/ComputerVisionandImaginginITS

across multiple states, seat belt usage rate has reached relatively high levels in recent years, reaching 86% in 2012 [3].

Although seat belt use in motor vehicles can significantly reduce the severity of a car accident, driver distraction remains the key cause of car accidents. The NHTSA considers distracted driving to include distractions such as: other occupants in the vehicle, eating, drinking, smoking, adjusting radio, adjusting environmental controls, reaching for an object in the car, and cell phone usage. Among all distractions, cell phone usage alone accounts for roughly 18% of car accidents caused by distracted drivers [4]. The studies in Refs. [5–7] found that the overall relative risk of having an accident for cell phone users when compared to non-cell-phone users averaged higher across all age groups. Studies also found that increasing cell phone usage correlated with an increase in relative risk. Several states have enacted regulations that prohibit drivers from using cell phones while driving due to the high number of accidents related to it. Thirty-nine states and the District of Columbia have at least one form of restriction on the use of mobile communication devices in effect [8].

Traffic congestion on busy commuter highways brings substantial financial cost and time loss to commuters, as well as contributing to air pollution, increased fuel consumption, and commuter frustration. Government officials and members of the transportation industry are seeking new strategies for addressing the problem. One mechanism to reduce the congestion on busy highway corridors is the introduction of managed lanes, that is, HOV and HOT lanes. Due to imposed limitations and fees, HOV/HOT lanes are often congestion free or much less congested than other commuter lanes. However, the rules of the HOV/HOT lane need to be enforced to realize the congestion reducing benefits. Typical violation rates can exceed 50–80%. To enforce the rules, current practice requires dispatching law enforcement officers at the roadside to visually examine incoming cars. Manual enforcement of HOV and HOT lanes can be a tedious and labor-intensive practice. Ultimately, it is an ineffective process with enforcement rates typically less than 10% [9]. Besides enduring environmental conditions of snow, darkness, sunlight reflections, and rain, law enforcement officers also have to deal with vehicles traveling at high speeds that may have darkened/tinted glass, reclining passengers, and/or child seats with or without children. As a result, there is a desire to have an automated method to augment or replace the manual roadside process.

Traditionally, enforcement of seat belt and cell phone regulations and counting the number of vehicle occupants are functions performed by traffic law enforcement officers that make traffic stops in response to roadside visual inspections. However, inspection by law enforcement officers faces many challenges such as safety, disruption of traffic, significant personnel cost and the difficulty of determining vehicle occupancy, and cell phone or seat belt usage at high speed. In response to the enforcement needs, in-vehicle systems and roadside solutions have been proposed for detecting seat belt usage, driver cell phone usage, and automatic counting of occupants in moving vehicles. The focus of this chapter is roadside-based sensing. We first briefly review systems that sense within the passenger compartment assuming particular compliance from occupants, where the systems can include vehicle instrumentation, smart phone applications, and cellular communications. Given the requirement on occupant compliance, they are not appropriate for general law enforcement.

4.2 Sensing within the Passenger Compartment

4.2.1 Seat Belt Usage Detection

References [10, 11] proposed two mechanical seat belt detection systems. In Ref. [10], the system includes a mechanism for detecting the state of a seat belt latch of an occupant; a control unit for generating an alarm signal when (i) an occupant detection mechanism detects the presence of an occupant within a seat and (ii) the latch-state-detection mechanism detects if the seat belt is not latched; and alarm unit for generating an alarm in response to the signal. In Ref. [11], the seat belt violation detection system includes sensors mounted to each seat belt, such as a retraction detector or alternatively a detection mechanism inside the seat belt buckle that determines if the seat belt is buckled, and a weight detector to determine if a seat belt is being worn by a passenger. Upon determination of a seat belt not being worn by a passenger, an alarm system activates an exterior visual seat belt violation indicator that may be observable by law enforcement personnel. Levy in Ref. [12] developed an in-vehicle vision-based seat belt detection system. The seat belt assembly includes strategically patterned indicators to improve the accuracy of image recognition. The indicators are known portions of the seat belt assembly that are provided with reflective portions. The indicators can include at least one of a nest indicator, an outer web indicator, an inner web indicator, a seat belt handle indicator, and a buckle indicator. The method uses an infrared (IR) illuminator and an image sensor located within the vehicle to capture an image of the seat belt assembly. An image analyzer matches identified indicators from the image sensor to a predefined set of indicators that characterize a particular status for the seat belt assembly to determine whether a seat belt assembly is buckled or not.

4.2.2 Cell Phone Usage Detection

For detecting driver cell phone usage, various methods that rely on cell phone signals have been proposed [13–19]. For example, Ref. [13] utilizes cell phone activity data from cell phone networks and vehicle operation data from vehicle monitoring systems to detects a vehicle is in motion. The vehicle operation data is then used to determine if any vehicles owned by a cell phone subscriber were moving during the use of an associated cell phone. Reference [15] describes a route-providing app that can also be used for cell phone usage control. The app determines the speed of the GPS coordinates and locks the cell phone when the user is moving at speeds greater than specific cutoffs. TransitHound [19] is a product that uses a sensor installed in a vehicle with an adjustable range to detect cell phone usage within that range. One of the limitations of these proposed approaches is the challenge in determining whether the driver or passenger(s) is the one engaged in the cell phone conversation or whether the usage is in hand-free mode or not. To solve these ambiguities, Refs. [14, 18, 20] proposed two alternative approaches. In Ref. [14], a phone-based driver detection system (DDS) was developed to determine whether the user in a car is a driver or passenger based on capturing and detecting key user movements during the car entry process, as well as specific sound patterns unique to vehicle post entry. Reference [18] used three-directional antennas, aimed toward the following three different locations: (i) front seat passenger side, (ii) head of a driver, and (iii) a location below the location at which the driver antennas are aimed on the driver side, respectively. A microprocessor processes the received signals and determines based thereon which antenna provides a strongest signal, which can be correlated to a most likely location of a cell phone being used by an occupant in the vehicle. Reference [20] used a combination of blue tooth signals and vehicle speakers to detect the cell phone usage by a driver.

4.2.3 Occupancy Detection

To comply with the mandated standard of having advanced or "smart" air bags in front seats of all new vehicles sold in the United States, car manufacturers have developed many in-vehicle technologies for occupancy detection. Various technologies, such as weight sensors [21–23], ultrasonic/radar sensors [24, 25], electromagnetic field based capacity sensors [26], and sensors for detecting physiological signals [27] or fingerprints [28], have been developed to detect the presence of an occupant in a vehicle. Although these technologies aimed at detecting occupants for appropriate air bag deployment, coupled with some form of communication, they can potentially be adapted to meet the need of managed lane enforcement. Owechko et al. [29] discussed an in-vehicle imaging-based solution using stereo cameras. Their sensor system consisted of a pair of vision sensors mounted near the rearview mirror and pointed toward an occupant. Multiple feature sets were extracted from synchronized image pairs to perform sensor fusion using a single vision sensor module. The four types of features utilized were: range (obtained using stereo vision), edges, motion, and shape. A classifier was trained for each feature type that classified the occupant into one of a small number of classes, such as adult in normal position or rear-facing infant seat. Each classifier generated a class prediction and confidence value. The class confidence outputs of the classifiers were combined in a vision fusion engine that makes the final decision to enable or disable the airbag.

4.3 Roadside Imaging

In recent years, traffic photo enforcement systems have been widely deployed in intelligent transportation systems for applications such as red light enforcement, unauthorized use of a bus lane, license plate recognition, and parking occupancy estimation. These camera-based enforcement systems may be mounted beside or over a road or installed in an enforcement vehicle to detect traffic regulation violations. Currently, the evidentiary packages including the captured photos or videos are typically manually reviewed by either an on-site law enforcement officer or human operator at a back-end office. The process is slow, labor intensive, and prone to human error. Therefore, there is a need for an automatic or semiautomatic solution that is robust and can handle the computational burden of high traffic volume.

Specifically concerning imaging of vehicle occupants from the roadside, various imaging systems proposed in the past included video, microwave, ultra-wideband radar, single-band near infrared (NIR), and multiband NIR. Studies in Ref. [30] have shown that NIR is the most promising roadside detection technology with the ability to address many challenges in vehicle occupancy detection such as windshield penetration and environmental conditions with good imaging resolution and fast image acquisition. Although recent studies on imaging into vehicles have focused on vehicle occupancy detection, many aspects of these systems and the conclusions of the studies can be extended to other types of passenger compartment violation detection such as seat belt usage and cell phone usage violations. Moreover, a comprehensive solution that can detect multiple types of violations using the same infrastructure can be cost effective.

4.3.1 Image Acquisition Setup

Figure 4.1 shows an illustration of an NIR image acquisition system as taught in Ref. [31]. The overall system includes front- and side-view camera triggers and camera systems to capture NIR images of both the front windshield and side view of approaching cars in HOV/HOT lanes. The exposure time of the cameras is set such that the image acquisition noise and motion blur is traded off to capture clear images of vehicles driving at a typical speed (30–80 mph). The front-view camera system is mounted on an overhead gantry, while the side-view camera is mounted along the road side. The camera triggers can be either induction loops installed beneath the road as in red light enforcement systems or laser break-beam triggers. The distance from the front-view camera to front-view trigger is on average 60-ft and the horizontal field of view of the camera is approximately 12 ft at that distance to accommodate the typical width of highway lanes in the United

Figure 4.1 Illustration of a gantry-mounted front-view and road-side side-view acquisition setup.

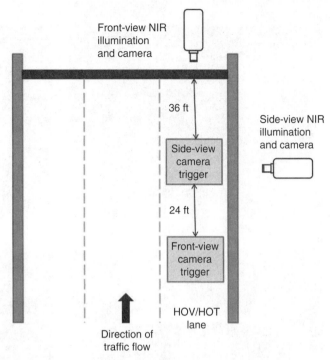

Front-view NIR illumination and camera

36 ft

Side-view camera trigger

Side-view NIR illumination and camera

24 ft

Front-view camera trigger

HOV/HOT lane

Direction of traffic flow

States. The distance between the side-view camera and the side-view trigger is set to 10 ft to capture side windows of a passing vehicle. The side-view trigger is placed 24 ft after the front-view trigger so that the system first captures the front windshield of an approaching vehicle and then captures a side view of the vehicle. Each camera system uses its own trigger, IR illuminator, and control and processing units. Although multiband NIR imaging systems can be employed for this purpose, the cost of multiband cameras alone can be quite high. Hence, a single-band NIR imaging system is still desirable. It is critical to filter out the visible light from the illuminator with a long pass filter (>750 nm) to reduce its visual impact on drivers. Reference [32] describes an alternative imaging and illumination system where the illumination is provided through the side window for a front-view gantry-mounted camera.

Figure 4.2 shows NIR images acquired by the HOV/HOT lane front- and side-view cameras of Ref. [31]. Note that there is a large variation in image intensity due to variations in windshield and window transmittance, illuminator state, sun position, cloud position, and other factors even though images were captured by an NIR camera [33]. A violation detection methodology based on these imaging conditions must be able to comprehend these challenges.

4.3.2 Image Classification Methods

An image classification approach has been shown to reliably detect passenger compartment violations under challenging imaging conditions. In this section, we present the classification method of Ref. [31]. A high-level view of the methodology of Ref. [31] for detecting passenger compartment violations is shown in Figure 4.3.

(a)

(b)

Figure 4.2 NIR images acquired by HOV/HOT lane front- and side-view cameras: (a) front-view images and (b) side-view images.

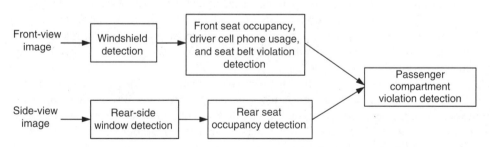

Figure 4.3 High-level view of detecting passenger compartment violations images in a classification-based approach
Source: Artan et al. [31]. Reproduced with permission of IEEE.

4.3.2.1 Windshield and Side Window Detection from HOV/HOT Images

There are many challenges in windshield detection such as (i) large variations in their shape, size, angle, and relative positions and distances to the imaging system; (ii) varying contrast between the windshield region and the vehicle body (e.g., against light vehicle exterior vs. dark vehicle exterior);

(iii) clutter on the dashboard and interior (e.g., variations in the appearance of vehicle occupants and objects visible through the front windshield). Several approaches have been proposed including geometric- and color-based detection [32, 34–37], using relationships to other objects on the vehicle [38], and more recently computer-vision-based approaches [31, 39]. The approach of Ref. [31] uses the front- and side-view images captured by an HOV/HOT image acquisition system as illustrated in Figure 4.1. The first step after image acquisition is localizing the front windshield and rear-side window from the front- and side-view images, respectively. This localization eliminates portions of the captured image irrelevant to passenger compartment violations while maintaining the relevant features. While front seat occupancy, driver mobile phone usage, and seat belt violations are detected from the localized front windshield image, the side-view image is used to detect rear seat occupancy. HOV/HOT occupancy violations are detected from processing the localized front- and side-view images and combining the front and rear seat occupancy information.

For windshield detection, Ref. [40], utilized a deformable parts model (DPM)-based object detection method, which jointly considers pose estimation and object detection tasks. This model forms a mixture of trees with a set of shared parts V, where each object part is considered as a node in the tree. The model then forms an optimization problem by jointly optimizing appearance and shape of the windshield parts. For a given image, this optimization problem is solved using dynamic programming to find the best configuration of object parts. Using the DPM framework, Ref. [31] generated a windshield model by positioning a set of landmarks around the windshield, the 13 landmark points are labeled manually in a set of training images similar to Figure 4.4 in the same sequence. In the implementation of Ref. [31], 10 positive and 20 negative images were used for training the front windshield model. Negative samples were selected from images that do not contain windshields in the scene. Rather than generating a mixture model, a single topological view was utilized due to the images being captured with a fixed camera and with vehicles always driving toward the camera along the same direction and angle with respect to the image plane. The windshield has a unique trapezoidal shape and does not show a significant variation across different vehicle types. A linearly parameterized tree-structured model was generated $T = (V, E)$, where V is the set of parts and E is the set of

(a)

(b)

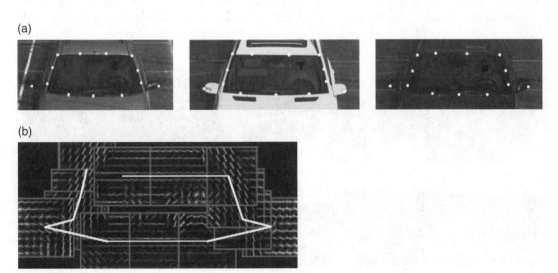

Figure 4.4 Windshield model learned using DPM-based method presented in Ref. [40]. (a) Training landmarks (light gray points) overlaid on top of the captured image and (b) Windshield model learned in training.
Source: Artan et al. [31]. Reproduced with permission of IEEE.

edges between parts. A windshield score for a particular configuration of landmarks $L = \{l_i; i \in V\}$ in a given image I is defined as follows:

$$S(I,L) = \text{App}_m(I,L) + \text{Shape}_m(L) = \sum_{i \in V}^{n} w_i \varphi(I, l_i) + \sum_{ij \in E}^{n} a_{ij} \, dx^2 + b_{ij} \, dy^2 + c_{ij} \, dx + d_{ij} \, dy. \qquad (4.1)$$

This score function is abridged from the general score function defined for mixtures of trees to accommodate a single tree structure [41]. In this function, $\varphi(I, l_i)$ represents the histogram of gradients (HoGs) features extracted at pixel location l_i [42], and $l_i = (x_i, y_i)$ stands for the pixel location of part i. The appearance evidence of each of the landmarks is represented in the *App* term and the evidence for the spatial locations of the landmarks with respect to each other is included in the *Shape* term. In Ref. [43], this model was viewed as a linear classifier with unknown parameters w_i and $\{a_{ij}, b_{ij}, c_{ij}, d_{ij}\}$, which is learned during training using a latent support vector machine (SVM). The training constructs a model by learning the appearance at each landmark point and the relationship between points as shown in Figure 4.4b.

For an incoming image I, we identify a list of candidate windshield areas by maximizing the score function Equation 4.2 over L using dynamic programming to find the best configuration of parts [40, 43].

$$S^*(I) = \max_{L} S(I,L). \qquad (4.2)$$

Similarly, a side-window model was developed using the DPM framework described before. For the side-window model, five mixtures were used corresponding to five different rear-side windows observed in the training data (e.g., rear-side windows for SUVs, sports cars, two different passenger cars, and vans) as shown in Figure 4.5a. Training utilized 50 positive images from five different side-window types (i.e., 10 images per side-window type) and 50 negative images. Landmark points (white points) overlaid on top of the side-view images corresponding to five mixtures are shown in Figure 4.5a. The training stage learned an appearance and shape model for each mixture component as shown in Figure 4.5b.

For some vehicles, front-side and rear-side windows resemble each other; and in these cases, the five-mixture side window detector, described earlier, causes false positives by detecting the front-side

(a)

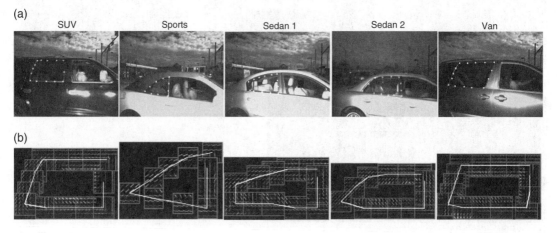

Figure 4.5 Side-window models learned using SVM-based method presented in Ref. [40]. (a) Training landmarks (light gray points) overlaid on top of the side-view images corresponding to five mixtures and (b) five mixture models learned in the training. *Source*: Artan et al. [31]. Reproduced with permission of IEEE.

rather than the rear-side window. To address this issue, Ref. [31] generated a DPM model for the detection of the B-pillar, which separates the front-side and rear-side windows. The B-pillar detector enabled capturing the side-window boundary more precisely and discarded false positives by comparing the location of the detected window and B-pillar. Assuming that x_w and x_b show the respective center of mass of the detected side window and of the B-pillar along the horizontal axis, respectively, a detected side window is validated if $(x_b - x_w) > t$ where t is a predefined threshold based on camera configuration and geometry. x_w and x_b are calculated from the detected landmark points as

$$x_{b,w} = \frac{\sum_{i=1}^{n} x_i}{n} \tag{4.3}$$

where n and x_i's represent the number and x-coordinates of the detected landmark points, respectively. Figures 4.6 (a–c) illustrate B-pillar sample images with training points overlaid. Figure 4.6d depicts the trained deformable part model obtained using 15 positive and 20 negative images.

Another approach is classification-based windshield detection. Li et al. [39] used the cascade AdaBoost classifier shown in Figure 4.7 to search for the windshield region. The cascade AdaBoost classifier is trained after the annotation and normalization. In their system, the maximum number of

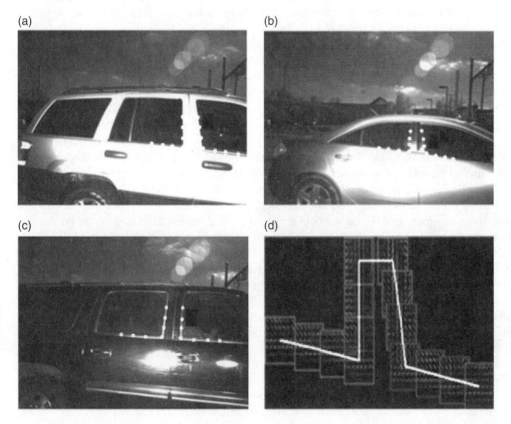

Figure 4.6 Parts (a–c) show a set of landmark points overlaid on three images for B-pillar detection. Part (d) illustrates the spatial shape model for tree structure learned in training.

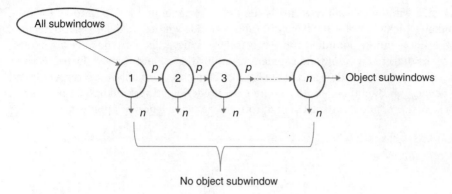

Figure 4.7 The structure of cascade classifier.

weak classifier of each stage is 30. The minimum detection rate is 0.995 and the maximum false-alarm rate is 0.5. A sliding window detector is used to detect the vehicle windows, and the size of detection windows is the same size as the positive samples. Images for detection are scanned by the detecting window from left to right, top to bottom. The detection operation is performed at multiple resolutions to enable detection of different size windows. Images for detection pass through every stage of the cascade classifier and the region that gets through all the stages is the desired target region.

Several other methods for detecting vehicle windows can be found in the literature. In Refs. [32, 34], the extraction of the vehicle windshield involves performing edge detection and Hough transform to detect the upper and lower horizontal lines of the windshield followed by employing a ±60° line detector mask and integral projection to detect the left and right borders of the windshield. Besides using edge information, there were also color-based approaches [35–37]. For example, the color-based approach in Ref. [37] used the mean shift algorithm, in which a color clustering is performed on the image to obtain regions with near-uniform color. After segmentation, only the region whose area is bigger than a threshold remains; the small regions are merged to the neighbor region whose color is most close. The mean shift algorithm of Ref. [37] was performed with the following steps: (i) choose the appropriate radius r of the search window, (ii) choose the initial location of the window, (iii) compute the mean shift vector, (iv) translate the search window by the shift amount, (v) repeat till convergence, and (vi) merge the segments and select dominant region. To improve the performance of the geometric- and color-based approach, Hao et al. [34] proposed the maximum energy method based on the observation that the color composition of the windshield region is more complex than the other parts of the vehicle due to its reflection and the background of vehicle interior. The method utilizes the candidate regions formed by the geometric- and color-based methods to define an energy function and select the windshield region with the maximal energy. Instead of searching for the windshield region directly in a scene, Xu et al. [38] identified other objects such as the headlights on a target vehicle and utilized a priori knowledge of the relative geometric relationships (e.g., distance and size) between the identified objects and the windshield.

4.3.2.2 Image Classification for Violation Detection
4.3.2.2.1 Defining Regions of Interest in Localized Windshield and Side Window Images
Following the method of Ref. [31] after front windshield localization, three regions of interest (ROIs) are defined on the image plane corresponding to the front seat occupancy, driver cell phone usage, and seat belt violation detections, respectively. These ROIs are illustrated in Figure 4.8. For front seat

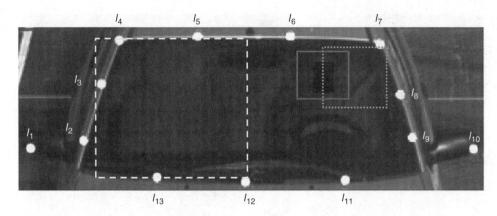

Figure 4.8 Regions of interest (ROI) defined for each case where white dashed rectangle shows ROI for front seat occupancy detection, solid rectangle for driver cell phone usage detection, and dotted rectangle for seat belt violation detection.

occupancy detection, the ROI is defined as the left half of the detected windshield as shown in the figure (i.e., dashed rectangle). Similarly, the ROI for driver mobile phone usage is defined as a rectangle around the driver's face (i.e., solid rectangle), and the ROI for seat belt detection is defined as the dotted rectangle in the figure.

The ROIs for occupancy and seat belt detection are calculated using the position of the detected windshield and landmarks l_i, $i = 1, 2,..., 13$. The center of the region for the seat belt detection (r, c), for example, is defined using landmarks l_6, l_7, and l_8, where r is set to the weighted average of the x-coordinates of l_6 and l_7 and c is calculated as the weighted average of y-coordinates of l_7 and l_8. The height of the region is determined based on the difference between y-coordinates of l_7 and l_8, and the width is determined based on the difference between x-coordinates of l_6 and l_7. Similarly, the location and size of the dashed rectangle corresponding to front seat occupancy detection is calculated using landmarks l_4, l_5, and l_{13}.

Defining the ROI for the driver mobile phone usage detection, on the other hand, requires face detection in the localized windshield image. Detecting the driver's face from the entire windshield image is challenging as the driver's face is often partially occluded by the sun visors, rearview mirror, or car roof. Figure 4.9 illustrates several instances of partial occlusions in HOV/HOT images. In order to detect faces in these cases while not increasing false-positive detections, Ref. [31] restricted the search space by first cropping the right half of the detected windshield image. Restricting the ROI for face detection provides flexibility for adjusting the detection threshold. Instead of setting a fixed threshold for face detection, Ref. [31] picked the window with the highest score calculated by the face detector as the driver's face, which enables detection of faces with partial occlusions whose score otherwise would not pass the detection threshold.

Unlike windshield images, the ROI for occupancy detection from the side-view images is the entire localized rear-side window as the passenger can be at any location in the localized image. The defined ROIs in the detected windshield image and the localized rear-side window image are further processed as described in Section 4.3.2.2.2 to identify violators.

4.3.2.2.2 *Image Classification*
For performing the image classification task, Ref. [31] considered local invariant descriptors that are aggregated into an image-level vector signature, which is subsequently passed to a classifier.

Figure 4.9 Examples for driver face images with partial occlusions.

Reference [31] used three locally aggregated descriptors bag of words (BoWs), vector of locally aggregated descriptors (VLADs), and Fisher vectors (FVs) due to their extensive usage and success in image classification and categorization tasks [44–46]. Among these descriptors, BoW has the oldest history and was the initiative for other locally aggregated descriptors [44]. From a set of training images, BoW first calculates dense scale-invariant feature transform (SIFT) features [47] and constructs a codebook of visual words that consist of K-centroids calculated by either K-means or Gaussian mixture model clustering algorithms applied on the calculated SIFT features. The dimension d for the SIFT feature vectors was 128. For an image in query I, local image descriptors $I = (x_1, x_2,..., x_N)$ are calculated and assigned to the closest cluster centers. Following the assignment step, a histogram of local descriptors is calculated and aggregated to generate the final image signature vector.

Similar to BoW, VLAD is a feature encoding method for image classification tasks that first constructs a vocabulary of visual words by clustering dense SIFT features for a set of training images. The visual vocabulary is generated by K-clusters calculated by either K-means or Gaussian mixture model clustering algorithms. Each cluster is represented by the cluster mean μ_k. For a query image, VLAD first calculates local image descriptors and assigns them to the closest cluster centroids. For the descriptors assigned to the same cluster, it calculates a total distance vector from the cluster mean as follows:

$$v_k = \sum_{x_i:N(x_i)=k}^{n} \left(x_i - \mu_k \right). \tag{4.4}$$

The final VLAD descriptor is formed by the concatenation of the d-dimensional distance vectors v_k for each cluster as $\varphi(I) = [v_1, v_2,..., v_k]$.

FVs have recently flourished as the probabilistic version of the VLAD and have been reported to achieve the best performance in several image classification and categorization tasks [45]. FVs have been proposed to incorporate generative models into discriminative classifiers [46]. Suppose $X = \{x_t, t = 1...T\}$ denotes the set of T local descriptors extracted from a given image. Assume that the generation process of local descriptors can be modeled by a probabilistic model $p(X|\theta)$, where θ denotes the parameters of the function. The gradient vector X is described as [46]

$$G_\theta^X = \frac{1}{T}\nabla_\theta \log p(x|\theta),$$

(4.5)

where the gradient of the log likelihood describes the contribution of the parameter θ to the generation process. Its dimensionality only depends on the number of parameters in θ.

A natural kernel on these gradient vectors is the Fisher kernel [46],

$$K(X,Y) = G_\theta^{X^T} F_\theta^{-1} G_\theta^Y,$$

(4.6)

where F_θ is the Fisher information matrix of $p(X|\theta)$,

$$F_\theta = E_{x\sim p}[\nabla_\theta \log p(x|\theta \nabla_\theta \log p(x|\theta)^T],$$

(4.7)

where F_θ^{-1} is symmetric and positive definite; it has a Cholesky decomposition $F_\theta^{-1} = L_\theta^T L_\theta$. Therefore, the kernel $K(X, Y)$ can be written as a dot product between normalized vectors (g_θ) shown in Equation 4.7.

$$g_\theta^X = L_\theta G_\theta^X.$$

(4.8)

Typically, g_θ^X is referred to as Fisher vector of X. Similar to the earlier work [48], we assume that $p(X|\theta)$ is a Gaussian mixture model (GMM): $p(x|\theta) = \sum_{i=1}^K w_i\, p_i(x)$. We denote $\theta = \{w_i, \mu_i, \sigma_i, i = 1, ..., K\}$ where w_i, μ_i, and σ_i are the mixture weight, mean vector, and variance matrix (assumed diagonal), respectively, of Gaussian pi. In this chapter, we only consider gradients with respect to the mean. We use the diagonal closed form approximation of the Fisher information matrix of Ref. [48] in which case the normalization of the gradient by L_θ is simply a whitening of the dimensions.

Let $\alpha_i(t)$ be the assignment of the descriptor x_t to the ith Gaussian,

$$\alpha_i(t) = \frac{w_i p_i(x_t|\theta)}{\sum_{j=1}^K w_j p_j(x_t|\theta)}.$$

(4.9)

Let g_i^X denote the d-dimensional gradient with respect to the mean μ_i of Gaussian i. Assuming that the x_t's are generated independently by $p(X|\theta)$, we obtain Equation 4.9 after mathematical derivations as follows:

$$g_i^X = \frac{1}{T\sqrt{w_i}}\sum_{t=1}^T \alpha_t(i)\left(\frac{x_t - \mu_i}{\sigma_i}\right).$$

(4.10)

The final vector g_θ^X is the concatenation of the g_i^X vectors for $i = \{1...K\}$ and is $K \times d$ dimensional. Experiments were performed for values ranging from $K = 32$ to $K = 512$.

After calculating locally aggregated image descriptors using either BoW, VLAD, or FV, Ref. [31] utilized a linear SVM to construct the classification model and perform the image classification for each case. The classifiers are trained using a set of positive and negative images. For occupancy detection, the positive set included images with one or more passengers and the negative set included images with no passengers.

4.3.3 Detection-Based Methods

4.3.3.1 Multiband Approaches for Occupancy Detection

Although most of the research on vehicle occupancy detection has been focused on the visible or very NIR spectrum range, multiband cameras have also been under development for more than a decade. Multiband techniques generally operate by selecting specific spectral bands that allow image-wise discrimination of skin, using illuminators and cameras that can provide sufficient signal in those bands, and applying a band fusing technique that makes the skin pixels apparent for segmentation. A key challenge is the selection of these discriminating spectral bands, along with a practical illuminator and sensor, where there is sufficient penetration through a windshield in any outdoor condition and at any time of the day while receiving enough reflected radiation from the objects of interest (i.e., human skin). A second challenge is computational, where a band fusing method must be developed that makes the skin pixels image-wise apparent. Once skin pixels are made apparent, the fused image can be processed to identify the number of occupants by applying morphological image processing operations or standard computer vision techniques to cluster the pixels to represent one or more occupants.

Concerning choice of spectral bands, several different wavelength bands are worthy of consideration. Visible spectral bands are ruled out because a visible flash illuminator at nighttime will distract drivers with potentially disastrous results. Also, visible range sensors perform poorly during foul weather conditions, such as rain and snow. Further, the appearance of vehicle occupants is quite variable in the visible range, depending on their physical characteristics, time of day, and the illumination conditions. This variability makes the machine vision task much more difficult [49].

The thermal IR band, including the mid wavelength 3–8 μm and long-wavelength IR 8–15 μm, initially seems attractive because it would distinguish the radiation from a warm body from the radiation emanating from the typical interior of a vehicle, and would not require an additional source of illumination. In practice, it is difficult to exploit thermal IR for HOV purposes because vehicle windshield glass severely attenuates EM radiation beyond 2.4 μm. An additional challenge arises due to the vehicle defogger, which can result in the thermal signature of the interior of the vehicle being dominated by the defogger heat [49].

The so-called NIR spectrum (0.7–2.4 μm) has been found to be most suitable for vehicle occupancy applications. A camera in this range can safely operate in both day and night, because a NIR illuminator will enhance the view of a scene in this band without disturbing a driver. The transmittance of typical vehicle windows in the NIR range is roughly 40%, so a camera is able to see through the vehicle's windshield and side windows. NIR spectral radiation is particularly effective in penetrating haze, so the camera can operate in adverse weather conditions. Considering detection of occupants, Ref. [49] recognized that humans consist of 70% water, and therefore exhibit spectral behavior very similar to water. Water absorbs NIR radiation heavily above 1.4 μm, and thus has low reflectance in this region. Human skin can be detected by examining spectral bands above and below 1.4 μm. Human skin has very high reflectance just below 1.4 μm wavelength, but very low reflectance just above 1.4 μm. If the NIR range is split into two bands, a lower band (0.7–1.4 μm) and an upper band (1.4–2.4 μm), then images of vehicle occupants will produce consistent signatures in the respective imagery. In the upper band images, humans will appear consistently dark, while in lower band images, humans will appear comparatively lighter.

References [49, 50] propose an occupancy detection system using two wavelength bands around 1.4 μm; the lower band covers 1.1–1.4 μm and the upper band spans 1.4–1.7 μm. In their system, one camera array was designed for the upper and lower bands, and the camera arrays were co-registered spatially and temporally relative to a scene and pixel transmission dynamics. The pairs of corresponding pixels from the two cameras were then fused with certain weights into a single pixel array by

$$P(i,j)_{\text{fused}} = P(i,j)_{\text{lower band}} - C \cdot P(i,j)_{\text{upper band}}, \tag{4.11}$$

where $P(i, j)$ is the pixel intensity at the pixel location i and j and C is the weighting factor. Given a higher reflectance in the lower band $P(i, j)_{\text{fused}}$ is generally positive and a threshold is then applied to the $P(i, j)_{\text{fused}}$ to determine whether a pixel corresponds to human skin or not. The threshold value was dynamically determined for each image according to the histogram of $P(i, j)_{\text{fused}}$ of the respective frame.

References [51, 52] describe a different spectral approach. A single camera with a single illuminator is used to capture images at N different infrared wavelength bands sequentially, and a fused image is generated by calculating the intensity ratios of images captured at the different wavelength bands. Humans are identified from the surrounding background via intensity ratios such as

$$P(i, j) = \frac{P_l(i, j) - P_m(i,j)}{P_n(i, j) - P_m(i,j)}, \tag{4.12}$$

where $P_{(l,m,n)}(i, j)$ is the pixel intensity at the pixel location i and j, and l, m, and n denote to the different spectrum bands from a multiband camera (e.g., 3). From the fused image $P(i, j)$, the number vehicle occupants can be identified based on pixel classification where the pixel intensities in the fused image are thresholded to create a binary image from which occupants faces are detected by applying morphological image processing operations. The number of vehicle occupants can also be identified using standard machine learning computer vision approaches from the fused image. Reference [50], for example, identifies the number of people from the fused image using a neural network classifier, where each class corresponds to different number of vehicle occupants.

Although no specific bands are given, Ref. [53] discloses a method to detect face of an occupant by designing the wavelength bands in a way that the lower and upper bands are sensitive to the eyebrows and eyes of a human.

4.3.3.2 Single Band Approaches

Another approach for passenger compartment analysis takes advantage of the distinct signatures of the objects of interest (empty seats, seat belts, passengers). For example, an unoccupied seat typically possesses images features such as long contiguous horizontal line segments and curved segments, and substantially uniform areas encompassed by these line segments. These features are not present in an occupied seat. On the other hand, human face features can typically be detected from an occupied seat. Hence, object detection–based approaches for vehicle occupancy detection rely on detecting empty seats or human face/body. Similar to the classification-based approaches, detection-based approaches can also benefit from a localized windshield area to reduce computation cost as well false detections.

Fan's et al. [54] approach for detecting empty seats was based on the observations that when a seat is occupied, long edges are partially (or fully) obscured by the seat's occupant. As a result, edges may be broken into shorter pieces or missing altogether. Occasionally, line segments representing a top edge of a seat remain intact when a shorter person occupies that seat. Edges associated with seats and headrests are typically "soft" with a relatively smooth transition. A typical seat and headrest edge profile is shown in Figure 4.10a. Figures 4.10b and c represent an edge with a sharp transition and a ridge edge, respectively. These two types of edges are more often seen at boundaries between adjoining

(a)   (b)   (c)

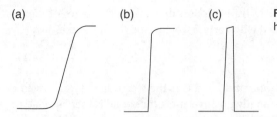

Figure 4.10 Edge profiles associated with seats and headrests. *Source*: US8611608 B2 [54].

fabrics of different colors and sewn seams. The edge profile is helpful in distinguishing seat edges from other edges. In their work, edge detection is performed using techniques such as horizontal edge detection followed by edge linking, or a Hough transform. One approach utilizes a complexity measure processor that incorporates a spatially sensitive entropy operator to calculate a local measure of complexity (entropy) for a given pixel in the image. This complexity measure is then used to reference a look-up table (LUT) to obtain a threshold value which, in turn, is used by the spatial bilateral filter to process pixels of the image. The detected line and curved segments are evaluated for their contiguousness. If those detected line segments are substantially contiguous, it is determined that the seat is unoccupied. Otherwise, it is determined that the seat is occupied.

In addition to edge features, pixels of the image associated with the seat and headrest areas can also be analyzed for color or intensity and a determination of whether the seat is occupied is based upon a uniformity of color or intensity in bounded areas. If the color or intensity of the area bounded by the line and curve segments associated with the seat and headrest is substantially uniform, then the seat is determined to be unoccupied.

Besides empty seat detection, one can also approach the problem as passenger detection, where features such as face and seat belt can indicate the presence of a person. Numerous presentations have been proposed for face detection, including pixel-based [55], parts-based [56], local edge features [57], Haar wavelets [58], Haar-like features [59], and more recently deformable part model based [40].

In addition to directly identifying a face or an empty seat, the presence of a seat belt is most likely an indication of the presence of a passenger. Various methods have been proposed [32, 34, 60, 61] based on the observations that seat belt has a 45° approximately in vehicle windshield images. For example, Ref. [61] employed the Hough transform to detect the seat belt after the Canny edge extraction. Given that the many straightlines are returned by Hough algorithm, the seat belt can be determined by two factors including the slope and the length of the line. Only when the slope of a line is in a neighborhood of 45° and the line length is greater than a certain threshold can the line be identified as a seat belt. The confidence is evaluated according to the slope and line length.

References

1 National Highway Traffic Safety Administration (NHTSA), *Lives saved calculations for seat belts and frontal air bags*. DOT HS 811206, U.S. Department of Transportation, Washington, DC, 2009.
2 Research Note: Traffic Safety Facts, National Highway Transportation System Administration, January, 2014.
3 M. Hinch, M. Solomon, and J. Tison, *The Click It or Ticket Evaluation, 2012*. No. DOT HS 811 989. 201
4 Distracted Driving 2009, Report, U.S. DOT National Highway Traffic Safety Administration, September 2010. http://www-nrd.nhtsa.dot.gov/Pubs/811379.pdf (accessed September 13, 2016).

5 C. Laberge-Nadeau, Wireless telephones and the risk of road crashes, *Accident Analysis and Prevention* 35(5): 649–660, 2003.

6 K. S. Lissy, J. T. Cohen, M. Y. Park, and J. D. Graham, Cellular phone use while driving: Risks and benefits, Phase 1 Report, Harvard Center for Risk Analysis and Harvard School of Public Health, Boston, MA, July 2000.

7 *Mobile phone use: A growing problem of driver distraction*, World Health Organization, Geneva, 2011, http://www.who.int/violence_injury_prevention/publications/road_trac/en/index.htm (accessed September 13, 2016).

8 J. K. Ibrahim, E. D. Anderson, S. C. Burris, and A. C. Wagenaar, State laws restricting driver use of mobile communications devices: Distracted-driving provisions, 1992–2010, *American Journal of Preventive Medicine* 40(6): 659–665, 2011.

9 S. Schijns and P. Mathews, A breakthrough in automated vehicle occupancy monitoring systems for HOV/HOT facilities, *Proceeding of the 12th HOV Systems Conference*, Houston, TX, April 20, 2005.

10 D. Yazdgerdi, Seat belt detection device, U.S. Patent application no. 10/081,687, August 28, 2003.

11 E. G. Rosenblatt, Seat belt violation alert system, U.S. Patent application no. 13/065,505, September 27, 2012.

12 U. Levy, Vision based seat belt detection system, U.S. Patent application no. 11/675,101, August 23, 2007.

13 J. C. Catten and S. McClellan, System and method for detecting use of a wireless device in a moving vehicle, U.S. Patent No. 7,876,205, January 25, 2011.

14 H. Chu, In-vehicle driver detection using mobile phone sensors, Graduation thesis, Duke University, April 2011.

15 A. Thiagarajan, J. Biagioni, T. Gerlich, and J. Eriksson, Cooperative transit tracking using smart-phones. *Proceedings of the 8th ACM Conference/Embedded Networked Sensor Systems (SenSys '10)*, Zurich, Switzerland, November 3–5, 2010.

16 S. H. Chu, V. Raman, J. Shen, A. Kansal, V. Bahl, and R. R. Choudhury, I am a smartphone and i know my user is driving, *Proceedings of the 6th International Conference on Communication Systems and Networks*, Bangalore, January 7, 2014.

17 http://bvsystems.com/Products/Security/TransitHound/transithound.htm (accessed September 13, 2016).

18 D. S. Breed and W. E Duval, In-vehicle driver cell phone detector, U.S. Patent no. 8,731,530, May 20, 2014.

19 https://www.bvsystems.com/product/transithound-in-vehicle-cell-phone-detector/

20 Advanced Vehicle Occupancy Verification Technologies Study: Synthesis Report, Texas Transportation Institute (TTI), August 2007 (prepared for the HOV Pooled Fund Study Group).

21 K. Jost, Delphi Occupant Detection for Advanced Airbags, Automotive Engineering International Online, October 18, 2000. https://trid.trb.org/view.aspx?id=709051 (accessed October 13, 2016).

22 N. Kuboki, H. Okamura, T. Enomonto, T. Nishimoto, T. Ohue and K. Ando, An occupant sensing system for automobiles using a flexible tactile force sensor, *Furukawa Review* 20, 89–94, 2001.

23 T. Gioutsos and H. Kwun, *The use of magnetostrictive sensors for vehicle safety applications*, Society of Automotive Engineers, New York, 1997.

24 A. P. Corrado, S. W. Decker, and P. K. Benbow, Automotive occupant sensor system and method of operation by sensor fusion. U.S. Patent No. 5,482,314. January 9, 1996.

25 J. G. Stanley, Occupant detection system. U.S. Patent No. 6,220,627. April 24, 2001.

26 B. George, H. Zangl, T. Bretterklieber, and G. Brasseur, A combined inductive–capacitive proximity sensor for seat occupancy detection, *IEEE Transactions on Instrumentation and Measurement* 59(5): 1463–1470, 2010.

27 O. A. Basir and B. Zhao, Occupant heartbeat detection and monitoring system. U.S. Patent No. 7,183,930. February 27, 2007.

28 Siemens system, as reported in "Candid Camera Security" in The Australian, p. 16., April 13, 2000.

29 Y. Owechko, N. Srinivasa, S. Medasani, and R. Boscolo, Vision-based fusion system for smart airbag applications, *IEEE Intelligent Vehicle Symposium* 1: 245–250, 2002.

30 B. L. Smith and D. Yook, Investigation of Enforcement Techniques and Technologies to Support High Occupancy Vehicle and High Occupancy Toll Operations, Final Contract Report VTRC 10-CR1, Virginia Transportation Research Council, Virginia, September 2009.

31 Y. Artan, O. Bulan, R.P. Loce, and P. Paul, Passenger Compartment Violation Detection in HOV/ HOT Lanes, submitted to IEEE Transactions on Intelligent Transportation Systems.

32 X. Hao, H. Chen, C. Wang, and C. Yao, A near-infrared imaging method for capturing the interior of a vehicle through windshield, *Proceedings of the 2010 IEEE Southwest Symposium on Image Analysis and Interpretation*, Austin, TX, May 23–25, 2010

33 N. Bouguila, R. I. Hammoud, and D. Ziou, Advanced off-line statistical modeling of eye and non-eye patterns, in R. I. Hammoud (ed.), *Passive eye monitoring*. Springer, Berlin, 2008, pp. 55–81.

34 X. Hao, H. Chen, C. Wang, and C. Yao, Occupant detection through near-infrared imaging, *Tamkang Journal of Science and Engineering* 14(3): 275–283, 2011.

35 Wang Y. and You Z., A fast algorithm for localization vehicle's window based on the mean of chromatism, *Computer Applications and Software* 21(1): 31–35, 2004, (In Chinese).

36 H. Fukui and T. Hasegawa, A study on detection of vehicles' positions using HSV color transformed template images, *IEEE Vehicular Technology Conference Fall* 2, 718–722, 2001.

37 X. Yuan, Y. Men, and X. Wei, A method of location the vehicle windshield region for vehicle occupant detection system, *Proceedings of International Conference on the Software Process*, Zurich, Switzerland, October, 21–25, 2012.

38 B. Xu and P. Paul, Windshield localization for occupancy detection, U.S. Patent 8971579, 2015.

39 W. Li, J. Lu, Y. Li, Y. Zhang, J. Wang, and H. Li, Seatbelt detection based on cascade AdaBoost classifier, *Proceedings of the 2013 6th International Congress on Image and Signal Processing (CISP)*, Hangzhou, December 16–18, 2013.

40 Z. Zhu and D. Ramanan, Face detection, pose estimation, and landmark localization in the wild, *Proceedings of the 2012 IEEE Conference on Computer Vision and Pattern Recognition (CVPR)*, Providence, RI, June 16–21, 2012. IEEE.

41 Y. Artan, O. Bulan, R. Loce, and P. Paul, Driver cell phone usage detection from HOV/HOT NIR images, *Proceedings of the IEEE Conference on Computer Vision and Pattern Recognition Workshops*, Columbus, OH, June 23–28, 2014.

42 N. Dalal and B. Triggs, Histograms of oriented gradients for human detection, *IEEE Computer Society Conference on Computer Vision and Pattern Recognition, 2005 (CVPR 2005)*, San Diego, CA, June 20–25, 2005, vol. 1. IEEE, pp. 886–893.

43 P. F. Felzenszwalb, R. B. Girshick, D. McAllester, and D. Ramanan, Object detection with discriminatively trained partbased models, *IEEE Transactions on Pattern Analysis and Machine Intelligence* 32(9): 1627–1645, 2010.

44 G. Csurka, C. Dance, L. Fan, J. Willamowski, and C. Bray, Visual categorization with bags of keypoints, *Workshop on statistical learning in computer vision, ECCV*, Prague, 2004, vol. 1, pp. 1–22.

45 H. Jégou, F. Perronnin, M. Douze, J. Sánchez, P. Pérez, and C. Schmid, Aggregating local image descriptors into compact codes, *IEEE Transactions on Pattern Analysis and Machine Intelligence* 34(9): 1704–1716, 2012.

46 T. Jaakkola and D. Haussler, Exploiting generative models in discriminative classifiers, *Advances in Neural Information Processing Systems* 11: 487–493, 1999.

47 A. Vedaldi and B. Fulkerson, Vlfeat: An open and portable library of computer vision algorithms, *Proceedings of the International Conference on Multimedia*, Firenze, Italy, October 25–29, 2010. ACM, pp. 1469–1472.

48 F. Perronnin and C. Dance, Fisher kernels on visual vocabularies for image categorization, *IEEE Conference on Computer Vision and Pattern Recognition, 2007 (CVPR'07)*, Minneapolis, MN, June 17–22, 2007. IEEE, pp. 1–8.

49 I. Pavlidis, P. Symosek, B. Fritz, M. Bazakos, and N. Papanikolopoulos, Automatic detection of vehicle occupants: The imaging problem and its solution, *Machine Vision and Applications* 11(6): 313–320, 2000.

50 I. Pavlidis, V. Morellas, and N. Papanikolopoulos, A vehicle occupant counting system based on near-infrared phenomenology and fuzzy neural classification, *IEEE Trans on Intelligent Transportation Systems* 1(2): 72–85, 2000.

51 Y. Wang, Z. Fan, and L. K. Mestha, Determining a number of objects in an IR image, U.S. Patent no. 8,587,657, November 19, 2013.

52 Y. Wang, Z. Fan, and L. K. Mestha, Determining a total number of people in an IR image obtained via an IR imaging system, U.S. Patent no. 8,520,074, August 27, 2013.

53 J. B. Dowdall and I. Pavlidis, Near-infrared method and system for use in face detection, U.S. Patent no. 7,027,619, April 11, 2006.

54 Z. Fan, A. S. Islam, P. Paul, B. Xu, and L. K. Mestha, Front seat vehicle occupancy detection via seat pattern recognition, US 8611608 B2, December 17, 2013.

55 H. Rowley, S. Baluja, and T. Kanade, Neural network-based face detection, *IEEE Transactions on Pattern Analysis and Machine Intelligence* 20: 23–38, 1998.

56 B. Heisele, T. Serre, and T. Poggio, A component-based framework for face detection and identification, *International Journal of Computer Vision* 74: 167–181, 2007.

57 F. Fleuret and D. Geman, Coarse-to-fine face detection, *International Journal of Computer Vision* 41: 85–107, 2001.

58 H. Schneiderman and T. Kanade, Object detection using the statistics of parts, *International Journal of Computer Vision* 56: 151–177, 2004.

59 P. Viola and M. Jones, Rapid object detection using a boosted cascade of simple features, *Proceedings of the 2001 IEEE Computer Society Conference on Computer Vision and Pattern Recognition, 2001 (CVPR 2001)*, Kauai, HI, December 8–14, 2001. pp. 511–518.

60 Image-based seat-belt detection, *2011 IEEE International Conference on Vehicular Electronics and Safety (ICVES)*, Beijing, China, July 10–12, 2011, pp. 161–164.

61 W. Li, J. Lu, Y. Li, Y. Zhang, J. Wang, and H. Li, Seatbelt detection based on cascade AdaBoost classifier, *Proceedings of the 2013 6th International Congress on Image and Signal Processing (CISP)*, Hangzhou, December 16–18, 2013.

5

Detection of Moving Violations

Wencheng Wu[1], Orhan Bulan[2], Edgar A. Bernal[3] and Robert P. Loce[1]

[1] *Conduent Labs, Webster, NY, USA*
[2] *General Motors Technical Center, Warren, MI, USA*
[3] *United Technologies Research Center, East Hartford, CT, USA*

5.1 Introduction

Law enforcement agencies and municipalities are increasing the deployment of camera-based road-way monitoring systems with the goal of reducing unsafe driving behavior. The most common applications are detection of violations for speeding, running red lights or stop signs, wrong-way driving, and making illegal turns. Other applications are also being pursued, such as detection of tailgating, blocking the box, and reckless driving. While some camera-based systems use the acquired images solely for evidentiary purposes, there is increasing use of computer vision techniques for automating the detection of violations.

Most applications in roadway computer vision systems involve analyzing well-defined and acceptable trajectories and speeds, which leads to clearly defined rules and detections. In some cases, the detections are binary, such as in red light enforcement or RLE (stopped or not), or divided highway driving (wrong way or correct way). Other applications require increased accuracy and precision, such as detecting speed violations and applying a fine according to the estimated vehicle speed. There are other deployed applications where the violation involves less definitive criteria, such as reckless driving.

The following sections present various applications, giving the motivation, system requirements, methodology, and effectiveness. The more common applications of speed and stop light will be described in detail, while the less common will be briefly noted.

5.2 Detection of Speed Violations

Studies have shown that there is a strong relationship between excessive speed and traffic accidents. In the United States in 2012, speeding was a contributing factor in 30% of all fatal crashes (10,219 lives) [1]. The economic cost of speeding-related crashes was estimated at $52 billion for 2010 [2]. In extensive review of international studies, automated speed enforcement is estimated to reduce injury-related crashes by 20–25% [3]. Hence, there is significant motivation to evolve and deploy speed enforcement systems.

Computer Vision and Imaging in Intelligent Transportation Systems, First Edition.
Edited by Robert P. Loce, Raja Bala and Mohan Trivedi.
© 2017 John Wiley & Sons Ltd. Published 2017 by John Wiley & Sons Ltd.
Companion website: www.wiley.com/go/loce/ComputerVisionandImaginginITS

A typical photo enforcement system for speed violations consists of (i) an imaging module, which provides visual confirmation of the violation; (ii) a speed measurement module; and (iii) a citation issuing module, which issues the citation to a violator based on the information collected from the imaging and speed measurement modules. Many technologies have been developed and deployed to real-world environments in all three modules. In this section, we focus the discussion on technologies applied to speed measurement using computer vision.

Common methods for speed measurement in transportation include the use of in-ground inductive loops, radar, lidar, and video cameras. There are several advantages to the use of a vision system over the use of inductive loops or radar/lidar, while presenting new challenges that need to be addressed. The disruption and expense of installing in-ground induction loops can be avoided. A vision system is needed to recognize the vehicle, so extending its capabilities to measure speed eliminates the cost and complexity of additional system components associated with radar and lidar. A high-quality vision system can include intelligence to perform additional functions, such as recognition of tailgating, reckless driving, and accidents; gather usage statistics; and serve as a general surveillance device. Conceptually, it is fairly simple for a vision system to provide some measure of speed of an object once the object of interest is properly detected, identified, and tracked. The issue is the accuracy and precision of the measurement. Although a significant body of research exists on applying computer vision technologies to traffic and traffic flow measurements, only a very small fraction of published research evaluates accuracy and precision of speed measurement of an individual vehicle, which is critical for speed enforcement applications.

5.2.1 Speed Estimation from Monocular Cameras

In order to yield accurate speed measurement of individual vehicles via computer vision, a first requirement is good performance of the vehicle detection and tracking methods. Much research has been conducted in object detection and tracking as fundamental building blocks for video processing technologies. An excellent survey can be found in Ref. [4]. Although many of these techniques are readily applicable to vehicle tracking for speed measurement, there is a distinct aspect that needs to be considered here. More specifically, common methods focus on a coarser concept of tracking. The objective of most trackers is to track the object "as a whole." The operation of a tracker is considered effective as long as it can track the object as it appears or reappears in the scene over time under various practical noises. For speed measurement, the tracking objective needs to be more refined: it is necessary to track a specific portion(s) of the object. Consider a simple example. If a tracker starts with a reference point located about the front of a vehicle, and as the vehicle moves leading to a change in perspective, ends with a reference point located about the rear of a vehicle, the tracked trajectory alone is not adequate to estimate speed with sufficient accuracy for most speed law enforcement applications. Consequently, a suitable tracker for an accurate speed measurement system adheres to one of the following: (i) directly tracking a specific portion(s) of the object to determine its trajectory or (ii) coarsely tracking the object as a common practice while applying additional processing to infer the trajectory of a specific portion(s) of the object indirectly.

Two common tracking approaches, cross-correlation tracking and motion-blob proximity association tracking, are presented here to illustrate the fine differences between tracking a specific portion of the object and coarsely tracking the object. The cross-correlation method tracks a specific portion of the object, while motion-blob proximity association only tracks the object coarsely.

Let $I(i, j; t)$ be the pixel value of an image frame at position (i, j) and time t and $v(i_t, j_t)$, $t = t_0 \sim t_1$ be the resulting trajectory of a tracker on a vehicle. Here, t_0 and t_1 are the start and end of the tracking of the vehicle, respectively. The first common tracking method to be described is cross-correlation matching approach. In this approach, first a region, $I_v(t_0) = I(i + i_{t_0}, j + j_{t_0}; t_0) \forall (i,j) \in R_v$, is identified and used as initial tracking template. In a typical setting, R_v is chosen as a rectangular region $(2m+1) \times (2n+1)$ centered around the centroid of the template (i_{t_0}, j_{t_0}). In the subsequent frames, the location that best matches the template $I_v(t_0)$ in the current frame $I(x,y;t)$ is then found using the following optimization criterion:

$$v(i_t, j_t) = \underset{(i_t, j_t)}{\arg\max} \frac{\sum_{i=-m}^{m} \sum_{j=-n}^{n} \left(I(i + i_t, j + j_t; t) \cdot I(i + i_{t_0}, j + j_{t_0}; t_0) \right)}{\sqrt{\sum_{i=-m}^{m} \sum_{j=-n}^{n} \left(I(i + i_t, j + j_t; t) \right)^2} \sqrt{\sum_{i=-m}^{m} \sum_{j=-n}^{n} \left(I(i + i_{t_0}, j + j_{t_0}; t_0) \right)^2}} \tag{5.1}$$

Conceptually, Equation 5.1 is equivalent to finding the location in the current image frame where its appearance is closest to the tracking template. The resulting $v(i_t, j_t)$ is the new center position, where the image content at time t is most similar to the template at time t_0 measured by cross-correlation. The tracker would repeat this process for the time duration where the tracked vehicle is in the scene and eventually yields the full trajectory $v(i_t, j_t)$, $t = t_0 \sim t_1$ of the vehicle. Since this approach utilizes appearance matching, an appropriate selection of template would warrant the tracker to track a specific portion of the vehicle to determine its trajectory. An excellent example of such a template is a portion of the license plate of the vehicle.

The second common tracking method to be discussed uses foreground or motion detection followed by a proximity association. In this approach, potential pixels with motion are identified through frame differencing.

$$M(x,y;t) = \begin{cases} 1 & \text{if } |I(x,y;t) - I(x,y;t-\delta)| > \eta \\ 0 & \text{otherwise} \end{cases} \tag{5.2}$$

The resulting binary image is then postprocessed with morphological filtering and connectivity analysis to extract motion blob(s), regions of potential moving objects. These candidate motion blobs may be further processed with thresholding on size, shape, aspect ratio, etc., to better detect blobs that are indeed from moving vehicles. Once the motion blob(s) are detected for the current frame, the tracker associates these blobs with those detected from past frame(s) based on measures such as proximity, smoothness of the overall trajectory, and coherence of the motions. The trajectory of the tracked vehicle would then be the collection of some reference points such as centroids of these associated blobs over time. Since the tracking of this approach is mainly based on motion blob rather than the appearance of object, it only tracks the object coarsely.

The schematic illustrations of the two tracking methods discussed earlier are shown in Figure 5.1 to help readers comprehend the fine differences. Figure 5.1a shows the first frame where the tracking starts. The dashed square is the template used by cross-correlation tracking to start the tracking, that is, $I_v(t_0)$. The triangle mark is the centroid of the template. The black outlined blob is the detected motion blob by motion-blob proximity association tracking. The solid-circle mark is the centroid of the motion blob. We set solid-circle and solid-triangle marks at the same location for illustration purpose. Note that the shadow of the vehicle was detected as part of the vehicle. Figure 5.1b shows the later frame where the tracking ends. The location of dashed square on this frame is found using Equation 5.1, and the solid-triangle mark is the centroid. As can be seen in Figure 5.1b, it is on the same location of the vehicle as before. The solid-circle mark in Figure 5.1b is the centroid of the

(a) (b)

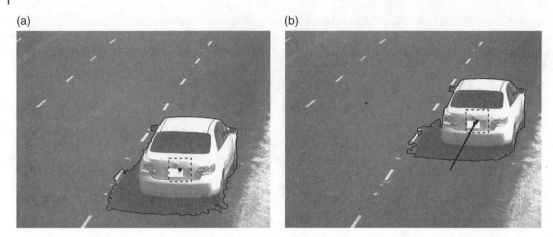

Figure 5.1 Schematic illustrations of cross-correlation and motion-blob proximity association tracking methods.

new black outlined blob, which is NOT on the same location of the vehicle as before. This is because the motion blob changes its shape a little bit due to noise and projective distortion. As a result, the pixel/distance traveled (solid and dash lines in Figure 5.1b) are slightly different by the two tracking methods (~5% difference). This difference is significant for the purpose of speed enforcement. The cross-correlation tracking method is preferred as discussed earlier.

In addition to accurate detection and tracking of vehicles, a computer vision–based speed measurement system requires (i) an accurate camera calibration strategy that produces a geometric mapping for translating image pixel positions to real-world coordinates [5–14], (ii) an understanding of the impact of tracked-feature height to speed accuracy [8, 14–16], and (iii) an accurate reference measurement system [17]. The geometric mapping is typically performed using a projective matrix transformation. More detailed discussions on these three requirements are presented in the following text.

Consider the work presented in Ref. [13], which introduces both the approach and potential pitfalls associated with *manual calibration methods*. In this work, the calibration is achieved by manually placing marks on the roadway, identifying image pixel locations that contain the marks, and then using the pixel location and mark location data to construct the camera calibration mapping. A couple of issues can arise with this approach. One consideration is that manually placing marks on the road may be impractical or costly, especially in high traffic areas. Second, both the placement and the identification of the location of the marks on the road need to be quite accurate. A systematic 10 cm combined error in the mark placement and pixel location for a 10 m spacing between marks would translate to a 1% bias error in subsequent speed measurements. Finally, the camera may move or change field of view over time (intentionally or unintentionally). Hence, camera recalibration may be needed periodically. Given these issues and constraints in *manual calibration methods*, it is preferred to use *model-based techniques*, which decompose entries of a projective matrix into functions of a set of parameters such as focal length, camera pose, etc., and estimate the parameters via scene analyses rather than the matrix entries directly from manually placed marks.

In camera calibration, the geometric mapping between pixel coordinates (i, j) to real-world planar coordinates $(x, y; z = z_0)$ can be characterized by a projective matrix as follows:

$$k \begin{bmatrix} x \\ y \\ 1 \end{bmatrix} = \begin{bmatrix} H_{11} & H_{12} & H_{13} \\ H_{21} & H_{22} & H_{23} \\ H_{31} & H_{32} & H_{33} \end{bmatrix} \begin{bmatrix} i \\ j \\ 1 \end{bmatrix}.$$

(5.3)

The $3 \times 3\ \boldsymbol{H}(z_0)$ matrix is known as the camera projective matrix for $z = z_0$. Different z would have a different projective matrix, but they are related across z's. If real-world coordinates are chosen such that its z-axis aligns with the camera optical axis, then the projective matrices for all z's can be described as follows:

$$
k \begin{bmatrix} x/z \\ y/z \\ 1 \end{bmatrix} = \begin{bmatrix} H_{11} & H_{12} & H_{13} \\ H_{21} & H_{22} & H_{23} \\ H_{31} & H_{32} & H_{33} \end{bmatrix} \begin{bmatrix} i \\ j \\ 1 \end{bmatrix}. \tag{5.4}
$$

Here, the projective matrix is the same for all z's. This is often not the case for speed enforcement applications since the plane of interest is the road surface, which is roughly on a 2D plane, but its normal direction rarely aligns with the camera optical axis. There is a conversion that one can use to relate the two coordinate systems. For a monocular camera, it is necessary to comprehend the z-position or depth position to have unique conversion between pixel coordinate and real-world coordinate. The task of camera calibration/characterization is now reduced to identifying $\boldsymbol{H}(z_0)$ for a given camera. A straightforward approach is to manually place reference markers on the plane $z = z_0$ whose (x, y) are known and whose (i, j) can be identified from the images. Due to noise and potential errors in (x, y) or (i, j) of the reference markers, a common practice is to use many more reference markers than the minimal number needed ($=4$) and the random sample consensus (RANSAC) process to derive a robust and accurate $\boldsymbol{H}(z_0)$. This type of approach is referred to as *manual calibration*, which directly estimates the projective matrix based on reference data without setting any constraint on the relationship among entries in the projective matrix.

A *model-based camera calibration method*, on the other hand, imposes meaningful structure/constraint to the projective matrix and derives the corresponding parameters via scene analysis. Figure 5.2 shows an example camera–road configuration discussed in Ref. [5] for deriving a model-based camera calibration method. It utilizes the identification of vanishing points along and perpendicular to the road travel direction and one additional piece of information, such as the height of the camera above the road. This method is briefly summarized here. Let $f, h, \phi,$ and θ be the camera focal length, camera height above the road, camera tilt angle, and camera pan angle, respectively. Assuming

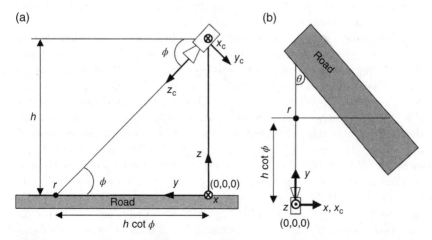

Figure 5.2 Illustration example for *model-based camera calibration method*: (a) left-side view of the scene and (b) top view of the scene. *Source*: Kanhere and Birchfield [5]. Reproduced with permission of IEEE.

that the camera has zero roll and has square pixels, it can be shown that the geometric mapping between pixel coordinates (i,j) to real-world coordinates on the road plane is

$$
k \begin{bmatrix} x \\ y \\ 1 \end{bmatrix} = \begin{bmatrix} h\sec\phi & 0 & 0 \\ 0 & -h\tan\phi & fh \\ 0 & 1 & f\tan\phi \end{bmatrix} \begin{bmatrix} i \\ j \\ 1 \end{bmatrix}.
\tag{5.5}
$$

The form is the same as Equation 5.3, but the entries of the projective matrix are more constrained. Although the matrix can be determined from actual measurements of f, h, ϕ, in practice it is difficult and inconvenient to measure these parameters directly. In Ref. [5], various known methods that utilize scene analysis to estimate these parameters indirectly are discussed and summarized. For example, in roadway transportation applications, it is feasible to determine a vanishing point (i_0, j_0) along the road travel direction and another vanishing point (i_1, j_1) perpendicular to the road travel direction from analyzing the image frame(s). The physical parameters f, ϕ, θ can then be determined by

$$
f = \sqrt{-\left(j_0^2 + i_0 i_1\right)}, \phi = \tan^{-1}\left(\frac{-j_0}{f}\right), \theta = \tan^{-1}\left(\frac{-i_0 \cos\phi}{f}\right)
\tag{5.6}
$$

If the height of the camera is also known, then one can recover the projective matrix Equation 5.5 using the detected pixel locations of the two vanishing points from scene analysis and Equation 5.6. The approach briefly reviewed here is only one of the many known model-based calibration methods. Note that there are several advantages to using model-based calibration methods over using manual approaches. First, model-based methods are more robust since fewer parameters need to be estimated. There is no need for any placement of reference marks on the road. The calibration can be easily updated and refined as the camera setting is changed or drifted over time since the parameters are derived through the scene analysis. In many road-side settings, the camera height is fixed but may be paned, zoomed, or tilted over time. The method of Equations 5.5 and 5.6 fits well for these scenarios.

Next, we review a few examples of model-based calibration methods from the perspective of impact on aspects of speed measurement, including accuracy. First, we discuss traffic-flow vision applications, where the goal is measurement of *average speed* and vehicle counting rather than law enforcement. In Refs. [6–12], the approaches taken focus on the use of vanishing points and/or heuristic knowledge for deriving the projective matrix transform. The vanishing point(s) are identified directly from the scene. Hence, they can be automatically updated as the scene changes, for example after pan, zoom, or tilt (PZT) operations. Furthermore, scene changes can be detected by analyzing the motion activity within the scene [7], which makes the calibration steps fully automated and dynamic. More specifically, in Ref. [6] the heuristic knowledge used includes a scale factor that varies linearly as a function of the traveling direction, which reduces the problem to a single dimension with known vehicle length distributions. The use of a known vehicle length distribution yields reasonable accuracy for average speeds over 20 s intervals (4% difference from inductive loop methods); however, the accuracy of individual vehicle speed estimates is quite poor. In Ref. [6], it is noted that the effect of shadows on centroid tracking is the main contributor for inaccuracies larger than 10%. In Ref. [8], lane boundaries and vanishing points are detected using a motion activity map. The histogram of average speed over 20 s intervals shows a bias of 4–8 mph compared to inductive loop measurements.

Note that unlike Ref. [6], where blob centroids are used for speed estimation, the pixel at the lowest row of the vehicle blob is used for speed calculation in Ref. [8]. Recently, Ref. [9] proposes a practical traffic camera calibration method from moving objects. This approach also follows the general concept of model-based camera calibration through the identification of vanishing points from the scene. The key difference of their approach comes from how the vanishing points are identified. None of the prior methods use the expected general orientations of moving vehicles or humans for identification of these vanishing points. The clear advantage of Ref. [9] is the sheer volume of available data since there are many more vehicles moving on the road than lane marks. However, there are also more challenges in order to get good quality data. Overall, Ref. [9] achieves good accuracy in a variety of traffic scenarios but no reported accuracy on how it translates to speed measurement accuracy.

Law enforcement is primarily concerned with the *speed of individual vehicles*, and here accuracy of the measurement becomes a critical concern. Accuracy requirements can be as tight as ±1 mph or ±1%. In Ref. [11], vanishing points and the assumption that the mean vehicle width is 14 ft are used to construct a camera calibration and resulting projective matrix transform. The reported inaccuracy of the estimated speed of an individual vehicle is below 10%, a figure somewhat below that achieved when lane boundaries are used for camera calibration [8]. Note that the improvement in speed estimation accuracy may not necessarily be due to differences in the calibration procedure; rather, it may be due to the use of a vehicle tracking method that is insensitive to shadows. In Ref. [12], the vanishing point is first detected from the road edges of the scene. The camera calibration mapping is then derived in a manner similar to the methods discussed earlier. The reported inaccuracy of the average speed of three test vehicles with 10 runs each is 4%. In Refs. [13–15, 18], the camera calibrations are all performed based on the known real-world coordinates of some form of landmarks. The reported inaccuracy of the speed estimates for individual vehicles ranges from 1.7, ±3, to ±5 km/h for five tested cars with speeds ranging from 13 to 25 km/h.

Consider an accuracy issue related to the height of a vehicle image feature being tracked and the dimensionality of the image acquisition scenario. As shown in Figure 5.3, a camera views a vehicle from an angle, and a tracking algorithm tracks one or more features (e.g., feature A and B) in the

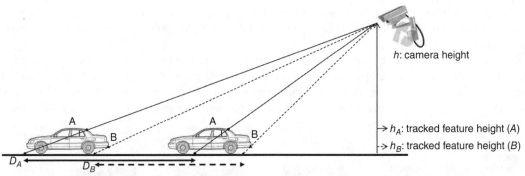

Observed frame-to-frame travel distance of image feature(s)
when projected to the ground
Note that perceived distance of tracked high feature A, D_A, is larger than that of tracked low feature B, D_B

Figure 5.3 Illustration of an accuracy issue related to tracked vehicle image feature height and the dimensionality of image acquisition.

acquired images. Speed on the road surface is the desired measure, while the feature being tracked is generally above the road at an unknown height. It is usually not possible to determine the height of the tracked feature because a single camera image is a 2D representation of a 3D phenomenon, which introduces mapping ambiguities. The calibration of 2D pixel locations to road locations assumes a given feature height, such as the road surface. Speed measurement based on tracked features at other heights will be inaccurate due to the discrepancy between assumed and real feature heights. The issue is less severe if vehicle speeds are calculated based on features that are the lowest edges or points of a motion blob (closest to the ground plane) [8, 14–16] rather than the centroids of a motion blob. Alternatively, if the height of the tracked feature is known or can be estimated, then the measured speed can be height corrected. Using the example in Figure 5.3, it can be shown that $D_A = \dfrac{h - h_B}{h - h_A} D_B$. This implies that the actual ground speed can be corrected from the tracked speed of the feature (A) with the factor $\left(1 - \dfrac{h_A}{h}\right)$. The higher the tracked feature, the more the reduction needed. The higher the camera height, the less the tracked height impacts the accuracy of the speed estimation without correction. As discussed in Section 5.2.2, the height of the tracked feature can be estimated through stereo imaging, which solves the dimensionality problem described here.

Finally, a typical accuracy requirement for speed enforcement systems can be as tight as ±1 mph or ±1%. It is thus necessary to have an accurate reference measurement system that is at least an order of magnitude more accurate and precise. An example of research on this topic is found in Ref. [17].

5.2.2 Speed Estimation from Stereo Cameras

Speed estimation from monocular imagery presents specific challenges that may be difficult to overcome. As discussed, vehicle detection is a key step in the process; methods for moving object detection from monocular video include foreground detection via background estimation and subtraction, temporal frame differencing, and flow analysis. Shortcomings associated with all three approaches include the "black hole" problem whereby it becomes difficult to detect an object in motion whose appearance closely matches that of the background; additionally, when the moving object casts a shadow, the shadow can be detected as part of the moving object itself, which may pose difficulties for subsequent tracking tasks. Even when object detection issues are overcome, ambiguities arising from the projection of the 3D world onto a 2D plane result in increased speed estimate uncertainties as described before.

Speed estimation based on binocular, depth-capable systems has been developed and deployed to overcome these limitations. Depth-capable systems rely on the use of two or more calibrated cameras with overlapping fields of view acquiring simultaneous images of the monitored scene. As illustrated in Figure 5.4, the speed estimation process from a stereo camera is similar to that from a monocular

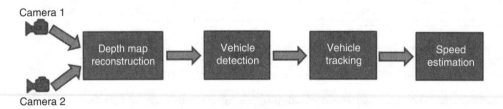

Figure 5.4 Speed estimation process in a stereo vision system.

camera, except that imagery from two cameras is processed jointly to obtain information of the scene including, but not limited to, the distance or range between the system and a feature in the scene, as well as real-world coordinates of objects in the scene. With the forthcoming discussion, it will become clear that the advantages brought about by depth-capable systems based on binocular imaging come at the cost of tighter design requirements. In particular, the accuracy of depth estimates depends on, among other factors, the placement of the system relative to the road being monitored, the resolution of the sensors in each of the monocular cameras in the system, and on the relative positioning between the cameras and the disparity between the assumed and real relative positioning. For instance, the monocular cameras comprising the system in Ref. [19] are located 100 cm from each other, have a 1.4 megapixel sensor each, and are housed in a structure made of carbon fiber, alloy, and polycarbonate to ensure high rigidity (torsion is rated at under 6.4 μrad) ensuring the relative positioning of the cameras is maintained under the most extreme environmental forces. The system is carefully calibrated at the factory, and error-checking routines are performed continuously in order to verify that the mutual geometric constraints between the monocular cameras are maintained so as to guarantee high precision depth (depth estimation errors are at most 3 cm) and speed (average speed estimation errors are under 1% in the 20–240 km/h speed range) estimates. The system is placed on a vertical pole mount at a slight angle from traffic, 15–25 m behind or ahead of the highway area being monitored.

It will also be appreciated with Sections 5.2.2.1–5.2.2.4 that the computational requirements of a depth-capable system are more demanding than those imposed by a monocular system, not only because two video streams are processed simultaneously but also because joint processing of the streams is required in order to extract depth estimates. This means that there is an additional processing overhead on top of the already doubled computational requirements, which places stricter constraints on the computational resources of the system, in particular when real-time monitoring of potentially busy multilane highways is desired. Ultimately, the ability to estimate the depth of objects in the scene leads to improved robustness to environmental and traffic conditions as well as increased capabilities to extract accurate vehicle trajectories, which in turn translates into added extensibility to other traffic enforcement-related applications such as vehicle classification, RLE, and forbidden turn monitoring.

5.2.2.1 Depth Estimation in Binocular Camera Systems

In the simplest scenario, the depth capable systems consist of two cameras, each defining coordinate systems x_1, y_1, z_1 and x_2, y_2, z_2 with origins O_1 and O_2, respectively [20]. Let us (approximately) model the cameras as pinhole cameras with each optical axis coinciding with their respective z-axis. Also, let the cameras have focal points located at F_1 and F_2 at distances f_1 and f_2 from their respective origins along their respective z-axis, as illustrated in Figure 5.5. Image planes π_1 and π_2 lie on planes $x_1 y_1$ and $x_2 z_2$, respectively.

Coordinates between the two coordinate systems are related by

$$\begin{bmatrix} x_2 \\ y_2 \\ z_2 \end{bmatrix} = R \begin{bmatrix} x_1 \\ y_1 \\ z_1 \end{bmatrix} + T, \tag{5.7}$$

where R is a 3 × 3 orthonormal matrix describing a rotation and T is a 3 × 1 vector representing translation. Since the cameras have overlapping fields of view, assume both cameras are imaging an object point P_0 located at $[x_{p0}, y_{p0}, z_{p0}]^T$ relative to O_1 and $[x'_{p0}, y'_{p0}, z'_{p0}]^T$ relative to O_2 in the 3D space that projects onto point P_1 on image plane π_1 and onto point P_2 on image plane π_2. Note that $[x_{p0}, y_{p0}, z_{p0}]^T$

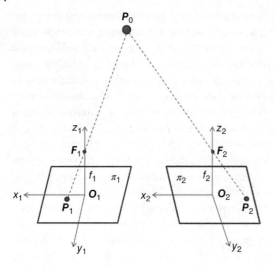

Figure 5.5 Depth capable system comprising two cameras.

and $[x'_{p0}, y'_{p0}, z'_{p0}]^T$ are related by Equation 5.7. The coordinates of P_1 relative to O_1 are $P_1 = [x_{p1}, y_{p1}, 0]^T$, and the coordinates of P_2 relative to O_2 are $P_2 = [x_{p2}, y_{p2}, 0]^T$ given that both P_1 and P_2 are on their respective image planes. Since P_0, F_1, and P_1 are collinear, and P_0, F_2, and P_2 are collinear as well, the relationships $P_1 = F_1 + \alpha_1(P_0 - F_1)$ and $P_2 = F_2 + \alpha_2(P_0 - F_2)$ hold for some scalars α_1 and α_2. Expanding each of these equations into their scalar forms, we get

$$\frac{x_{p0}}{x_{p1}} = \frac{y_{p0}}{y_{p1}} = \frac{f_1 - z_{p0}}{f_1} \tag{5.8}$$

$$\frac{x'_{p0}}{x_{p2}} = \frac{y'_{p0}}{y_{p2}} = \frac{f_2 - z'_{p0}}{f_2} \tag{5.9}$$

Assuming the focal lengths of the cameras, the rotation matrix R, and the translation vector T, P_1, and P_2 are known, Equations 5.7, 5.8, and 5.9 can be used to compute P_0 in the 3D world. Note that this assumes a correspondence between P_1 and P_2 has been established. A depth map of the scene can be reconstructed by combining 3D coordinates of points for which correspondences are found. At other locations, depth information can be inferred via interpolation techniques.

Given two or more images acquired with two or more cameras with at least partially overlapping fields of view, the correspondence problem refers to the task of finding pairs of points (one in each image) in the images that can be identified as being projections of the same points in the scene. Correspondences can be found via local correlations, or via feature extraction and matching. Assume that the location of an image point P_1 on image plane π_1 is known and that we need to find the location of the corresponding point P_2. Instead of searching across the whole image plane π_2 for corresponding point P_2, the size of the search space can be significantly reduced by noting that P_2 must be located along the line segment defined by the intersection between π_2 and the plane defined by the points P_1, F_1, and F_2. This is referred to as the epipolar constraint.

5.2.2.2 Vehicle Detection from Sequences of Depth Maps

Once a sequence of depth maps of the scene is available, vehicles traversing the scene can be detected with a high degree of confidence. As mentioned earlier, object detection in monocular video sequences is usually addressed by detecting differences in appearance (e.g., intensity or color) between the foreground object and a background model, which may lead to erroneous detections when the object's appearance closely matches that of the background. In the case of a stereo vision system, in addition to scene appearance, the depth of various points in the scene relative to the camera system is known. Since foreground objects are usually located between the background and the camera system, more robust foreground object segmentation can be achieved by augmenting background appearance models with range data.

More specifically, consider modeling the attributes of the background via a statistical model such as a distribution comprising a mixture of K Gaussian components, as proposed in Ref. [21]. According to this approach, a historical statistical model for each pixel is constructed and updated continuously with each incoming frame at a predetermined learning rate. Foreground detection is performed by determining a measure of fit of each pixel value in the incoming frame relative to its constructed statistical model: pixels that do not fit their corresponding background model are considered foreground pixels. Instead of modeling attributes related to the appearance of the background only, the statistical models can be seamlessly extended to capture the behavior of the background in terms of its depth relative to the scene, as proposed in Ref. [22]. This process is illustrated in Figure 5.6. Let $F_{i,j}$ denote the ith video frame, acquired by camera j, where i represents a discrete temporal index, and $j \in \{1, 2\}$. Let F_i denote the augmented ith video frame including color and depth information obtained by fusing frames $F_{i,1}$ and $F_{i,2}$. F_i will generally be smaller in spatial support, but possibly larger in terms of number of channels/bits per channel than $F_{i,1}$ and $F_{i,2}$ since depth information can only be extracted from scene locations that fall in the common field of view of the cameras. Background estimation is achieved by estimating the parameters of the distributions that describe the historical behavior in terms of color and depth values for every pixel in the scene as represented by the augmented video frames F_i. Specifically, at frame i, what is known about a particular pixel located at (x, y) in F_i is the history of its values $\{X_1, X_2, \ldots, X_i\} = \{F_m(x, y), 1 \le m \le i\}$, where X_m is a vector containing color and depth values, that is $X_m = [R_{Xm}, G_{Xm}, B_{Xm}, d_{Xm}]$ in the case of RGB cameras. The value of

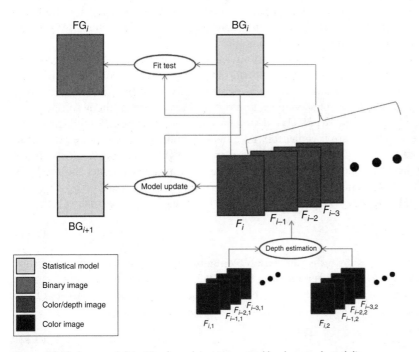

Figure 5.6 Foreground detection based on augmented background modeling.

d_{Xm} can be estimated via Equations 5.7, 5.8, and 5.9, while the values of R_{Xm}, G_{Xm}, and B_{Xm} can be estimated, for example, by averaging the RGB values of the corresponding pixels used to estimate d_{Xm}. For pixels for which a correspondence wasn't found, but which are still within the reduced field of view associated with F_i, RGB and depth values can be estimated via interpolation.

The recent history of behavior of values of each pixel can be modeled as a distribution consisting of a mixture of K Gaussian components, so that the probability of observing the current value is

$$P(X_m) = \sum_{k=1}^{K} w_{km} \Phi(X_m, \mu_{km}, \Sigma_{km}),$$

(5.10)

where w_{km} is an estimate of the weight of the kth Gaussian component in the mixture at time m, μ_{km} is the mean value of the kth Gaussian component in the mixture at time m, Σ_{km} is the covariance matrix of the kth Gaussian component in the mixture at time m, and $\Phi(\cdot)$ is the Gaussian probability density function. Sometimes, a reasonable assumption is for the different pixel attributes to be uncorrelated, in which case $\Sigma_{km} = \sigma_{km}I$, for a set of scalars σ_{km}, $1 \leq k \leq K$, where I is the $K \times K$ identity matrix. Let BG_i denote the ith augmented background model, that is, an array of pixel-wise statistical models of the form of Equation 5.10 including color and depth information. Then FG_i, the ith foreground binary mask indicating the pixel locations associated with a detected foreground object can be obtained via comparison between BG_i and F_i. Foreground detection is performed by determining a measure of fit of each pixel value in the incoming augmented frame F_i relative to its constructed statistical model which is stored in BG_i and has the form of Equation 5.10. In the simplest implementation, as a new frame comes in, every pixel value in the augmented frame F_i is checked against its respective mixture model so that a pixel is deemed to be a background pixel if it is located within T_1 standard deviations of the mean of any of the $K - 1$ color components, and its estimated depth is within T_2 standard deviations of the depth component with the largest mean, where T_1 and T_2 are predetermined thresholds. The former condition checks for consistency in the color appearance between the current pixel and the background model, and the latter verifies that the objects in the current frame are as far as possible from the camera system, relative to what has been observed in the past.

Since scenes are usually dynamic and in order to maintain accurate foreground detection, the background model BG_i can be updated to obtain a background model BG_{i+1} after incoming augmented frame F_i is processed. Note that FG_{i+1} will subsequently be determined via comparison between BG_{i+1} and augmented frame F_{i+1}. If the current pixel value is found not to match any of the K components according to the fit test described before, the pixel is identified as a foreground pixel; furthermore, the least probable component in the mixture is replaced with a new component with mean equal to the incoming pixel value, some arbitrarily high variance, and a small weighting factor. If, on the other hand, at least one of the distribution components matches the incoming pixel value, the model is updated using the value of the incoming pixel value. Updating of the model is achieved by adjusting the weight parameters for all components (i.e., for all k, $1 \leq k \leq K$) in Equation 5.10 according to

$$w_{k(i+1)} = (1 - \alpha) w_{ki} + \alpha M_{ki},$$

(5.11)

where α is the learning rate, M_{ki} is an indicator variable equaling 0 for every component except the matching one, in which case $M_{ki} = 1$. After applying Equation 5.11 to update the mixture weights, the weights are renormalized to sum up to 1. Once the weights are updated, the parameters of the distribution k_0 that was found to match the new observation are updated according to

$$\mu_{k_0(i+1)} = (1 - \rho) \mu_{k_0 i} + \rho X_i$$

(5.12)

$$\sigma_{k_0(i+1)}^2 = \left(1-\rho\right)\sigma_{k_0i}^2 + \rho\left(X_i - \mu_{k_0(i+1)}\right)^T\left(X_i - \mu_{k_0(i+1)}\right)$$ (5.13)

where X_i is the value of the incoming pixel and $\rho = \alpha\Phi\left(X_i \mid \mu_{k_0i}, \sigma_{k_0i}^2\right)$. The parameters of the components that didn't match are left unchanged.

5.2.2.3 Vehicle Tracking from Sequences of Depth Maps

Tracking can be performed once the vehicle in motion is detected. Most of the existing video-based object trackers are optimized for monocular imagery. Recall that one of the requirements for the output of a tracking algorithm to provide robust speed-related information is for it to convey positional data of highly localized portions of the vehicle being tracked. In view of this requirement, and given the fact that salient scene points are extracted when solving the correspondence problem, the authors of Ref. [23] proposed a method for efficiently tracking features extracted for stereo reconstruction purposes. Assuming the stereo system cameras satisfy the epipolar constraint, the displacement of corresponding features across frames can be represented via a 3D vector. For example, in the case where the cameras are vertically aligned, feature displacement across video frames can be uniquely represented by two horizontal offsets, one for each camera, and a shared vertical offset, $D = (d_1, d_2, d_y)$. This assumption achieves a reduction of one degree of freedom relative to the more general approach of independent tracking of the features in their respective video feeds. Let us assume features $T_1(x_1)$ and $T_2(x_2)$ from windows centered at locations x_1 and x_2 of the point being tracked are extracted from frames acquired by cameras 1 and 2, respectively, for frame i, as illustrated in Figure 5.7. For simplicity, assume that the extracted features are pixel values within the window.

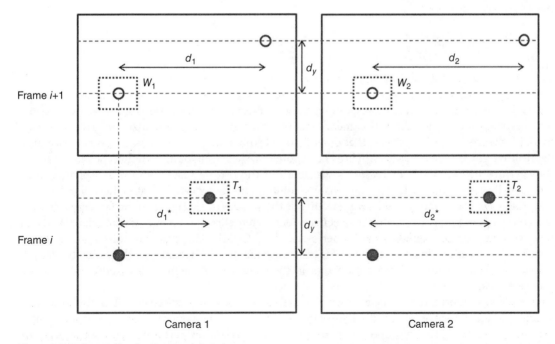

Figure 5.7 Point tracking in the stereo domain.

In the scenario considered by Lucas and Kanade [24], the task of tracking the point consists of finding its displacement between frames i and $i + 1$, which can be expressed as finding warping parameters $\boldsymbol{p}^*$ that satisfy

$$\boldsymbol{p}^* = \operatorname{argmin}_p \sum_{j=1}^{2} \sum_x \left[I\left(W_j\left(\boldsymbol{x}_j; \boldsymbol{p}\right)\right) - T_j\left(\boldsymbol{x}_j\right) \right]^2, \tag{5.14}$$

where $I(W_j(\boldsymbol{x}; \boldsymbol{p}))$ is an image resulting from warping frame $i + 1$ acquired by camera j at location $\boldsymbol{x}$ according to warping parameters $\boldsymbol{p}$. For simplicity, assume that the frame rate of the acquired video is large enough relative to the apparent motion of the object so that $\boldsymbol{p}$ can be well approximated by displacements (d_j, d_y) for $j = 1, 2$. Equation 5.14 then becomes

$$\left(d_1^*, d_2^*, d_y^*\right) = \operatorname{argmin}_{\left(d_1, d_2, d_y\right)} \sum_{j=1}^{2} \sum_x \left[I\left(W_j\left(\boldsymbol{x}_j; \left(d_j, d_y\right)\right)\right) - T_j\left(\boldsymbol{x}_j\right) \right]^2, \tag{5.15}$$

where $I(W_j(\boldsymbol{x}_j; (d_j, d_y)))$ is the image window displaced by (d_j, d_y) from pixel location $\boldsymbol{x}_j$ in frame $i + 1$ acquired by camera j, for $j = 1, 2$. The expression in Equation 5.15 can be solved iteratively by assuming a current estimate of (d_1, d_2, d_y) is available and solving for increments $(\Delta d_1, \Delta d_2, \Delta d_y)$ as demonstrated in Ref. [23].

5.2.2.4 Speed Estimation from Tracking Data

Tracking information for features extracted in and around the identified vehicle consists of a temporal sequence of 3D coordinates $\boldsymbol{P}_s^{(1)}, \boldsymbol{P}_s^{(2)}, \ldots, \boldsymbol{P}_s^{(i)}$, where $\boldsymbol{P}_s^{(k)}$ denotes the coordinates at frame k of feature s being tracked. Pair-wise instantaneous speed estimation computation for frame k is then performed by computing

$$v_s^{(k)} = r \left\| \boldsymbol{P}_s^{(k)} - \boldsymbol{P}_s^{(k-1)} \right\|, \tag{5.16}$$

where r is the frame rate of the image capture system.

There are multiple sources of error that affect the instantaneous speed estimation process described before. For one, the location of corresponding features is established relative to a discrete set of coordinates determined by the discrete camera sensors, which gives rise to quantization errors intrinsic to digital systems. It can be shown that quantization errors result in uncertainties on the estimated depth or range of the features for which a correspondence has been found, and that the magnitude of the depth estimation error increases as the depth of the feature relative to the sensor increases. Additionally, for fixed focal length and sensor resolution, the larger the separation between the cameras, the larger the range of depths supported, but the smaller the accuracy in the feature matching; this conflict gives rise to a trade-off between how accurately correspondences can be established and how accurately range estimation can be performed [25]. Clearly, the parameters of a stereo imaging system affect depth estimation capabilities, and consequently, the accuracy of the estimated speed. Careful selection of the system parameters is then required to achieve the required accuracy specifications.

In order to address errors associated with instantaneous speed measurements, Ref. [26] proposes acquiring a set of measurements at different times and performing regression on the multiple measurements. According to the proposed approach, linear regression is performed on a set of range or distance measurements; the slope of the resulting linear model indicates whether the vehicle is approaching or receding from the camera system. A mean speed estimate can be obtained from the slope estimate. Additionally, a confidence level represented by the R^2 value of the regression process can be computed, which in turn reflects the confidence of the estimated speed. Estimates with confidence level below certain predetermined thresholds may be discarded.

5.2.3 Discussion

In summary, although individual vehicle speed is a straightforward output from most computer vision systems, there is an accuracy gap for single camera systems. While stereo cameras [19, 27] for photo enforcement are becoming widely available, there are very few scientific publications on calibration and practical accuracy of 3D systems. There are also potential issues with a lack of accurate reference measurements. In addition to 3D solutions, another common approach to photo enforcement of speed has been through use of radar/lidar for speed and a camera for vehicle identification and evidence recording [28].

5.3 Stop Violations

Driver violations at intersections are the major cause of fatal accidents in urban traffic. According to the US Department of Transportation (USDOT), nearly half of all traffic accidents and 20% of all fatal crashes occur in close proximity to an intersection [29]. This statistic has remained substantially unchanged since the past decade despite significant efforts by transportation agencies for improved intersection designs and sophisticated applications of transportation engineering [29]. To combat this persistent trend, municipalities and law enforcement agencies are employing automated stop enforcement technologies at red lights and stop signs, as well as for stopped school buses.

5.3.1 Red Light Cameras

The most common violation at intersections is running a red light. Several studies found in the literature show the prevalence and severity of stop light violations [30–32]. One study conducted in Arlington, Virginia, reported a red light violation occurs on an average of every 20 min at each intersection [33]. The rate of violations significantly increases at peak hours, and rises as high as one per 5 min, which in turn causes a high number of traffic accidents, with associated damaged property and lost lives [33].

Prevention of intersection-related crashes to improve public safety and reduce property damage has led to the adoption of two primary approaches [34], namely optimizing signal light timing and stop light enforcement via red light cameras (RLCs). Several studies claim that longer yellow light duration reduces red light running violations [35, 36], and increased yellow light duration with an all-red interval can reduce the number of accidents [37, 38]. The Institute of Transportation Engineers (ITE) has provided a standard for minimum yellow light duration Y, given by Equation 5.17 and computed in Table 5.1.

$$Y = t + \frac{1.47v}{2(a+Gg)}, \tag{5.17}$$

where

Y = length of yellow interval (s)
t = perception-reaction time (use 1 s)

Table 5.1 Minimum yellow light intervals given approach speeds for straight road and 0 grade.

Approach speed (mph)	Yellow interval (s)
25	3.0 (rounded up)
30	3.2
35	3.6
40	4.0
45	4.3
50	4.7
55	5.0
60	5.4
65	5.8

v = approach speed (mph)
a = deceleration rate in response to the onset of a yellow indication (use $10\,\text{ft/s}^2$)
g = acceleration due to gravity ($32.2\,\text{ft/s}^2$)
G = fractional grade, with uphill positive and downhill negative

Most studies examined increasing Y by 0.5–1.5 s over the ITE minimum.

RLC systems, also known as RLE systems, are deployed extensively throughout the world. The efficiency of RLCs on reducing number of red light running has been reported in several studies [30, 32, 39, 40]. In Ref. [39], the efficiency of RLCs was evaluated in New York City, Polk County, Florida, Howard County, Maryland, and it is reported that enforcement cameras yielded 20% reduction in violations in New York City while showing promising results for the other municipalities. Similarly, Ref. [40] reports that RLCs reduce stop light violations on average up to 50% based on a study performed on international RLCs.

5.3.1.1 RLCs, Evidentiary Systems

Common RLC systems consist of the following three modules: (i) violation detection, (ii) evidentiary photo/video capturing, and (iii) control unit [41, 42]. Figure 5.8 shows an illustration of a typical RLC system. The violation detection module uses in-ground induction loops to estimate the speed of an approaching vehicle before the stop bar at the intersection. The speed estimate is used to conjecture whether the vehicle will run through the red light. The cameras are solely utilized to capture evidentiary images and videos, and vehicle identification information (i.e., license plate number) of violators is extracted from the evidentiary images by human operators [42].

The violation detection module relies on two magnetic induction loops buried under the pavement and in communication with the controller unit to estimate vehicle speed as it approaches the stop bar. The speed estimation can be as simple as dividing the distance between induction loops by the time difference that vehicle is detected by each loop. More complicated regression models can be also employed based on the road geometry and historical data. When a vehicle activates the first and second induction loops within a time threshold (e.g., estimated speed above threshold) when the light is red, then a violation signal is sent to trigger the evidentiary cameras. The cameras and control system record information about the violation event, such as date, time, estimated speed, license plate, and lane of violation. This auxiliary information can be combined with an image or images and used in a citation document.

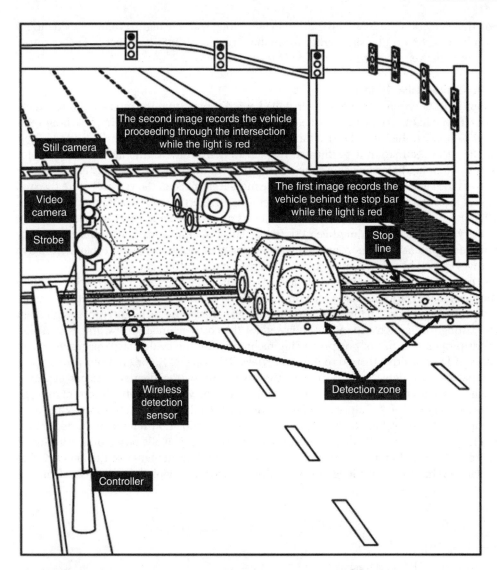

The second image records the vehicle proceeding through the intersection while the light is red

Still camera

Video camera

Strobe

The first image records the vehicle behind the stop bar while the light is red

Stop line

Wireless detection sensor

Detection zone

Controller

Figure 5.8 Red light enforcement system as described in Ref. [43]. *Source*: App. No 14/278196.

As Figure 5.8 shows, current RLE systems typically include two digital cameras. The first (i.e., upper camera in Figure 5.8) is a relatively high-resolution (e.g., 4080 × 2720 pixels and above) still image camera that takes two evidentiary pictures (Shot A and Shot B) intended to show a violator before the stop bar and at the intersection. The license plate of the violator is acquired from high-resolution evidentiary images captured by the first camera. The high-resolution cameras might also have NIR capabilities to enable night-time functionality with an external IR illuminator. NIR cameras, however, capture images in grayscale, which can be restrictive in some evidentiary settings. Therefore, in most RLC systems, RGB cameras are used with a visible light illuminator to provide night-time functionality. The illuminator is flashed behind the violator to view the rear license plate and to prevent disrupting the driver. The second camera (i.e., the lower camera in Figure 5.8) operates at lower resolution (e.g., 896 × 596 pixels) and captures a video of the incident given a trigger from the violation detection module. The captured video along with the high-resolution evidentiary photos are transferred to a processing center for manual review. A citation ticket is mailed to the registered owner of the vehicle once a violation is manually reviewed.

5.3.1.2 RLC, Computer Vision Systems

Induction loops in RLC systems are typically installed at a distance from the intersection for safety reasons, so that traffic lights can be automatically switched based on the speed of approaching vehicles to prevent T-bone crashes [44]. The performance of the violation detection module is adversely impacted as the speed estimation is performed further away from the traffic lights, especially for vehicles abruptly stopping before reaching the intersection. In these cases, the violation detection module identifies the vehicle as a violator based on the measured speed of the vehicle before the stop bar, and triggers the enforcement camera. In one operating RLC system, for example, it is reported that only 20% of the detected cases were actual violators and the remainders were false positives detected by the violation detection module [43]. In current RLC implementations, the evidentiary photos and videos of potential violators are manually reviewed to avoid issuing improper citations to the false positives caused by vehicles making a sudden stop at the traffic light.

Most of the computer vision and video analytics algorithms in the literature focus on the violation detection module [43, 45–50]. The proposed computer vision and video analytics algorithms can be divided into two main groups. The methods in the first group propose techniques to complement and support existing RLC systems to reduce the need for the costly manual review process [43, 45]. These methods propose the use of postprocessing techniques on the evidentiary images/videos to reduce number of false positive that go to human review. Reference [43], for example, proposes an algorithm to reduce false detections of nonviolators using the two evidentiary images captured before the vehicle reaches the stop bar (e.g., Shot A) and when the vehicle is at the intersection (e.g., Shot B) as shown in Figure 5.9. A feature matching algorithm is applied between the Shot A and Shot B images using speeded up robust feature (SURF) points and a violation/nonviolation decision is made based on the attributes of the matched features. The decision is made based on a criterion of finding "coherent clusters of matched features" that comply with the attributes of the matched features on a violating vehicle as shown in Figure 5.10. More specifically, a violation is detected from the

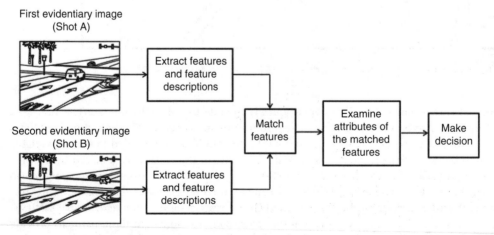

Figure 5.9 High-level overview of the red light violation detection system proposed in Ref. [43] from evidentiary photos. *Source*: App. No 14/278196.

(a)   (b)

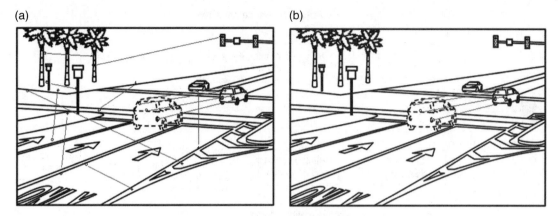

Figure 5.10 Feature matching in evidentiary red light images from Ref. [43]. Matched SURF features in Shot A and Shot B images (a), and coherent cluster of matched features after eliminating feature pairs irrelevant to red light violation (b). *Source*: App. No 14/278196.

evidentiary images if a coherent cluster of matched features is detected satisfying the following conditions:

- Length of lines connecting matched feature pairs within the cluster are longer than a specific threshold T.
- Angle lines connecting matched feature pairs within the cluster are within a specified angular interval around the road direction.
- Matched feature pairs within the coherent cluster should all start before the stop bar and end in the intersection.

Reference [45] proposes another computer vision method to support existing RLC systems by reducing false positives. The method analyses the evidentiary video clips to automatically identify and distinguish violators from nonviolators. The following steps are included in the method: (i) defining ROIs in the video (i.e., virtual detection zones before traffic lights for vehicle detection); (ii) detecting the state of traffic lights; (iii) when a vehicle is detected in the ROI and the traffic light is red, extracting a set of attributes/features from the vehicle and tracking it across frames; and (iv) processing and analyzing the trajectories of the tracked vehicles to identify a violation. When coupled with the method in Ref. [43], it is reported that a significant reduction can be achieved in automatically reducing false positives of current RLC systems [45].

The second group of computer vision methods eliminate the need for induction loops by using video cameras to identify violations in real time [46–52]. Video-based violation detection provides a nonobtrusive approach that can reduce construction and maintenance costs associated with the installation of induction loops in the pavement. Figure 5.11 illustrates an overall block diagram of the video-based red light violation detection system proposed in Ref. [48]. The system consists of two main modules: (i) traffic light sequence detection (TLSD) module that determines the spatial coordinates of the lights and estimates light sequence from video, and (ii) vehicle motion estimation (VME) module to detect a red light running violation. The position of traffic lights in a video can be determined by performing edge detection and finding the closed loops corresponding to the shape of the traffic signals within the edge map. This, however, is a challenging task given that outdoor scenes are uncontrolled and can be quite complex. A better way to find the position of traffic lights is based on detection of colored regions in video. This detection requires multiple frames as different

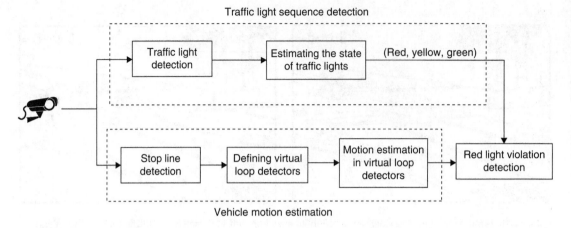

Figure 5.11 Block diagram for the video-based red light violation detection system based on Ref. [48]. *Source*: Yung and Lai [48]. Reproduced with permission of IEEE.

lights turn on at different times. Colored regions can be detected in various color spaces. Reference [48] performs color segmentation for red, yellow, and green regions in the Hue-Saturation-Value (HSV) color space based on the following criteria:

$$\text{Red region} = \left\{ I(x,y) : \left| I_H(x,y) \right| < t_h \ \text{ and } \ \left| I_V(x,y) \right| > t_v \right\}$$

$$\text{Yellow region} = \left\{ I(x,y) : \left| I_H(x,y) - \frac{\pi}{3} \right| < t_h \ \text{ and } \ \left| I_V(x,y) \right| > t_v \right\}$$

$$\text{Green region} = \left\{ I(x,y) : \left| I_H(x,y) - \frac{2\pi}{3} \right| < t_h \ \text{ and } \ \left| I_V(x,y) \right| > t_v \right\},$$

where $I(x,y)$ represents the current frame, $I_H(x,y)$ and $I_V(x,y)$ represent the hue and value channels of the current frame. t_h and t_v are threshold values to account for color variations in outdoor environments. Reference [48] reports typical values for the thresholds t_h and t_v as $\pi/6$ and 0.7, respectively. The location of a traffic light in the video is then determined by finding red, yellow, and green regions that are spatially related (i.e., in sequence from top to bottom) and having similar sizes. After finding the location of traffic lights in the video, they are continuously monitored, as well as tracked for possible movement, to determine the sequence of the lights as the video is streamed. In the VME module, the first step is detecting the stop line in the video. This detection is performed from a single background frame where the stop line is visible. The background frame can be constructed at the initialization using conventional background subtraction methods [48]. The stop line is then detected by applying a Hough transform on the constructed background image where the a priori information about the orientation of the stop line is assumed to be precalculated based on the traffic flow along the street. The stop line is expected to be orthogonal to the traffic flow, which can be estimated by the predominant motion vectors calculated in the video over a period of time.

After detecting the stop line, a prohibited zone is defined beyond the stop line toward intersection. When the lights are red, no moving car is expected to be seen in the prohibited zone moving along

Figure 5.12 Three situations in the cross junction. The VLDs cannot distinguish right turns (situation 3) from the red light violations (situations 1 and 2). *Source*: Luo and Qin [51]. Reproduced with permission of IEEE.

the road direction. In order to detect moving vehicles, a number of virtual loop detectors (VLDs) are defined in the prohibited zone. The number of VLDs depends on both the size of the prohibited zone and the size of the VLDs. Following assignment of VLDs, a reduction process is performed to eliminate redundant VLDs based on the mean and standard deviation of motion vectors across a small set of frames with vehicle motions. The loop detectors that fall into any of the following categories are eliminated:

- VLD with average motion vector not aligned with the road orientation
- VLD with average motion vector magnitude lower than the mean magnitude over all VLDs
- VLD with standard deviation of the motion vector magnitude higher than the average standard deviation of motion vector magnitude over all VLDs

Given a reduced number of VLDs and a signal from the TLSD, the motion of each VLD is estimated by using a block matching algorithm between two consecutive frames. The block matching algorithm tries to find the best match in the target frame (i.e., the first frame) corresponding to the block in the reference frame (i.e., the second frame) in terms of mean absolute difference (MAD). A violation decision is made based on the calculated motion vectors for all VLDs and the signal from the TLSD. If motion along the road direction is detected in more than half of the VLDs with magnitudes larger than half of the mean magnitudes of motion vectors for a given lane, the algorithm makes a decision that the vehicle is in violation.

While VLDs detect red light runners by mimicking the function of induction loops, alone they are not able to distinguish legal right-turning vehicles from the red light violators, which in turn causes false positives for right-turning vehicles. Figure 5.12 illustrates three trajectories that a vehicle can follow at a junction. The first two trajectories correspond to red light runners, while the third trajectory is for the vehicles making a legal right turn. Tracking-based approaches have been proposed in the literature to address this issue [45, 49–52]. Similar to the VLD method, these methods also use video cameras to detect the traffic light sequence without a direct connection to traffic light controller. Apart from traffic light and stop line detection, these methods typically include the following three main steps to detect red light runners: vehicle detection, vehicle tracking, and trajectory analysis. Reference [51], for example, proposes a tracking-based method where vehicle detection is performed using motion analysis. Another common way to detect vehicles is using

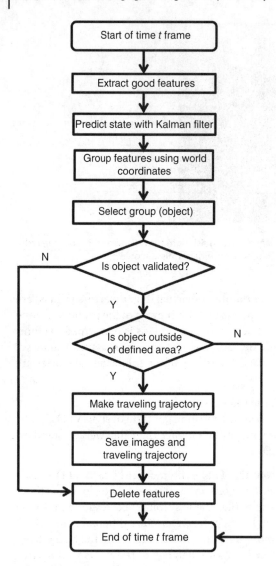

Figure 5.13 Flowchart of the tracking algorithm proposed in Ref. [49]. *Source*: Lim et al. [49]. Reproduced with permission of IEEE.

background subtraction where the background image is adaptively updated using the current frame and background image [49]. Background subtraction can be performed in conventional red, green, and blue channels [49] or in the hue channel after converting RGB image to HSV color space [52]. Once a vehicle is detected in the prohibited region using either motion analysis or background subtraction, the detected vehicle is first checked for correspondence with a vehicle already being tracked. This check ensures that only one tracker is assigned to a vehicle to prevent additional computation burden in tracking. If the detected vehicle is not in the list of vehicles already being tracked, a set of attributes/features are extracted from the detected vehicle. The extracted attributes/features depend on the type of the tracker used. After extracting a set of features/attributes from the detected region, they are tracked across the video. The tracking is performed as long as the vehicle stays within the field of view of the camera. Tracking techniques such as mean shift tracking, contour tracking, Kalman filtering, KLT tracking, and particle filtering can be employed. Figure 5.13, for example, shows the flowchart of the tracking algorithm used in Ref. [49]. The tracking algorithm specifically uses Kalman filtering where the state $p(t)$ at time t is defined as the 4D vector including the center and size of the vehicle. The state transition between time t and $t+1$ is expressed as follows:

$$p(t+1) = p(t) * \varphi(t,t+1) + w(t).$$

Here, $\varphi(t,t+1)$ and $w(t)$ are a 4×4 identity state transition matrix and a noise term, respectively.

The output of the tracking algorithm is a sequence of x–y coordinates $\{(x_1,y_1),(x_2,y_2),\ldots,(x_n,y_n)\}$ that shows the pixel position of the vehicle in the image plane at each time instant where n denotes the number of moving points. Ideally, it is desired that the sequence is continuous along the y-axis in a way that for every y_i there is a corresponding x_i so that violation detection can be determined using the continuity of the trajectory. This continuity in the sequence, however, is not always possible due to variations in vehicle speed, occlusion, and camera geometry. References [50] and [51] perform a cubic spline interpolation in order to complete the missing points in the sequence and provide continuity in the trajectory. For a viewing geometry similar to Figure 5.12, the violation decision is made based on the following criteria:

- As y_i values in the sequence increase, the extent of the change in the corresponding x_i values is less than a predefined threshold T; the vehicle is concluded as going straight, which in turn indicates a red light violation.
- As y_i values in the sequence increase, the corresponding x_i values also increase and the extent of the increase is larger than a predefined threshold T; the trajectory is determined to be that of a left-turning vehicle and indicates a red light violation.
- As y_i values in the sequence increase, the corresponding x_i values reduce and the extent of the reduce is larger than a predefined threshold T; the trajectory is determined to be that of a right-turning vehicle.

Other alternative methods for trajectory analysis have been also considered in the literature for red light violation detection [45, 49]. Reference [49], for example, determines a red light violation using template matching techniques from the calculated trajectories. Reference [45] proposes a technique that investigates start and end points of the calculated trajectory, where a violation is detected if any of the calculated trajectories has a component in an ROI defined after the traffic lights. In the case of a point tracker, a coherent cluster of trajectories, all having a component in the ROI, can be required to declare a violation.

5.3.2 Stop Sign Enforcement Systems

Nearly 700 thousand annually police-reported motor vehicle crashes and one-third of all intersection crashes in the United States occur at stop signs. Approximately one-third of intersection crashes involve injuries, and more than 40 percent of all fatal crashes occur at stop sign–controlled intersections [53]. Stop sign cameras are a relatively new tool applied to reducing stop sign violations, and resulting accidents [54–56]. The deployed stop sign camera systems are based roughly on RLC platforms, but they differ in the violation–detection mechanism. Stop sign compliance requires a complete stop (zero velocity), which cannot be detected reliably with induction loops. One proposed system uses a laser speed detector to measure velocity operating within a particular distance range for a given lane [57]. Reference [55] describes a system using a combination of high-resolution still images with full-motion video for irrefutable evidence.

Stop sign violation detection based solely on computer vision has been proposed in the literature. Reference [58] proposes a video-based method using motion vectors. The motion vectors can be read from the live, incoming video stream, or it can be read from a partially decompressed video file that may have been sent to a server for processing. Key steps in the method include the following: (i) capture video using the camera directed at the target stop area; (ii) determine motion vectors from the incoming video stream; (iii) detect the presence of a vehicle that moves across the target area by using a cluster of the motion vectors that are above a threshold in length; (iv) determine whether the detected vehicle stopped within the target area by using a cluster of motion vectors and a motion vector length threshold; (v) if a violation is detected, provide a signal of violation; and (vi) optionally, frames that capture the start, middle, or end of the violation event can be encoded as reference frames to facilitate future searches or rapid retrieval of evidentiary imagery.

In one version of the method proposed in Ref. [58], a block-based approach is used to generate the motion vectors. Motion vectors in block-based approaches describe motion between matching blocks across adjacent frames. The current frame is divided into a number of blocks of a predetermined size, for example, 16×16 pixels. For each reference block of $m \times n$ pixels in a reference frame, a current frame is searched for a target block that is most similar to the reference block. A search window can be defined around the location of the reference block in the reference frame. The search for the best-matching block in a corresponding region of the search window can be conducted using a process, such as, a full extensive search, a binary search, a three-step search, spiral search algorithms, and a combination of these.

The search is achieved by computing a similarity measure between each reference block and potential target blocks in the current frame. Displacements are computed between the select reference block and the target blocks. The displacement can computed using a mean-squared error (MSE) or MAD as follows:

$$\text{MSE}\left(d_1, d_2\right) = \frac{1}{mn}\sum\left(B\left(k,l,j\right) - B\left(k+d_1, l+d_2, j-1\right)\right)^2, \tag{5.18}$$

$$\text{MAD}\left(d_1, d_2\right) = \frac{1}{mn}\sum\left|B\left(k,l,j\right) - B\left(k+d_1, l+d_2, j-1\right)\right|, \tag{5.19}$$

where (d_1, d_2) is the vector that describes the relative displacement between reference and target blocks, and $B(k,l,j)$ denotes the pixel located on the kth row and lth column of the $m \times n$ block of pixels in the jth frame; $(j-1)$th frame is the reference frame, and jth frame is the current frame. Because both MSE and MAD measure how dissimilar two blocks are, the similarity measure can be defined as the reciprocal or the negative MSE or MAD. As mentioned earlier, this vector-based method can be used on partial uncompressed video that may have been sent to a server for off-line processing. Compression/decompression algorithms, such as the H264 and MPEG4 algorithms, can be used for extracting the motion vectors. If the standard compression block size and motion vectors are not sufficiently accurate, at minimum they can be used to screen for potential violators that are more deeply analyzed.

While the previous violations involve stopping at an intersection, school buses present a stop sign scenario that can occur in various locations along a roadway. The potential for injury to a child is quite high when vehicles drive past a school bus that is loading and unloading children. Although the buses activate flashing red lights and deploy a stop sign, it is estimated that 50,000 vehicles illegally pass a school bus each day in New York State alone [59]. Cameras are being deployed to aid in the enforcement of school bus signs. While some enforcement systems are using manual review of video to detect violators, others are proposing computer vision solutions with significantly reduced labor and cost [60]. In Ref. [60], the video is partitioned into segments each corresponding to a single bus stop, moving vehicles are detected using techniques such as frame differencing or analysis of motion vectors, frames are tagged where moving vehicles are detected, license plates are localized, and automatic license plate recognition (ALPR) is performed.

5.4 Other Violations

Driver violations at intersections are the major cause of fatal accidents in urban traffic. According to the USDOT, nearly half of all traffic accidents and 20% of all fatal crashes occur in close proximity to an intersection [29]. This statistic has remained substantially unchanged since the past decade despite significant efforts by transportation agencies for improved intersection designs and sophisticated applications of transportation engineering [29]. To combat this persistent trend, municipalities and law enforcement agencies are employing automated stop enforcement technologies at red lights and stop signs, as well as for stopped school buses.

5.4.1 Wrong-Way Driver Detection

According to the US National Transportation Safety Board [61], wrong-way driving is "vehicular movement along a travel lane in a direction opposing the legal flow of traffic." Although wrong-way driving events occur relatively infrequently (accounting for only 3% of accidents on high-speed divided highways), they can result some of the most serious types of accidents that occur on highways. Consequently, accidents resulting from wrong-way driving are much more likely to result in fatal and serious injuries than other types of accidents. As a result, considerable effort has been put into developing technologies that enable prevention and detection of wrong-way drivers. While detection systems based on microwave and magnetic sensors, as well as on Doppler radar, have been proposed [62], we focus on systems relying on video acquisition and processing.

In one instance, the authors of Ref. [63] exploited the fact that a wrong-way driving event is an anomalous event and, consequently, detectable by a system that is aware of expected patterns of motion associated with traffic on a highway. Based on this assumption, they propose a system that undergoes a learning stage in which optical flow is computed for video sequences in which vehicles are traversing the highway in a lawful manner. The normal direction of motion for each lane in the scene is represented by a mixture of Gaussians aimed at modeling the statistical behavior of the orientation parameter in the optical flow field. Once a model is constructed, optical flow is computed from frames in the incoming video feed. A determination is made that an object is traveling in the wrong direction when the difference between the direction of the flow associated with the object and that learned by the model is larger than 2.57 times the standard deviation of the closest matching Gaussian component in the model. In order to minimize false alarms, a verification stage ensures that a wrong-way notification is only issued when an anomalous motion pattern is detected more than n times across m consecutive frames, for two positive integers n and m. In Ref. [62], the feasibility of a thermal imaging system aimed at detecting entry of wrong-way vehicles onto the highway system was tested. The proposed system relied on a long-range thermal camera with dual detection zones that estimated the direction of travel of vehicles entering and exiting the highway based on the sequence of and the temporal intervals between the triggered detections. The performance of the proposed system was shown to outperform that of competing systems based on regular RGB imaging, which failed to detect wrong-way vehicles at night when the vehicle headlights were off.

5.4.2 Crossing Solid Lines

Unsafe lane changes account for a high percentage of the total accidents that occur on the road, second only to speeding. There are several scenarios for unsafe lane changes, a common one being overtaking a vehicle by crossing a solid line. The corresponding accidents are often quite severe due to the head-on nature of the collisions. Crossing a solid line is a moving violation in many jurisdictions worldwide. One photo enforcement system described in the literature is known as *Police Eyes* [64]. Police Eyes operates by detecting moving blobs and their intersection with a violation region. Key steps in the Police Eyes systems are the following:

1) *System Initialization*: An operator specifies a violation region and the processing area on an initial image indicated by manually clicking on image points.
2) *Image Acquisition from IP Cameras*: Images are acquired from two IP cameras continuously. A low-resolution image from one camera is used to detect blobs and to identify violations. A high-resolution image from the second camera is used to identify the vehicle. The images and video clips and can be later used for evidentiary purposes and to reject nontrivial cases such as avoidance of hazards.
3) *Update a Background Model and Background Subtraction*: The background model is initialized using a single frame and then updated for every new frame. A Gaussian mixture model is used for every pixel in the image. The number of Gaussian components is constantly adapted per pixel. Frame differencing from the background model produces a foreground image.
4) *Shadow Detection*: Shadow pixels must be removed from the foreground image to avoid false violation detections. Shadow pixels are identified using a combination of normalized cross-correlation between the foreground region and the corresponding background pixels, along with RGB vector distances between the foreground pixels and underlying background pixels.
5) *Blob Extraction*: Foreground blobs are extracted from the foreground image through connected component analysis after performing morphological operations on the foreground image to remove noisy blobs. The base profile is extracted for each remaining blob. The base of a blob is identified as the set of lowermost pixels of the external contour of the blob.
6) *Violation Analysis*: Analysis of the region of intersection of the base profile of each blob with the violation area is used to detect violations.

References

1 Traffic Safety Facts, 2012 Data, Speeding, National Highway Traffic Safety Administration, NHTSA's National Center for Statistical Analysis, Washington, DC, DOT HS 812 021 (2014).
2 The Economic and Societal Impact of Motor Vehicle Crashes, NHTSA's National Highway Traffic Safety Administration, National Center for Statistical Analysis, Washington, DC, DOT HS 812 013, (2010) (Revised).
3 L. Thomas, R. Srinivasan, L. Decina, and L. Staplin, Safety effects of automated speed enforcement programs: critical review of international literature, *Transportation Research Record: Journal of the Transportation Research Board*, 2078, 117–126, 2008, Transportation Research Board of the National Academies, Washington, DC.

4 A. Yilmaz, O. Javed, and M. Shah, Object tracking: a survey, *ACM Computing Surveys*, 38(4), 1–45, 2006.

5 N. K. Kanhere and S. T. Birchfield, A taxonomy and analysis of camera calibration methods for traffic monitoring applications, *IEEE Transactions on Intelligent Transportation Systems*, 11(2), 441–452, 2010.

6 D. J. Dailey, F. W. Cathey, and S. Pumrin, An algorithm to estimate mean traffic speed using uncalibrated cameras, *IEEE Transactions on Intelligent Transportation Systems*, 1(2), 98–107, 2000.

7 S. Pumrin and D. J. Dailey, Roadside camera motion detection for automated speed measurement, in *Proceedings of the IEEE 5th International Conference on Intelligent Transportation Systems*, Singapore, September 3–6, 2002, pp. 147–151.

8 T. N. Schoepflin and D. J. Dailey, Dynamic camera calibration of roadside traffic management cameras for vehicle speed estimation, *IEEE Transactions on Intelligent Transportation Systems*, 4(2), 90–98, 2003.

9 Z. Zhang, T. Tan, K. Huang, and Y. Wang, Practical camera calibration from moving objects for traffic scene surveillance, *IEEE Transactions on Circuits and Systems for Video Technology*, 36(5), 1091–1103, 2013.

10 F. W. Cathey and D. J. Dailey, A novel technique to dynamically measure vehicle speed using uncalibrated roadway cameras, in *Proceedings of IEEE Intelligent Vehicles Symposium*, Las Vegas, NV, June 6–8, 2005, pp. 777–782.

11 N. K. Kanhere, S. T. Birchfield, and W. A. Sarasua, Automatic camera calibration using pattern detection for vision-based speed sensing, in *Transportation Research Board Annual Meeting*, 87th Annual Meeting, Washington, DC, January 13–17, 2008, pp. 30–39.

12 J. Wu, Z. Liu, J. Li, C. Gu, M. Si, and F. Tan, An algorithm for automatic vehicle speed detection using video camera, in *Proceedings of 2009 4th International Conference on Computer Science and Education*, Nanning, July 25–28, 2009.

13 L. G. C. Wimalaratna and D. U. J. Sonnadara, Estimation of the speeds of moving vehicles from video sequences, in *Proceedings of the Technical Sessions*, 24, March 2008, pp. 6–12, Institute of Physics, Colombo, Sri Lanka. http://www.ipsl.lk/index.php/technical-sessions/18-publications/technical-sessions/58-volume-24-2008 (accessed on October 14, 2016).

14 L. Grammatikopoulos, G. Karras, and E. Petsa (GR), Automatic estimation of vehicle speed from uncalibrated video sequences, in *Proceedings of the International Symposium on Modern Technologies, Education and Professional Practice in Geodesy and Related Fields*, Sofia, November 2005, pp. 332–338.

15 A. G. Rad, A. Dehghani, and M. R. Karim, Vehicle speed detection in video image sequences using CVS method, *International Journal of the Physical Sciences*, 5(17), 2555–2563, 2010.

16 E. A. Bernal, W. Wu, O. Bulan, and R. P. Loce, Monocular vision-based vehicular speed estimation from compressed video streams, in *Proceedings of the 16th International IEEE Conference on Intelligent Transportation Systems for All Transport Modes*, The Hague, October 6–9, 2013.

17 P. Bellucci, E. Cipriani, M. Gagliarducci, and C. Riccucci, The SMART project: speed measurement validation in real traffic condition, in *Proceedings of the 8th International IEEE Conference on Intelligent Transportation Systems*, Vienna, September 16–16, 2005.

18 Z. Tian, M. Kyte, and H. Liu, Vehicle tracking and speed measurement at intersections using video-detection systems, *ITE Journal*, 79(1), 42–46, 2009.

19 Available: http://project-asset.com/data/presentations/P_011-4.pdf (accessed on September 19, 2016).

20 J. Battle, E. Mouaddib, and J. Salvi, Recent progress in coded structured light as a technique to solve the correspondence problem: a survey, *Pattern Recognition*, 31(7), 963–982, 1998.

21 C. Stauffer and W. Grimson, Adaptive background mixture models for real-time tracking, in *Proceedings of the IEEE Computer Society Conference on Computer Vision and Pattern Recognition*, Fort Collins, CO, June 23–25, 1999.

22 G. Gordon, T. Darrell, M. Harville, and J. Woodfill, Background estimation and removal based on range and color, in *Proceedings of the IEEE Computer Society Conference on Computer Vision and Pattern Recognition*, Fort Collins, CO, June 23–25, 1999.

23 K. Ni and F. Dellaert, Stereo tracking and three-point/one-point algorithms: a robust approach in visual odometry, in *Proceedings of the IEEE International Conference on Image Processing*, Atlanta, October 8–11, 2006.

24 S. Baker and I. Matthews, Lucas-Kanade 20 years on a unifying framework, *International Journal of Computer Vision*, 56(3), 221–255, 2004.

25 R. Balasubramanian, S. Das, S. Udayabaskaran, and K. Swaminathan, Quantization error in stereo imaging systems, *International Journal of Computer Math*, 79(6), 671–691, 2002.

26 L. Hardin and L. Nash, Optical range and speed detection system. U.S. Patent No. 5,586,063, December 17, 1996.

27 Available: http://www.imagsa.com/main/images/datasheet/Atalaya3D_Speed.pdf (accessed on September 19, 2016).

28 Available: http://www.foxnews.com/leisure/2011/11/11/russian-super-speed-camera-can-issue-thousands-tickets-per-hour/ (accessed on September 19, 2016).

29 The National Intersection Safety Problem, Brief Issues #2, U. S. Department of Transportation, Federal Highway Administration, FHWA-SA-10-005.

30 R. A. Retting, A. F. Feldman, C. M. Farmer, and A. F. Williams, Evaluation of red light camera enforcement in Fairfax, Va., USA, *ITE Journal*, 69, 30–35, 1999.

31 R. A. Retting, R. G. Ulmer, and A. F. Williams, Prevalence and characteristics of red light running crashes in the United States. *Accident Analysis and Prevention*, 31, 687–694, 1999.

32 T. Supriyasilp, D. S. Turner, and J. K. Lindly, *Pilot Study of Automated Red Light Enforcement*. University Transportation Center for Alabama (*UTCA*) *Report* 470, September 30, 2003.

33 S. E. Hill and J. K. Lindly, *Red light running prediction and analysis*. Ph.D. dissertation, University of Alabama, 2002.

34 R. A. Retting, S. A. Ferguson, and C. M. Farmer, Reducing red light running through longer yellow signal timing and red light camera enforcement: results of a field investigation, *Accident Analysis and Prevention*, 40, 327–333, 2008.

35 J. A. Bonneson and H.J. Son, Prediction of expected red-light running frequency at urban intersections. *Transportation Research Record: Journal of the Transportation Research Board*, 1830, 38–47, 2003.

36 J. A. Bonneson and K. H. Zimmerman, Effect of yellow-interval timing on the frequency of red light violations at urban intersections. *Transportation Research Record: Journal of the Transportation Research Board*, 1865, 20–27, 2004.

37 P. Guerin, *City of Albuquerque Yellow Light Timing Change and All-Red Clearance Interval Timing Change Effectiveness Study Final Report*. Prepared for: The City of Albuquerque, Department of Municipal Development and the Office of the Mayor, September 2012. http://isr.unm.edu/reports/2012/city-of-albuquerque-yellow-light-timing-change-and-all-red-clearance-interval-time-change-effectiveness-study-final-report..pdf (accessed on October 14, 2016).

38 Q. Yang, L. D. Han, and C. R. Cherry, Some measures for sustaining red-light camera programs and their negative impacts, *Transport Policy*, 29, 192–198, 2013.

39 D. M. Smith, J. McFadden, and K. A. Passetti, Automated enforcement of red light running technology and programs: a review, *Transportation Research Record: Journal of the Transportation Research Board*, 1734, 29–37, 2000.

40 R. A. Retting, S. A. Ferguson, and A. S. Hakkert, Effects of red light cameras on violations and crashes: a review of the international literature, *Traffic Injury Prevention*, 4, 17–23, 2003.

41 K. A. Passetti and T. H. Hicks, *Use of automated enforcement for red light violations*. Texas A&M University, College Station (1997).

42 R. W. Lock, Image recording apparatus and method. U.S. Patent No. 8,390,476. March 5, 2013.

43 O. Bulan, A. Burry, and R. P. Loce, Short-time stopping detection from red light camera evidentiary photos. U.S. Patent App. No 14/278196, filed May 15, 2014

44 M. Glier, D. Reilly, M. Tinnemeier, S. Small, S. Hsieh, R. Sybel, and M. Laird, Method and apparatus for traffic light violation prediction and control. U.S. Patent App. No. 09/852487, published May 9, 2002.

45 O. Bulan, A. Burry, and R. P. Loce, Short-time stopping detection from red light camera videos. U.S. Patent App. No 14/278218, filed May 15, 2014.

46 O. Fucik, P. Zemcik, P. Tupec, L. Crha, and A. Herout, The networked photo-enforcement and traffic monitoring system Unicam, in *Proceedings of the 11th IEEE International Conference and Workshop on Engineering of Computer-Based Systems*, Brno, May 27, 2004, pp. 423–428.

47 Camea, Red light monitoring, http://www.camea.cz/en/traffic-applications/enforcement-systems/red-light-violation-detection-unicamredlight-2/ (accessed December 3, 2014).

48 N. H. C. Yung and H. S. Lai, An effective video analysis method for detecting red light runners, *IEEE Transactions on Vehicular Technology*, 50(4), 1074–1084, 2001.

49 D. W. Lim, S. H. Choi, and J. S. Jun, Automated detection of all kinds of violations at a street intersection using real time individual vehicle tracking, in *Proceedings of the Fifth IEEE Southwest Symposium on Image Analysis and Interpretation*, Sante Fe, NM, April 9, 2002, pp. 126–129.

50 M. Heidari and S. A. Monadjemi, Effective video analysis for red light violation detection, *Journal of Basic and Applied Scientific Research*, 3(1), 642–646, 2013.

51 D. Luo, X. Huang, and L. Qin, The research of red light runners video detection based on analysis of tracks of vehicles, in *Proceedings of the IEEE International Conference on Computer Science and Information Technology*, Singapore, August 29 to September 2, 2008, pp. 734–738.

52 K. Klubsuwan, W. Koodtalang, and S. Mungsing, Traffic violation detection using multiple trajectories evaluation of vehicles, in *Proceedings of the 4th International Conference on Intelligent Systems Modelling and Simulation (ISMS)*, Bangkok, January 29–31, 2013. IEEE.

53 Insurance Institute for Highway Safety, 37, No 9, October 26, 2000. http://safety.fhwa.dot.gov/intersection/conventional/unsignalized/case_studies/fhwasa09010/ (accessed on October 14, 2016).

54 TheNewspaper.com: Driving Politics. Stop Sign Ticket Cameras Developed: New Stop Sign Cameras Will Issue Automated Tickets for Boulevard Stops. http://www.thenewspaper.com/news/17/1742.asp, September 5, 2007.

55 http://www.bypass.redflex.com/application/files/5814/4902/9070/redflexstop.pdf (accessed October 14, 2015).

56 M. Trajkovic and S. Gutta, Vision-based method and apparatus for monitoring vehicular traffic events. U.S. Patent 6,442,474, August 27, 2002.

57 M. Phippen and D. W. Williams, Speed measurement system with onsite digital image capture and processing for use in stop sign enforcement. U.S. Patent No. 6,985,827. January 10, 2006.

58 E. Bernal, O. Bulan, and R. Loce, Method for stop sign law enforcement using motion vectors in video streams. U.S. Patent App. No. 13/613174, March 13, 2014.

59 OPERATION SAFE STOP, An Educational Campaign by the New York Association for Pupil Transportation. Prepared by NYAPT through the National Transportation Safety Administration under a grant from the Governor's Traffic Safety Committee. http://slideplayer.com/slide/1508781/ (accessed on October 14, 2016).

60 A. Burry, R. Bala, and Z. Fan, *Automated processing method for bus crossing enforcement*. U.S. Patent App. No. 13/210447, published February 21, 2013.

61 US National Transportation Safety Board, Wrong-Way Driving. Special Investigative Report 12/01. US National Highway Traffic Safety Administration, 2012. pp. 1–77.

62 S. Simpson, Wrong-way Vehicle Detection: Proof of Concept, Report to the Arizona Department of Transportation Research Center, FHWA-AZ-13-697, March 2013.

63 G. Monteiro, M. Ribeiro, J. Marcos, and J. Batista, A framework for wrong-way driver detection using optical flow, *Proceedings of the 4th International Conference on Image Analysis and Recognition*, Montreal, August 22–24, 2007.

64 R. Marikhu, J. Moonrinta, M. Ekpanyapong, M. Dailey, and S. Siddhichai, Police eyes: real world automated detection of traffic violations, in *Proceedings of the 2013 10th International Conference on Electrical Engineering/Electronics, Computer Telecommunications and Information Technology (ECTI-CON)*, Krabi, May 15–17, 2013.

6

Traffic Flow Analysis

Rodrigo Fernandez[1], Muhammad Haroon Yousaf[2], Timothy J. Ellis[3],
Zezhi Chen[3] and Sergio A. Velastin[4]

[1] Universidad de los Andes, Santiago, Chile
[2] University of Engineering and Technology Taxila, Taxila, Pakistan
[3] Kingston University, London, UK
[4] Universidad Carlos III de Madrid, Madrid, Spain

6.1   What is Traffic Flow Analysis?

6.1.1   Traffic Conflicts and Traffic Analysis

Traffic is the movement of people and goods in the public space. Some people can ride a car, a public transport vehicle, a bicycle, or travel on foot. This movement will generate interactions between people: in one moment two individuals who are circulating may coincide at the same point. These interactions give rise to the so-called traffic conflicts [1].

Traffic conflict will occur when two or more people intend to use the same resource of the transport system simultaneously: a section of a road, an area within a junction, a space inside a public transport vehicle. Depending on how and why traffic conflicts occur, they can be classified into converging conflicts or directional conflicts. Converging conflicts take place in one portion of a road, traveling in the same direction but at different speeds. Directional conflicts occur in an intersection due to the various maneuvers made by vehicles.

Traffic flow can be studied from the point of view of traffic conflicts. Its study can be divided into two approaches. The simplest one is uninterrupted traffic flow analysis, which studies converging conflicts at road links, such as highways, rural roads, or viaducts not disrupted by junctions or entry/exit ramps. Another, more complex alternative, is analysis of interrupted traffic flow, which considers directional conflicts, such as those found at road junctions. We shall discuss uninterrupted traffic flow analysis hereafter; however, when necessary, a brief reference to traffic signal analysis is made.

Uninterrupted flow is characterized by the absence of stops along a road due to junctions or ramps. However, stops may occur because of interactions between vehicles traveling in the same direction. The study of traffic in these conditions was historically the first to be carried out. There are two approaches to the problem: the fluid-dynamic model in which traffic is considered as a continuous stream and its analysis is based on average steady-state variables; and the car-following model, which studies the interaction between each pair of vehicles and extrapolates that analysis to the whole traffic stream. For simplicity, henceforth we discuss fluid-dynamic traffic analysis.

Computer Vision and Imaging in Intelligent Transportation Systems, First Edition.
Edited by Robert P. Loce, Raja Bala and Mohan Trivedi.
© 2017 John Wiley & Sons Ltd. Published 2017 by John Wiley & Sons Ltd.
Companion website: www.wiley.com/go/loce/ComputerVisionandImaginginITS

A section of road is not necessarily a straight line. There may be horizontal and vertical curves; in addition, there may be more than one traffic lane and they have some width. However, from a viewpoint of a section of 1000 m long, traffic can be represented along one spatial axis. Moreover, traffic takes place during a time period; therefore, a time axis is also necessary to "see" the traffic. Then, traffic takes place in a space–time "window." Figure 6.1 shows how to observe the traffic of a group of vehicles traveling at different speeds on a section of length L and during a time period T, where the slopes of the time–space trajectories of vehicles are their speed [2].

In Figure 6.1, the slope of the time–space trajectories of the vehicles is their speed; the more vertical the curve, the higher the speed. Also, it shows two points where converging conflicts are solved in different ways. Point A represents an overtaking. At point B, the faster vehicle adjusts its speed to the slower one. In the figure, traffic can be observed from two viewpoints. In a time observation, the observer stands at a point s_0 and sees what happens throughout the period T. In this case, $n = 5$ vehicles. A space observation takes place at one instant of time t_0 and the observer takes a look of the whole section L, for example, from an aerial photograph, observing $n' = 4$ vehicles. Steady-state variables of the traffic stream are obtained depending on the type of observation; these are explained next.

6.1.2 Time Observation

Flow (q): It is also called flow rate or volume. It is the number n of vehicles passing the point s_0 during the period T. It is usually expressed in vehicles per hour (veh/h).

$$q = \frac{n}{T} \tag{6.1}$$

Time mean speed (v_t): It is the arithmetic average of the instantaneous velocities (time–space trajectories) of the n vehicles that pass the point s_0 during the time period T. It is usually expressed in kilometers per hour (km/h) or meters per second (m/s).

$$v_t = \frac{1}{n}\sum_i v_i \tag{6.2}$$

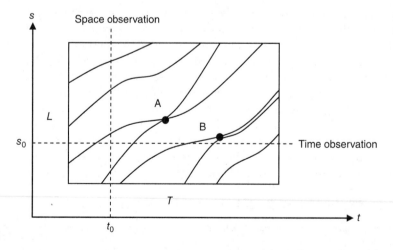

Figure 6.1 Time–space traffic diagram.

6.1.3 Space Observation

Concentration (k): It is also called density. It is the space equivalent of the flow rate. It is the number, n', of vehicles observed at time t_0 in entire section L. It is measured in vehicles per kilometer (veh/km).

$$k = \frac{n'}{L} \tag{6.3}$$

Space mean speed (v_s): It is the speed based on the average time taken to each vehicle to travel section L, in (km/h) or (m/s).

$$v_s = \frac{L}{\bar{t}} = \frac{L}{1/n' \sum_i t_i} = \frac{1}{1/n' \sum_i (t_i/L)} = \frac{1}{1/n' \sum_i (1/v_i)} \tag{6.4}$$

Consequently, the average space speed is the harmonic mean of the instantaneous speeds of vehicles on the section L.

6.1.4 The Fundamental Equation

The relationship between the average steady-state variables flow (q), concentration (k), and space mean speed (v_s) is called the "fundamental equation of traffic flow" and it can be demonstrated [2] that it is given by the following equation:

$$q = kv_s \tag{6.5}$$

The equation explains the behavior of an uninterrupted traffic stream. Note that the time mean speed does not appear in the equation. However, it has been first demonstrated by Wardrop [3] that the relationship between time and space means is the following:

$$v_t = \frac{1}{v_s}\left(v_s^2 + \sigma_s^2\right) = v_s + \frac{\sigma_s^2}{v_s}. \tag{6.6}$$

Here, σ_s^2 is the variance of the space mean speed. That is, the time mean speed is greater than the space mean. To be equal, the variance of space speeds should be 0. Hence, the importance of how the speed is measured to calculate the appropriate means. On highways, if the traffic composition is more or less the same, the variance of speeds is not too large. On the contrary, on urban roads, speeds are quite different.

The fundamental equation establishes a relationship between three variables of traffic flow. As such, it defines a plane that describes their behavior. However, it is possible to reduce this three-dimensional (3D) description establishing relationships between pairs of these variables.

6.1.5 The Fundamental Diagram

It has been postulated that the relationship between v_s and k is the following linear function, where v_f is the free-flow speed and k_j is the jam concentration. The free-flow speed is the speed chosen by a driver traveling alone on a link, and the jam concentration is that at which a traffic stream stops because vehicles are too close.

$$v_s = v_f\left(1 - \frac{k}{k_j}\right) \tag{6.7}$$

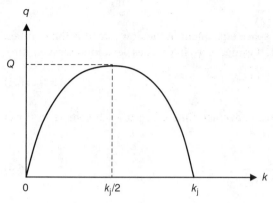

Figure 6.2 Flow concentration function.

Assuming that $v_s(k)$ is a linear function, from the fundamental equation $q = kv_s$, the following relationship between flow and concentrations $q(k)$ can be found:

$$q = v_f k - \frac{v_f}{k_j} k^2 \tag{6.8}$$

That is, the function $q(k)$ is an inverted parabolic function which crosses the k-axis at $k = 0$ and $k = k_j$. It is easily seen that the vertex of the parabola is located at $k = k_j/2$ (Figure 6.2).

An important property of the flow is derived from the $q(k)$ function. That is, there is a maximum value for the flow. This maximum is the capacity Q of a road section, which is defined as "the maximum number of vehicles per unit of time that can pass a point, under the prevailing traffic conditions." Therefore, the graphic $q(k)$ is called the "fundamental diagram." As the vertex of the parabola is located at $k_j/2$, then it follows that

$$Q = \frac{v_f k_j}{4}. \tag{6.9}$$

Note that Q depends on parameters k_j v_f defined in the $v_s(k)$ function. Therefore, the capacity depends on (i) physical characteristics of the road (e.g., geometric design and road surface); (ii) characteristics of the driver (e.g., age, personality, and physical and psychological conditions); (iii) vehicle features (e.g., power, acceleration, and maneuverability); (iv) environmental conditions (e.g., light, weather, and environment); and (v) traffic composition (e.g., length and proportion of different type of vehicles).

For the aforementioned reasons, in the definition of capacity the statement "under the prevailing traffic conditions" is included. Therefore, under different circumstances the value of the capacity may change. Consequently, the capacity is not an absolute value even for the same section of a road.

6.1.6 Measuring Traffic Variables

As mentioned previously, the variables that describe traffic behavior are *flow*, *speed*, and *concentration*. However, from the viewpoint of traffic analysis in both highways and urban roads, the most important ones are flow, capacity, and speed. The flow rate represents the demand of traffic through a given device such as a road link, junction, ramp, and bottleneck. Similarly, capacity is the throughput that those traffic devices can offer. On the other hand, speed (or inversely, the delay) represents the quality of the traffic as a function of the rate between flow and capacity which is called the "degree of saturation" $x = q/Q$ [1].

The simplest data for traffic analysis is the flow of different vehicles passing a point on either a road link or at a point such as junction, ramp, or bottleneck. Vehicle counting is divided into two classes: user type and movement type. User-type categories normally used in traffic counts are cars and light goods vehicles (LGVs), medium goods vehicles, heavy goods vehicles (HGVs), buses and other public transport, motorcycles, bicycles, pedestrians, and passengers. On the other hand, movement-type categories are through movements, right turns, left turns, and others (e.g., "U" turns) [4].

6.1.7 Road Counts

Traffic flows can be recorded either manually, using paper forms and tally counters, or by portable electronic devices [5]. This is a simple method, but requires many man-hours. Alternatively, data can be obtained using automatic detectors such as pneumatic or piezoelectric sensors which count the number of axles, instead of vehicles, passing a point on a time period. However, manually sample counts are necessary to classify vehicles and verify the accuracy of these detectors.

Today, the use of inductive-loop detectors is common either on top or under the road surface, which count and classify vehicles according to their magnetic field. The advantage of the loop detectors is that counts may be sent to a traffic control unit (TCU) via any data transmission system.

The use of image processing from video cameras is becoming popular because they "take the field to the office," so data can be reprocessed if any doubt arises. Cameras may be public (e.g., belong to the TCU or the police), or they are located on the field by the traffic engineering team. The advantage of this method is that flows of all type of vehicles may be obtained from the image instead of using time-consuming manual analysis.

6.1.8 Junction Counts

At junctions, counts must record the directional traffic conflicts. Therefore, it is necessary to count the maneuvers of all vehicles that cross the junction. This task can be carried out either manually or automatically [4].

Methods are the same as those described before. The main difference is that counters (people or detectors) must be placed just downstream of the stop-line of each approach to the junction, because individual traffic streams are identified more easily at this point. The use of video cameras and visual- or image processing should always be considered. In such a case, cameras should be placed at elevated locations to get an appropriate view of the junction; for example, buildings or structures.

There are other measurements for modeling purposes that should be carried out at junctions. Among others are saturation flows at traffic signals (i.e., throughput during the green time); queue lengths; and queuing times. Traditionally, these data have been collected by human observers because they require selection and discrimination of traffic events (e.g., the instant at which a vehicles stops), but video recordings with subsequent human processing are becoming popular among traffic engineers [4, 5].

At complex junctions, large roundabouts or highway interchanges the use of any type of counter that may be expensive in terms of time and resources. In this case a "plate survey" may be required [5]. This survey consists of recording some part of the plate number (e.g., the last two digits) of each vehicle entering and leaving the junction, as well as the time when they enter and leave the junction. This information may be recorded by any means—from paper forms to electronic data devices—and then downloaded into computer software that matches the plate numbers of vehicles entering and leaving the junction. An improvement of the method is the application of image-processing techniques for recording plate numbers and letters automatically.

6.1.9 Passenger Counts

This measurement consists of counting the number of passengers using both public and private vehicles. It can be carried out from either outside the vehicle, in the case of private vehicles, or inside the vehicles, in public transport. The reader can find more details in Chapter 4.

In the case of public transport, counts can be carried out by either boarding/alighting counts at stops along a route, or they can be conducted on-board each vehicle. As counting passengers on-board an overcrowded public transport vehicle may be difficult or time consuming, they are usually counted from outside by assigning occupancy categories to the number of passengers inside the vehicle as shown in Table 6.1 [4]. Then, according to his criterion, the observer assigns an occupancy category to each public transport vehicle.

Another way to counting public transport passengers is at terminal points (e.g., bus stations or terminal bus-stops), in which the amount of all passengers boarding and alighting vehicles at the terminal is manually recorded [5].

Recently, the use of smartcards as a mean of fare payment may provide information on the number of boarding passengers to public transport vehicles. The disadvantage is that in some systems, such as buses or some metros, the smartcard is not required to disembark the vehicle. As a consequence, the information of the number of alighting passengers is missing. In this case, a survey of the alighting passengers at bus-stops or stations must be conducted; this survey usually involves manual counts.

6.1.10 Pedestrian Counts

Pedestrian counts are needed to footway design (width, length, alignment) as well as to provide facilities at crossing points. In the case of crossing facilities, pedestrian counts are required to evaluate the need to install some form of pedestrian crossing (zebra crossing, traffic signal, etc.). For this purpose, pedestrian counts will be carried out over some length on either side of the proposed crossing point.

Whatever the objective, pedestrian counts are more difficult to perform than vehicle counts, because people tend to walk and wait in platoons. For this reason, the use of video cameras and either visual- or image processing is normally required.

6.1.11 Speed Measurement

As stated before, the average speed may be measured in two ways. One way is to measure individual speeds with a radar speedometer and average arithmetically. The other is to measure travel times of individual vehicles and get the average over a given road length. The first case would give the time mean speed v_t and the space mean speed v_s in the second one. However, it is possible to convert one into the other by the Wardrop equation. Time mean speed is used for accident analysis or for setting speed limits. Space mean speed, on the other hand, is applied in traffic modeling and management.

Table 6.1 Occupancy of a 12-m bus.

Category	Occupancy level	No. of passengers
A	Overcrowding	100
B	More than half aisle with standing passengers	70
C	Less than half aisle with standing passengers	50
D	More than half of seats busy	30
E	Less than half of seats busy	15

In highways, rural and arterial road speeds can also be measured automatically with a pair of consecutive inductive-loop detectors connected to the TCU that manage the road. This can also be done by image-processing techniques from the video cameras belonging to the TCU.

Finally, a rather old technique for measuring journey speeds is the "moving observer" method [5]. In this method, a test vehicle travels along a road section at the average speed of the traffic platoon. At each run, the travel time of the test vehicle is recorder either manually or automatically. After a number of, say 10, runs a good estimation of the average speed is obtained. In order to do that, the driver must pass the same number of vehicles that pass him/her.

Having considered the main variables that are relevant to the traffic engineer, we now look at ways in which computer vision technology may be used to measure some of these variables or other traffic-related parameters.

6.2 The Use of Video Analysis in Intelligent Transportation Systems

6.2.1 Introduction

Intelligent transportation systems (ITS) have been widely proposed with improvements in computational technologies and the increasing number of multipurpose camera installations. Researchers have paid considerable attention to exploring innovative computer vision–based solutions for traffic flow analysis. Traffic flow analysis covers a wide range of applications for ITS, for example, vehicle detection, vehicle counting, vehicle classification, traffic density measurement, traffic flow estimation, traffic speed estimation, and traffic rules violation monitoring and incident detection. To propose solutions for such applications, already installed surveillance cameras are usually preferred to enhance the usefulness of existing infrastructure. However, surveillance cameras are usually of poor quality and the height and direction of the camera also play an important role in video analytics. In an open and cluttered environment, occlusion, illumination, shadows, nonvehicle objects, and varying weather conditions make traffic flow analysis a challenging task. Section 6.2.2 deals with what has become a common framework for traffic flow analysis systems and outlines the main state-of-the-art techniques. Application domains and challenges faced especially in developing and underdeveloped countries are considered later in Sections 6.2.3 and 6.4.

6.2.2 General Framework for Traffic Flow Analysis

Here, we show a typical framework (Figure 6.3) for a computer vision–based traffic flow analysis system (see Section 6.3 for a more detailed example). Online or offline video/image data are the input to the framework. Vehicle detection is the core element for any application in traffic flow analysis, requiring a *region of interest* (ROI) or image features that must be extracted to represent the objects in the scene. The ROI/feature extraction phase transforms the input visual data into a mask/silhouette/feature vector. The output of ROI/feature extraction is then passed to a *classification stage* to detect the identified vehicle or other objects. The classification stage is highly dependent on training data. Detected objects are analyzed and used for different applications in traffic flow analysis. With reference to Figure 6.3, the following are the typical stages involved in video-based traffic flow analysis:

Video data: This required as an input to any computer vision–based system for traffic flow analysis. This visual data can be processed in both online or offline mode. To capture a useful video dataset, cameras need to be properly installed at the appropriate location, height, and angle toward the road. Parameters for installation of cameras may vary for different application domains, that is, urban, motorways, and tunnels.

ROI/feature extraction: Frames are extracted from the video sequences and passed to the ROI/ feature extraction block. This block converts the input frame into a ROI/silhouette/feature vector, which is then passed to the classifier. According to the literature, ROI/feature extraction is categorized into two types, that is, foreground estimation/segmentation or an object recognition–based approach. A pictorial representation of ROI/feature extraction is presented in Figure 6.4.

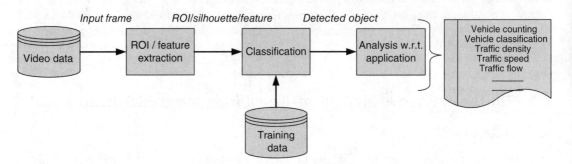

Figure 6.3 Typical/general framework for traffic flow analysis.

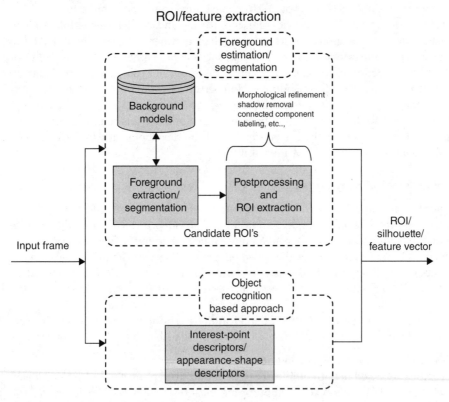

Figure 6.4 ROI/feature extraction.

6.2.2.1 Foreground Estimation/Segmentation

Foreground estimation/segmentation is a key step in many computer vision applications. This step is concerned with the detection of changes or potential objects in the image sequence. Foreground represents the objects that are not stationary in the scene for a period of time. In ITS, foreground detection ultimately aims to identify potential objects that can be declared as vehicles after a due classification process. In the dynamic and real-time environment of an ITS, foreground estimation becomes more challenging due to noise, illumination changes, weather conditions, and a cluttered environment. Generally, foreground estimation can be performed using two approaches, that is, temporal differencing and segmentation.

Temporal differencing utilizes the motion/change as a clue for the foreground estimation. Simply, temporal differencing is used to find changes in the scene by taking the difference between the image frame and some reference frame. A reference frame can be the previous image frame or a model frame that estimates the static background pixels of the scene. The differencing operation returns patches of the scene that result from significant change over a certain time period.

Frame differencing: Frame differencing computes the pixel-by-pixel difference between two adjacent image frames. Postprocessing steps (e.g., thresholding and morphological operation) are used to establish well-defined foreground objects. For example, frame differencing is used in Ref. [6] for vehicle detection to observe street parking violations. Although computationally very fast, frame differencing is not robust to noise, illumination changes, and periodic movements of other unwanted objects that can occur in real-time traffic scenes.

Background subtraction: A more refined form of temporal differencing, called *background subtraction* uses a background model as a reference image. Various approaches have been proposed in the literature [6]. An example is to take the temporal average of pixels over a sequence of image frames, but that is unresponsive to changes in illumination and unable to cope with slow-moving or stationary objects, resulting in unwanted patches detected as foreground.

Keeping in view the gradual changes in time and to improve robustness compared to averaging, a single temporal Gaussian model can be used for every pixel of the background. Observations from several consecutive frames are used to compute a mean and a variance image for the background model. For a pixel in the current frame, the pixel value is compared with the Gaussian distribution of the background model, and the pixels that deviate from the background model are labeled as foreground pixels. A single Gaussian background model is used in Ref. [7] for traffic surveillance to address occlusions and background variations. However, a single Gaussian is not adequate for traffic videos because foreground pixels can exhibit significantly different values over short time periods (due to fast-moving vehicles of different colors, shadows, and specular reflectances).

Gaussian mixture model (GMM): A significant improvement in background modeling is achieved by using statistical models to estimate the background color of each pixel, that is, GMM. In GMM, every pixel in the input frame is evaluated against the background model by comparing it with every Gaussian in the model in order to find a match. If a matching Gaussian is found, the mean and variance of the matched Gaussian are updated; otherwise, a new Gaussian with the mean equal to the input pixel color and some initial variance is introduced into the mixture. Each pixel is classified based on whether the matched distribution represents the background process. GMMs have been widely used for vehicle detection in traffic flow analysis [8–10]. In Ref. [11], a recursively updated GMM with a multidimensional smoothing transform is proposed to segment the foreground based on vehicle color. The temporal dimension of the transform is used to deal with noise associated with adverse imaging conditions such as rain and camera shake. As edges are not sensitive to sudden illumination changes, an edge-based GMM is proposed for moving vehicle detection in Ref. [12]. To cope with the slow and sudden illumination changes in outdoor traffic scenes, a self-adaptive GMM is proposed in Ref. [13]. This dynamic learning-based model shows a better foreground estimation results as compared to other techniques. A key disadvantage of the GMM for foreground detection is the higher computational complexity as compared to static and averaging background models. A much more detailed description of the GMM algorithm for background/foreground segmentation is presented in Chapter 9.

Others: In Ref. [14], a background subtraction scheme based on histogram-based filtering and region properties is proposed for vehicle detection and tracking. The algorithm generates reliable and actual background instances under different traffic conditions. A fuzzy background subtraction is proposed in Ref. [15] for moving vehicle detection. The algorithm employs the fusion of color components and texture features, claiming high accuracy and dealing with the illumination changes and shadow.

6.2.2.2 Segmentation

The objective of foreground segmentation is to partition the scene into perceptually similar regions, that is, vehicles, road, and pedestrians in a traffic flow analysis application. Two major evaluation parameters for foreground segmentation include the criteria of a good partition and a scheme for achieving better partitioning. The *mean shift algorithm* proposed by Comaniciu et al. [16] can be widely seen in the literature for traffic flow analysis. The mean shift algorithm is proposed in Ref. [17] for the simultaneous detection and tracking of vehicles in the video sequence. The algorithm shows efficient results for the detection of several vehicles and segmentation even under occlusion. The mean shift algorithm has also been used for segmentation and tracking of moving objects on an autonomous vehicle testbed platform in Ref. [18]. *Graph cut segmentation* is another common method to extract objects of interest from the background. In traffic flow analysis, the graph cut algorithm can lead to a robust segmentation as the object (vehicles) shapes are well defined. In Ref. [19], several types of graph cuts are investigated for segmenting traffic images. Different variations of graph cuts are explored in detail and compared with others to choose the best for traffic detection. Several weighting schemes based on grayscale value differences, pixel variances, and mean pixel values are investigated for vehicle segmentation. The method was applied to video sequences of traffic under various lighting conditions and locations. A graph cut algorithm is also used for the purpose of vehicle tracking in Ref. [20].

6.2.2.3 Shadow Removal

Shadow removal plays a critical role in foreground estimation/segmentation in traffic flow analysis. Moving shadows in outdoor environments are associated with both moving objects and other objects in the scene. Vehicle shadows will be adjacent to and follow vehicles. Tall static objects such as trees, signboards, poles, and buildings will cast shadows that move slowly with the movement of the sun. Therefore, an efficient shadow removal technique is required to ensure reliable estimation of foreground objects. Shadow removal for different color models (RGB, HSV, and YCbCr) is evaluated for vehicle detection in Ref. [7]. To deal with real-time sudden illumination changes and camera vibration in urban traffic videos, a background GMM and shadow removal method is proposed in Ref. [21]. The method showed a high accuracy rate for vehicle detection even in challenging weather conditions. In Ref. [8], shadow removal of moving vehicles is carried out based on the differences of brightness distortion between vehicle shadows and moving vehicles relative to background, indicating improved results for vehicle detection even in the case of a sudden change in illumination. A computationally efficient algorithm using conditional random fields is proposed by Wang [22] for real-time vehicle detection with the removal of cast shadows.

6.2.2.4 Morphological Operations

Foreground estimation often results in silhouettes that are incomplete or distorted because of shadows, occlusion, illumination, and weather conditions. To deal with such artifacts, morphological operations are employed to recover vehicle-like shapes. Connected component labeling, opening, closing, convex hull, and region filling are typically used as a post-processing step on the silhouettes to improve the representation. Due to the more structured nature of highway traffic, morphological operations are also proposed independently for vehicle detection in Refs. [23–25].

6.2.2.5 Approaches Based on Object Recognition

Another way of classifying objects in traffic scenes is to use object recognition–based approaches. The quality of foreground estimation/segmentation is impacted by traffic density, varying illumination and weather conditions, and the range of vehicle types in the scene. Therefore, object recognition approaches provide a more organized and supervised way of detecting vehicles and other objects in traffic videos. Object recognition–based approaches generally employ appearance-based feature extraction and classification to detect the objects or part of the objects. These approaches can be categorized into two groups: interest-point feature descriptors and appearance shape–based feature descriptors. The following Sections 6.2.2.6–6.2.2.8 provide more insight into the range of object recognition approaches for traffic flow analysis.

6.2.2.6 Interest-Point Feature Descriptors

Interest points (also referred to as keypoints) are image positions from where features can be extracted. The selection of interest points can be based upon appearance-based algorithm (e.g., Harris corners and scale-invariant feature transform or SIFT) or a dense selection scheme. Once the interest points are selected, they are required to be described using some feature representation, that is, a feature descriptor. Feature descriptors are typically a vector representation that is used for both training and testing a classifier. Therefore, describing an interest point using features is a key toward better detection of the objects in this process.

SIFT [26] has been proved a powerful feature descriptor for object recognition. A feature based on edge points and modified SIFT descriptors is proposed in Ref. [27] to provide a rich representation of vehicle classes. In Ref. [28], groups of SIFT features are matched between successive image frames to enable vehicle tracking.

Speeded-up robust feature (SURF) descriptor [29] was introduced to deal with the relatively higher computational complexity and the lack of robustness against different transformations of the SIFT descriptor. In Ref. [30], a combination of SURF and edge features are used for vehicle detection in blind spot areas and showed reliable results for detection of sedan vehicles. SURF is unable to detect symmetrical objects, which is important in traffic videos. A new symmetrical SURF descriptor is proposed in Ref. [31] for vehicle detection and recognition of vehicle make and model.

Histogram of oriented gradients (HOGs) descriptor [32] was initially proposed for human detection in videos, but it is also used in various object detection and recognition applications. HOG features are used for vehicle detection in Ref. [33], where the orientation is determined using multiplicative kernel leaning. Support vector machine (SVM) is used for HOG feature classification to detect vehicles. In Ref. [34], on-road multivehicle tracking is carried by modeling a vehicle as a deformable object. Local (HOG) and global (latent SVM) features of vehicles are combined to model a vehicle as a deformable object. A study of feature combinations (HOG, PCA, Gabor filters) for vehicle detection is investigated and reported in Ref. [35].

6.2.2.7 Appearance Shape–Based Descriptors

Object detection and recognition can benefit from modeling objects by their appearance and shape features, capturing the spatial relationship among the object parts. Haar-like features are proposed for vehicle detection in real-time vehicle detection and tracking [36] using a stationary camera under various challenges. In Ref. [37], a hybrid approach using Haar-like features (HOG and SVM) is proposed for robust vehicle detection and verification. A region tracking–based algorithm is proposed in Ref. [38] for vehicle detection in nighttime traffic videos, addressing a challenging issue in traffic flow analysis. In Ref. [39], an online adaptive boosting (AdaBoost) approach is employed using Haar-like features for real-time vision-based vehicle detection. Similarly, a combination of Haar-like features and an AdaBoost classifier is proposed in Ref. [40] for on-road vehicle recognition and tracking in highway traffic videos.

6.2.2.8 Classification

Classification is the process of assigning a class label to an unknown instance/object. To recognize a particular object, the classifier needs information about an object, which is available in terms of features (a.k.a. feature vector). To classify an unknown object, the classifier needs to be trained initially on a representative set of feature vectors. For example, a classifier can be used to differentiate between vehicles and pedestrians, as well as for categorizing vehicles into different classes. Following segmentation, ROIs can be represented using combinations of region, contour, and appearance-based features. In case of high-dimensional data, dimensionality reduction techniques can be employed such as principal component analysis (PCA) and linear discriminant analysis (LDA) to simplify the computation of the classifier. The following discussion is focused on the classifiers that have been applied to the task of traffic flow analysis.

The nearest-neighbor (NN) classifier is the simplest nonparametric classifier. Class labels are assigned on the basis of a simple distance (Euclidean, City-block, Mahalanobis, etc.) calculation between the feature vector of an unknown instance and every feature vector of the training data. A much improved version of the NN classifier is the k-nearest neighbor (KNN) algorithm, which assigns the class according to a majority vote of the k closest vectors, which is more robust than the NN. KNN is employed for vehicle classification and tracking in Refs. [41, 42] for highway traffic videos. The study in Ref. [43] presents segmentation of vehicle detection data for improved traffic flow prediction based upon KNN. A key disadvantage of NN and KNN is that they require several distance calculations and do not scale very well for large training sets in terms of computational complexity and memory requirements.

SVM has been extensively used for vehicle detection and classification in ITS. It is employed for vehicle detection in Ref. [44] (EOH + SVM), [45] (HOG + SVM), and [29] (HOG + SVM). A fuzzy SVM is applied over shape, texture, and gradient histogram features in Ref. [46] for vehicle detection and classification. While only a binary classifier, it can be used to tackle multiclass problems; however, for multiclass classification it can be computationally expensive.

Boosting: AdaBoost is another useful algorithm for vehicle classification that can also be used in conjunction with other learning algorithms. The AdaBoost algorithm with Haar-like features has been used for real-time vehicle detection and tracking in Refs. [35, 36]. A much needed requirement in ITS, both vehicle and pedestrian detection, are carried out by using Haar-like features and an AdaBoost classifier in Ref. [47]. AdaBoost classifiers are fast, simple, and versatile, but are vulnerable to uniform noise and may result in classifiers that are too weak and complex and lead to overfitting.

Random forest: Random forest classification is an ensemble learning algorithm that consists of many decision trees at training time and output the class that is the mode of the classes returned by individual trees. A vehicle classification scheme is proposed by Dalka and Czyzewski [48] to classify vehicles into sedans, cars, and trucks using various classifiers including the random forest classifier. In Ref. [49], vehicle detection from aerial images is carried out on HOG feature descriptor using NN, decision tree, random forest, and SVM classifiers. In this scheme, rural traffic videos are targeted for applications in surveillance, traffic monitoring, and military operations. A model-based vehicle pose estimation and tracking scheme using a random forest is also proposed in Ref. [50]. Random forest is one of the accurate learning algorithms on large datasets, but they have been observed to overfit for some datasets with noisy classification.

6.2.2.9 Analysis

The framework for traffic flow analysis may return the detected vehicles, nonvehicle objects, vehicle types and models. The outcome can be employed in various applications in ITS, such as vehicle detection, vehicle counting, automatic number plate recognition, vehicle tracking, vehicle classification, traffic density measurement, traffic flow estimation, traffic speed estimation, surveillance, and traffic rules violation monitoring and incident detection.

6.2.3 Application Domains

Traffic flow monitoring and analysis applications need to be developed for road environments where different conditions, such as traffic density and speed, vary for each environment. To make the analysis more explicit, three application domains are considered: urban, highways/motorways, and tunnels. Table 6.2 shows the comparison of potential challenges in traffic flow analysis for these application domains. These are affected to differing degrees by the presence of nonvehicle objects, shadows, occlusion, illumination changes, and the density of traffic. However, under free-flowing conditions, typical on highways and through tunnels, there may be fewer problems with occlusion and shadows associated with high traffic densities, simplifying the video analysis task.

Extensive research work is reported in the literature (as discussed in Sections 6.2.2.1–6.2.2.8) regarding traffic flow analysis in urban and highway traffic videos. In addition to that, a multishape descriptor composed of HOG, edge, and intensity-based features is proposed for urban traffic by Chen and Ellis [51] for classification of vehicles into four categories, that is, car, van, bus, and motorcycle. The work was further enhanced in Ref. [17] by proposing a new background GMM and shadow removal technique to deal with sudden changes in the illumination in the urban conditions. In Ref. [52], vehicle detection is carried out to track them in highway surveillance videos using a background mixture model with a binary classifier, resulting in higher detection accuracy for various quality of crowded and uncrowded videos. A comprehensive review of video analytics systems for urban and highway traffic videos is presented in Ref. [53], highlighting algorithms and performance. Keeping in

Table 6.2 Comparison of real-time challenges in traffic flow analysis for different application domains.

Application domains	Real-time challenges						
	Nonvehicle objects	Shadow	Occlusion	Impact of weather	Illumination changes	Traffic congestion	Traffic flow analysis
Urban	Yes	Yes	Yes	Partial	Yes	Yes	Complex
Highways	Partial	Partial	Partial	Yes	No	No	Moderate
Tunnels	No	Partial	Partial	No	Yes	No	Moderate

view the poor imaging conditions and lighting from the vehicles in tunnels, a vehicle detection and tracking scheme is proposed in Ref. [54] for tunnel surveillance. The same was extended as well for vehicle classification in Ref. [55] for tunnel surveillance using multicamera videos. The scheme employs Haar-like features and a cascade AdaBoost classifier to validate the results for three tunnel videos. In Ref. [56], a combination of foreground masks, optical flow at corner points, and image projection profile with a Kalman filter are used as clues for real-time vehicle tracking in tunnels.

We have seen so far that advances in computer vision in recent years show real promise, and in Section 6.3 we give an example of vision algorithms applied to traffic flow analysis using standard closed-circuit television (CCTV) cameras. Due to space limitations, we focus on vehicle detection, without considering pedestrians or cyclists.

6.3 Measuring Traffic Flow from Roadside CCTV Video

Roadside CCTV cameras are deployed to enable remote monitoring of critical sections of the road network. The monitoring is typically performed by trained operators, whose role is to assess the traffic flow rates, identifying and responding to the impact of events that adversely affect the flow and may lead to congestion, such as accidents, roadworks, adverse weather, and high traffic volumes. In urban environments, traffic flow is also affected by the road layout (junctions, roundabouts) and signals (traffic lights, pedestrian crossings) that cause the traffic to slow or come to a halt. As outlined earlier in this chapter, the primary measurements for estimating traffic flow are based on counting the number of vehicles over a period of time and estimating vehicle velocities. From these measurements, other measures can be derived such as vehicle density, average flow rate, and measures of congestion and queuing. Traffic flow patterns vary according to various influences: time of the day, day of the week, time of the year, public and personal holidays, school holidays, major events, and so on. The challenge of traffic flow analysis is to capture and support data to model these varying patterns and manage the traffic signaling and route planning systems in order to optimize the flow and minimize congestion. In this section, we consider examples of practical and algorithmic solutions to measuring the properties of traffic streams based on vehicle speed, density, and flow. The methods analyze the video feeds by detecting individual vehicles and estimating their speed by tracking them over successive image frames. Vehicle-type classification is made with respect to a set of standard classes that can be used to profile the proportion of vehicle types and their changing volumes over time. The remainder of this section describes a system developed for the detection, segmentation, classification, and estimation of flow statistics of traffic from CCTV cameras deployed in an urban environment. We also describe methods for scene calibration and traffic lane detection that are used to support a more detailed analysis of the flow patterns.

6.3.1 Video Analysis Framework

Figure 6.5 presents the video analysis system partitioned into four stages. The first stage learns a background model of the scene (i.e., pixels that do not change significantly over time) to deal with illumination changes and objects that (temporarily) become stationary, using an adaptive background estimation process. Detection is then based on subtracting the current image from the most appropriate background model, segmenting objects that are moving through the static background, suppressing shadows and "mending" holes in the binary silhouette using a morphological operation (see Figure 6.6a). Vehicles are sampled within a detection zone that has been manually located in the image and aligned to the road surface (Figure 6.6b). It is located in the foreground of the image to

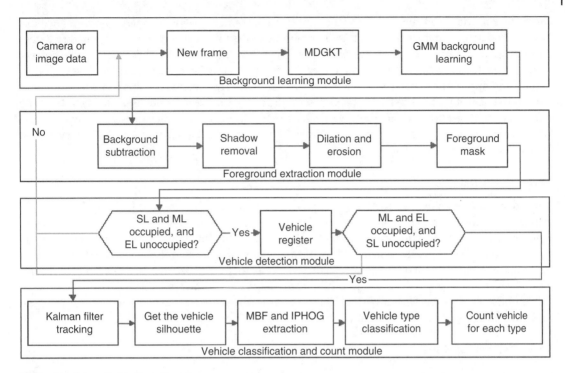

Figure 6.5 System's data flow diagram.

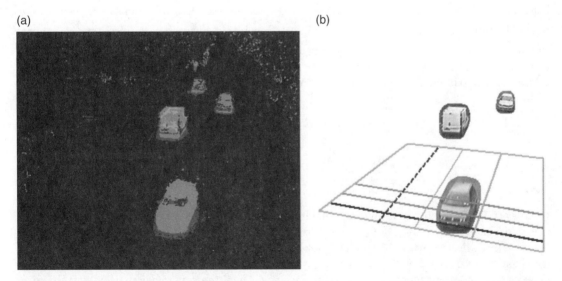

Figure 6.6 Vehicle segmentation result: (a) background subtraction results: modeled background (black), foreground object (white), shadow (light gray), or reflection highlights (dark gray); and (b) foreground image (morphologically dilated), detection zone (the dark lines are lane separators, while the dashed black line designates the bus stop area), and the vehicle detection lines (dark, gray, and light gray).

ensure that the detected vehicles are at their most visible (and hence detectable). The zone includes three virtual detection lines that are used to minimize errors in vehicle counting by applying a detection gating logic that counts vehicles only once when they intersect with these lines in the correct sequence. Detected vehicles are tracked over five image frames by a Kalman filter. A set of size-, shape-, and appearance-based image features are extracted from each blob and assigned to one of four classes (car, van, bus, and motorcycle/bicycle) using an SVM classifier. A majority vote of the classification of each tracked blob instance over the five frames determines the final assigned class type. Vehicle velocities are estimated over the five image frames, based on the distance traveled on the ground plane.

6.3.2 Vehicle Detection

The video data we analyze in the following experiments have been captured from roadside CCTV cameras that are controlled by operators. Although the cameras are pan–tilt–zoom (PTZ), they are typically deployed in a fixed-view configuration, and return to a preset viewpoint subsequent to any manually controlled roaming. Hence, the first stage of our image analysis utilizes vehicle detection based on temporal differencing, constructing a statistical model that describes the state of each background pixel. A widely used approach, originally proposed by Stauffer and Grimson [57], models the multimodal background distribution with a GMM. The method can reliably cope with slow lighting changes, repetitive motions associated with clutter (e.g., wind-blown leaves), and long-term scene changes.

6.3.3 Background Model

Video analysis begins with a noise reduction step that is applied at each image frame using a spatiotemporal Gaussian filter ($3 \times 3 \times 3$). After filtering, background pixels are modeled as a Gaussian mixture (please see Chapter 9 for more details on GMMs) using a recursive updating computation that suits online operation [58], estimating the mixture parameters and automatically selecting the number of components for each pixel. Assuming an adaptation period T (e.g., 100 frames), if $x^{(t)}$ is the value of a pixel at time t, $X_T = \left\{ x^{(t)}, x^{(t-1)}, \ldots, x^{(t-T)} \right\}$. For each new sample, we update the model and re-estimate the density. These samples may contain values that belong to both the background (BG) and foreground (FG) object, so the estimated density is denoted as $\hat{p}\left(x^{(t)} | X_T, \mathrm{BG} + \mathrm{FG}\right)$ [58]. Using a GMM with a maximum of K components (normally between 3 and 7, and in our case, set to a value 4),

$$\hat{p}\left(x^{(t)} \mid X_T, \mathrm{BG} + \mathrm{FG} \right) = \sum_{m=1}^{K} w_m \cdot \eta\left(x^{(t)} ; \mu_m, \Sigma_m \right) \tag{6.10}$$

$$\eta\left(x^{(t)} ; \mu_m, \Sigma_m \right) = \sum_{i=1}^{K} \frac{w_{i,t}}{\left(2\pi \right)^{d/2} \left| \Sigma_{i,t} \right|^{1/2}} e^{-\frac{1}{2}\left(x^{(t)} - \mu_{i,t} \right)^T \Sigma_{i,t}^{-1} \left(x^{(t)} - \mu_{i,t} \right)}, \tag{6.11}$$

where each Gaussian distribution is parameterized by the estimated mean value (μ_m), the estimated covariance matrix (Σ_m), and nonnegative estimated mixing weights (w_m), normalized (i.e., sum to 1) for the mth GMM at time t. We use the simplifying assumption that each channel of the color space is independent of the others (so Σ_m is a diagonal matrix) and that the red, green, and blue channels pixel values are independent, with the same variance. Then, the covariance matrix takes the form $\Sigma_m = \sigma_m I \Sigma_m = \sigma_m I$, where I is a 3×3 identity matrix, which avoids a costly matrix inversion at the expense of some loss of accuracy.

GMMs are computationally intensive, sensitive to sudden changes in global illumination and require carefully tuned parameters. The original algorithm uses a fixed learning rate to respond to illumination changes. A low learning rate may yield an inaccurate model with a large variance that will have low detection sensitivity, while a high learning rate updates the model too quickly and slow moving objects are absorbed into the background model. Slow moving or stationary objects are a common occurrence for urban traffic and will be "absorbed" into the background model; when they finally move, there is a lag in detection, resulting in poor segmentation. To address this transient stop–start behavior, we introduce a self-adaptive GMM [59] that remembers the original background, invoking this model and reacting faster when the traffic begins to move again. This uses a global illumination change factor g between the learnt background I_r and the current input image I_c [13] computed from

$$g = \text{median}_{s \in S} \left(\frac{I_{c,s}}{I_{r,s}} \right), \tag{6.12}$$

where s excludes pixels for which $|q_i - \mu_i| > \sigma_i$, where μ_i and σ_i are the mean and standard deviation of the pixel-wise ratio $q_i = I_{c,i}/I_{r,i}$, which biases the measure in favor of pixels that change significantly. This value is used in the detection step for deciding the best background model, and in the computation of the brightness and chromaticity distortion applied for shadow and highlight detection. The factor g keeps track of how the global illumination changes, while a counter (c_m) keeps track of how many data points have contributed to the parameter estimation of that Gaussian. Each time the parameters are updated, a learning rate β_m is recalculated based on the basic learning rate α and the current value of the accumulative counter, c_m. Given a new data sample $x^{(t)}$ at time t the recursive update equations are as follows:

$$w_m = (1-\alpha)w_m + \alpha \left(o_m^{(t)} + c_T \right) \tag{6.13}$$

$$\beta_m = \frac{\alpha(l + c_m)}{c_m} \tag{6.14}$$

$$\mu_m = \mu_m + o_m^{(t)} \left(\frac{\beta_m}{w_m} \right) \delta_m \tag{6.15}$$

$$\sigma_m^2 = \sigma_m^2 + o_m^{(t)} \left(\frac{\beta_m}{w_m} \right) \left(\delta_m^T \delta_m - \sigma_m^2 \right) \tag{6.16}$$

$$c_m = c_m + 1. \tag{6.17}$$

Here, l is a constant, $x^{(t)} = [x_1, x_2, x_3]^T$, $\mu_m = [\mu_1, \mu_2, \mu_3]^T$, and $\delta_m = x^{(t)} - \mu_m$ for a three-channel color image. Instead of the time interval T, a constant α defines an exponentially decaying envelope that is used to limit the influence of the old data. c_T is a negative prior evidence weight [58], which means that the model is accepted to exist only if there is enough evidence from the data for the existence of this class. This will suppress the components that are not supported by the data and ensures that components with negative weights are discarded. For a new sample, the ownership $o_m^{(t)}$ is set to 1 for the "close" component with the largest w_m and the others are set to 0. A "close" sample to a component corresponds to a Mahalanobis distance (MD) from the component that is less than a threshold distance, d_c. The squared MD from the mth component is calculated as follows:

$$D_m^2 \left(x^{(t)} \right) = \hat{\delta}_m^T \hat{\Sigma}_m^{-1} \hat{\delta}_m. \tag{6.18}$$

Here, $\hat{\delta} = g \cdot x^{(t)} - \mu_m$. If there are no "close" components, a new component is generated with $w_{m+1} = \alpha$; $\mu_{m+1} = x^{(t)}$; $\sigma_{m+1} = \sigma_0$; $c_{m+1} = 1$, where σ_0 is an appropriately initialized variance value. If the maximum number of components K is reached, the component with the smallest w_m is discarded. After each weight update in Equation 6.13, the weights are renormalized such that $\sum_{\forall k} w_k = 1$.

The algorithm uses an online clustering algorithm that can be updated on a frame-by-frame basis in real time. Usually, intruding foreground objects are represented by additional clusters with small weights w_m. Therefore, the background model can be approximated by the M largest clusters as follows:

$$\hat{p}\left(x^{(t)} \mid X_T, \text{BG}\right) \sim \sum_{m=1}^{K} w_m \cdot \eta\left(x^{(t)}; \mu_m, \Sigma_m\right). \tag{6.19}$$

Reordering the components by descending weights w_m, we have

$$M = \text{argmin}_b \left(\sum_{i=1}^{b} w_i > \left(1 - c_f\right) \right), \tag{6.20}$$

where c_f is a measure of the maximum portion of the data that should be accounted for by the foreground objects without influencing the background model, and b is the number of Gaussian background models that satisfy the condition. This approach has three significant benefits: (i) if an object is allowed to become part of the background, it does not destroy the existing background model. The original background values remain in the GMM if the object remains static for long enough, and its weight becomes larger than c_f. If the object then moves, the distribution describing the previous background still exists with the same estimated mean and variance; (ii) from the dynamic learning rate update Equation 6.14, it can be seen that if the background changes quickly, the value of c_m will become smaller and the new learning rate β_m will increase, so the background model will update quickly. The model will quickly achieve a good estimate of the mean and variance. Maintaining a dynamic learning rate for each Gaussian component will improve convergence and approximation of a smaller data cluster. Otherwise, if the background is stable, as more data samples are included in its parameter estimation, β_m will approach the basic learning rate α, while still maintaining the same temporal adaptability, because the weights update Equation 6.13 still uses the basic learning rate; and (iii) from Equation 6.18, we can see that the MD calculation will compensate for the global illumination change, making the MD insensitive to sudden illumination changes [13].

Another challenge to the video analysis is the identification of shadows cast by objects moving through the scene. It is desirable to discriminate between objects and their shadows and to identify the type of features that could be utilized for shadow detection. The GMM is susceptible to both global and local illumination changes, such as shadows, and highlights reflections from specular surfaces (e.g., shiny car bodies) that can cause the failure of consequent processes, for example, tracking and classification, because they change according to object pose, camera viewpoint, and the location and strength of the illumination source. The shadow suppression algorithm used in our system is based on the algorithm proposed by Horprasert [60] that is stable in RGB color space. This has been modified to cope with sudden changes in the global illumination by incorporating the global illumination change factor g described by Equation 6.12.

The distorting effect of shadow and highlight in RGB space is decomposed into two components, a brightness and chromaticity distortion, and computes an adjustment to re-estimate the chromaticity according the brightness variation. The ith pixel's brightness distortion B_i is a scalar value that brings the observed color closer to the expected chromaticity line. For the ith pixel value of $I_i = \left[I_{R_i}, I_{G_i}, I_{B_i}\right]^T$ in RGB space, the estimated mean is $E_i = \left[\mu_{R_i}, \mu_{G_i}, \mu_{B_i}\right]^T$. The color channels are rebalanced by scaling the color values using the pixel standard deviation $\sigma_i = \left[\sigma_{R_i}, \sigma_{G_i}, \sigma_{B_i} \right]$, and the distortion of the brightness B_i and chromaticity CD_i are computed as follows:

$$B_i = \frac{g \left(I_{R_i} \mu_{R_i} / \sigma_{R_i}^2 + I_{G_i} \mu_{G_i} / \sigma_{G_i}^2 + I_{B_i} \mu_{B_i} / \sigma_{B_i}^2 \right)}{\left(\mu_{R_i} / \sigma_{R_i} \right)^2 + \left(\mu_{G_i} / \sigma_{G_i} \right)^2 + \left(\mu_{B_i} / \sigma_{B_i} \right)^2} \tag{6.21}$$

$$CD_i = \sqrt{\left(\frac{\left(g I_{R_i} - B_i \mu_{R_i} \right)}{\sigma_{R_i}} \right)^2 + \left(\frac{\left(g I_{G_i} - B_i \mu_{G_i} \right)}{\sigma_{G_i}} \right)^2 + \left(\frac{\left(g I_{B_i} - B_i \mu_{B_i} \right)}{\sigma_{B_i}} \right)^2} \tag{6.22}$$

A foreground pixel is then classified according to the following:

$$\begin{cases} \text{Shadow} & CD_i < \gamma_1 \text{ and } \gamma_2 < B_i < 1 \\ \text{Highlight} & CD_i < \gamma_1 \text{ and } B_i > \gamma_3 \end{cases} \tag{6.23}$$

Here, $\gamma_1 (0 < \gamma_1 < 1)$ is a threshold to distinguish between chromaticity values of the GMM-learnt background and the current image frame. If there is a case where a pixel from a moving object in the current image contains a very low RGB value, then this dark pixel will always be misclassified as a shadow, because the value of the dark pixel is close to the origin in RGB space and all chromaticity lines in RGB space meet at the origin. Thus, a dark color point is always considered to be close or similar to any chromaticity line. We introduce a second threshold, $\gamma_2 (0 < \gamma_2 < 1)$, to address this problem. Similarly, a third threshold, $\gamma_3 (\gamma_3 > 1)$, is applied to the normalized brightness distortion, in order to detect highlights.

Figure 6.6a shows the result of image segmentation that applies the earlier classification steps to segment foreground objects from background by image subtraction, using a background estimated using the GMM. Shadow and highlight pixels have also been labeled, though for the vehicle blobs we are more concerned with the shadow pixels, which distort the object size and shape parameterization if included in the final blob representing the object (vehicle). Hence, the shadow pixels are interpreted as background, while the highlight pixels are assigned to the foreground object.

6.3.4 Counting Vehicles

To ensure that vehicles are only counted once as they transit the detection zone, gate detection logic is employed. This comprises three detection line:, designated StartLine (SL), MiddleLine (ML), and EndLine (EL). These line detectors are sensitive to misdetection as a consequence of the ragged edge of a vehicle boundary. To minimize this effect, the detectors have a finite width to ensure a stable detection of the vehicle when it intersects the line (a width of five pixels was used in the experiments described later). The separation between detector lines depends on the average traffic speed, and was set to 30 pixels in the experiments described here. A detection line is considered occupied if the proportion of pixels intersecting the line is above a threshold (30% of the lane width); otherwise, it is unoccupied. This threshold is chosen as a trade-off between detecting small vehicles (i.e., bicycles and motorbikes) while being insensitive to small blobs associated with noise. A vehicle is determined to be "present" only when both SL and ML are occupied and EL is unoccupied (for traffic moving toward the camera, that is, the two rightmost lanes in Figure 6.6b). A vehicle is said to be "leaving" when ML and EL are occupied and SL unoccupied. A vehicle is counted only when it changes from the "present" state to the "leaving" state. This is reasonable in congested situations and for stationary traffic. Lines SL and EL are swapped to account for vehicles moving away from the camera (e.g., leftmost lane in Figure 6.6b).

6.3.5 Tracking

The output of the vehicle detection step is a binary object mask (blob) that is tracked using a constant velocity Kalman filter model. The state of the filter is the location and velocity, $s = [c_x, c_y, v_x, v_y]^T$ of the blob centroid, and the measurement is an estimate of this entire state, $y = \hat{s} = [\hat{c}_x, \hat{c}_y, \hat{v}_x, \hat{v}_y]^T$. The data association problem between multiple blobs is addressed by comparison of the predicted centroid location with the centroids of the detections in the current frame. The blob whose centroid is closest to the predicted location is chosen as the best match. The track class label is computed at each frame, but the final label is assigned by a majority voting scheme, which considers the entire track in making a decision of class type, rather than employing a single frame that could be corrupted by different sources of noise. Figure 6.7 shows the results of vehicle labeling with track identifiers.

6.3.6 Camera Calibration

Estimating true (road) velocities requires a calibration process to convert pixel coordinates into real units (e.g., meters). Note that a similar requirement is encountered in Chapter 5. The commonest method of performing camera calibration employs the use of a predefined target (calibration) object that is sampled at different locations in the scene. The target defines a measuring rod that is selected as an object of known dimensions. It should be easily detectable and present a consistent projection into the image. Automatic image analysis algorithms can then be used to detect and measure the object size in pixels, and then translate these into real units. For high accuracy, the measuring rod should be aligned with and in contact with the plane surface that is to be calibrated (in this case, the road).

Researchers have used various calibration targets; some are static and "native" to the scene, such as lampposts or traffic signs, which tend to have standard dimensions, or objects (e.g., people) that move through the scene (which may require an assumption about their average size, such as vertical height). For the experiments described here, we utilize road markings for calibration that satisfies our target constraints—that is, they are aligned with (painted onto) the road surface and their real dimensions can be easily measured (or are known from well-prescribed regulations that govern their dimensions). While we could directly measure the locations of significant points on the line (e.g., the location of the endpoints, or where lines intersect or cross), making such surveys requires the road to be temporarily closed. Instead, we extract coordinate data from Google Earth imagery, comparing

Figure 6.7 Kalman filter track labeling results.

(a)

(b)

(c)

Figure 6.8 (a) Google Earth image with manually labeled locations, (b) calibration reference image (middle), and (c) zoomed portion of (b). Circles and index numbers indicate the corresponding points and the asterisks indicate the reprojected points. The average reprojection error was 0.97 pixels for this view.

manually located road marking coordinates visible in the CCTV camera image with corresponding points visible in the Google Earth view of the scene.

Figure 6.8 shows a camera image and the Google Earth view. The mouse is used to select corresponding locations in the two images (indicated by the numbered circles), and Heikkila's method [61] is used to compute the camera calibration parameters. The reprojection of the points is indicated by the asterisks, with average errors of one to two pixels across different cameras. In the United Kingdom, the lane markings are prescribed to be 10–15 cm in width and are clearly visible in the

Figure 6.8a, indicating a ground plane resolution of similar scale. A potential drawback with this source of calibration coordinate data arises if the road markings are changed, as updates of Google Earth imagery are infrequent (~1–3 years).

6.3.7 Feature Extraction and Vehicle Classification

Each blob above a minimum size (applied to eliminate noise) is represented by a set of morphological and appearance-based features based on (i) 13 shape features comprising measures of size and shape from the binary silhouette, encompassing bounding box (width, height, and area), circularity (dispersedness, equivdiameter), ellipticity (length of major and minor axis, eccentricity), and shape-filling measure (filled area, convex area, extent, solidity) [51]; and (ii) a multiscale pyramid of HoGs (PHOGs [62]). The PHOG descriptors represent local shape properties, capturing the distribution of intensity gradients within a region, and their spatial layout by tiling the image into multiple resolutions. The descriptor consists of HoGs over each image subregion at each resolution level of the detection bounding box, normalized for each level. The resulting feature vector is made up of 202 elements (13 morphological and 189 PHOG for three levels of resolution). Figure 6.9 depicts an object illustrating the first level of tiling and three levels of histogram resolution. Defining stable class types is challenging because some instances of a class (e.g., MPVs and small vans) may be very similar

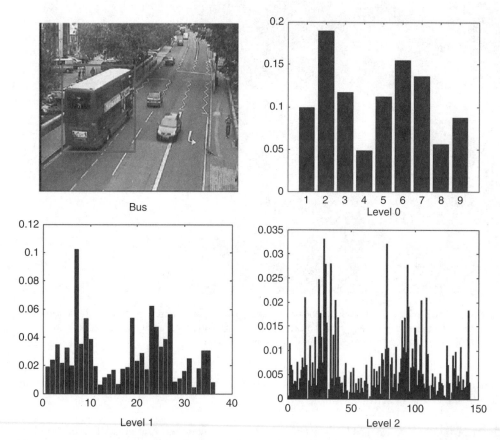

Figure 6.9 An input image and the shape spatial pyramid representation of HOG for a bus over three spatial scales: 9 features (level 0), 36 features (level 1), and 144 features (level 2).

in appearance (especially from particular viewpoints) and the classification may be ambiguous, resulting in higher error rates. We combine tracking with classification and use a majority voting over several frames to improve the classification performance. Classification uses SVM with a polynomial kernel, constructing a multiclass classifier using a one-versus-all strategy, in which the mth classifier constructs a hyperplane between class m and the $N-1$ other classes. The SVM classifier is trained on a set of labeled exemplars for which the vehicle silhouette has been manually delineated and labeled. An effective training set typically requires a large quantity of training data that is representative of the object classes under consideration. However, manual annotation of the object silhouette is slow and laborious, so we have developed an approach that eliminates the need to manually extract many silhouettes and minimizes the number of manual label annotations required [59].

6.3.8 Lane Detection

We learn the location of road and traffic lanes by detecting vehicles and analyzing their pattern of movement based on vehicle motion trajectories. While individual trajectories may be unreliable for inferring such information, the collective behavior averaged over a large number of observations allows more robust estimation associated with both the geometry of the scene and typical patterns of motion. The detected vehicle blobs are fitted with one of four wireframe models (depicted in Figure 6.10a) that represent a broad categorization of the vehicle types that commonly use these urban roads—lorries and buses (including double deckers), vans, motor cars, and motorcycles/bicycles. The wireframe models are projected into the image, scaled, and oriented according to the ground-plane resolution and road direction. A grid search of the pose parameters is made to determine the best-fit of each model to the blob's silhouette. Figure 6.10b shows a typical result of the model fit.

Trajectories are typically constructed using a single point to represent an object instance in each frame. The blob centroid coordinate is widely used and is quite adequate for tracking, but it is an inaccurate feature to locate the vehicle with respect to the ground plane (i.e., the road surface). This is because the centroid position depends on the size and orientation of the object (vehicle) and also on the position of the camera. For tall vehicles (e.g., a bus or lorry), the blob centroid represents a point located at approximately half of the vehicle height, which might be 2–3 m above the ground and its projection (along the optical axis of the camera) can be well beyond the vehicle location on the ground plane. This biases the lane detection algorithm, as the projection of each vehicle's centroid is displaced away from the camera, and in some cases, well beyond the actual lane boundary.

A more suitable point is one on (or close to) the ground plane, such as the centroid of the base face of a wireframe model fitted to the detected blob, as shown in Figure 6.11. This shows the difference between the centroid location of the ROI and the centroid of the base face of a wireframe model of a vehicle.

We use this representation as a reliable object feature to construct the trajectory of the vehicle through the scene. An agglomerative clustering algorithm is used to group similar trajectories using a similarity distance measure that can cope with trajectories of different lengths, overlap, and numbers of sample points [63]. A low-degree polynomial approximates the lane prototype (the centerline), computed from a refined set of trajectories. Vehicle type combined with lane usage statistics can be used to identify special lane designations (e.g., bus lanes) and atypical trajectories (e.g., lane changing and U-turns) can be identified as departures from the lane prototype, as measured by their similarity distance. Results of trajectory clustering and lane delineation are shown in Figure 6.12. The projected lane widths are obtained using the 3D lane model according to the lane width (using the standard value of lane width for urban all-purpose roads (3.65 m) in the United Kingdom [64]).

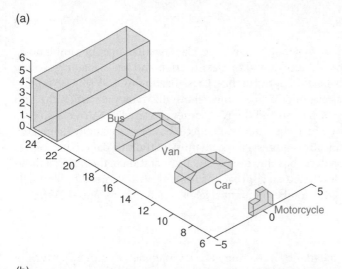

(a)

(b)

Figure 6.10 3D wireframe models with true size (a) and samples of their projections on the ground plane (b).

Figure 6.11 Alternative point-based representation of a vehicle's location.

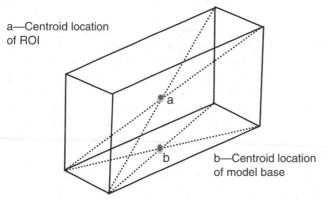

a—Centroid location of ROI

b—Centroid location of model base

6.3.9 Results

The experimental data is taken from 5 h of video recorded from a single pole-mounted roadside camera in daytime on a busy road leading into the local town center. Referring to Figure 6.12, the road comprises two lanes heading into town (lanes 2 and 3), the rightmost of which is a designated bus lane with traffic restricted to busses, taxis, and cycles during certain periods of the day. The left-most lane (lane 1) heads out of town, and has a bus stop in the near foreground. There is a pedestrian crossing midway in the scene that is operated by pedestrian requests. The capture rate is 25 frames per second and the image is RGB of size 352×288. Weather conditions ranged from dry and overcast (~1 h), to periods of light and heavy rain, after which the road surface was wet and shiny. A total of 7456 vehicles were manually observed over this period (car: 6208, van: 725, bus: 330, motorcycle: 186; and lorry: 7).

Figure 6.12 Semantic lane labeling, camera 2: (a) shows initial trajectories clustering results; (b) presents the refined results obtained using random sample consensus (RANSAC), the black "+" shows the fitted center line of each traffic lane; and (c) segmented and labeled lanes, boundaries, and the direction of traffic flow (arrows) derived from tracking results.

Table 6.3 Hourly counts (and %) by vehicle classification type over a 5-h period during the middle of the day.

Type/hour	1	2	3	4	5	All
Car	1010 (77%)	1062 (75%)	1271 (81%)	1414 (83%)	1192 (83%)	5949
Van	175 (13%)	182 (13%)	165 (10%)	125 (7%)	87 (6%)	734
Bus	77 (6%)	139 (10%)	104 (7%)	104 (6%)	101 (7%)	525
Cycle	41 (3%)	24 (2%)	25 (2%)	44 (3%)	46 (3%)	180
Other	(3)	(9)	(12)	(15)	(7)	(46)
All	1303	1407	1565	1687	1426	7388

Table 6.4 Lane usage (over a 5 h period) of each vehicle type.

Lane/type	Car	Van	Bus	Cycle	Other	All
1	3667 (81%)	449 (10%)	264 (6%)	112 (2%)	(19)	4492
2	2245 (84%)	284 (11%)	107 (4%)	29 (1%)	(21)	2665
3	37 (16%)	1 (0%)	154 (65%)	39 (16%)	(6)	231
All	5949	734	525	180	(46)	7388

Table 6.3 shows the results of vehicle counting over a 5-h period for the four main classes of vehicles. Automatic detection using the self-adaptive GMM found a total of 7388 vehicles (car: 5949, van: 734, bus: 525, motorcycle: 180; the proportion of lorries present was too few to consider), representing an overall detection rate (DER) of over 97%. As can be seen, cars dominate the traffic volume. Table 6.4 shows the vehicle count data associated with each traffic lane. The larger proportion of traffic is detected in lane 1, associated with vehicles heading out of town at the end of the working day. The restricted use of lane 3 for bus, cycle, and taxi traffic is also clearly depicted. Figure 6.13 provides a graphical representation of traffic flow densities over 10-min intervals.

Vehicle speed is estimated by computing the ground-plane distance covered by each vehicle tracked over five consecutive image frames (i.e., a period of 0.2 s at 25 fps video sampling rate). A vehicle's ground-plane location is computed using the feature point described in Section 6.3.8. Figure 6.14 shows the distribution of speeds for each vehicle type over the period.

6.4 Some Challenges

We conclude this chapter by highlighting some of the challenges facing those developing computer vision tools to assist traffic flow management. The largest growth in vehicular traffic is taking place in developing countries and emerging economies. Without proper management and the use of advanced ITS techniques, this is leading to congestion, pollution, and increased road fatalities. However, state-of-the-art techniques and technologies which are available for traffic flow analysis might not be very useful in developing countries such as Pakistan, India, and Africa. These developing countries neither have a well-managed and developed road infrastructure, nor do they have

Figure 6.13 Plot of vehicle densities for each vehicle type sampled in 10-min periods.

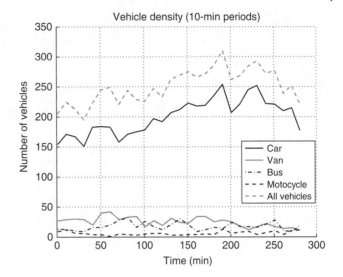

Figure 6.14 Plot of vehicle speed distributions for each vehicle type.

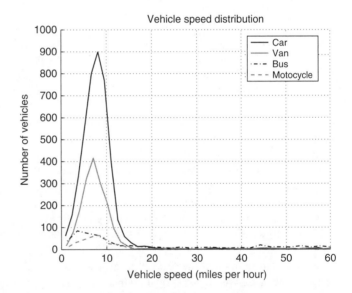

enough resources to implement technologies related to traffic flow analysis. Some of the key reasons as to why new technologies are difficult to deploy in developing countries are given next.

Lack of infrastructure and resources: Many metropolitan regions and large cities in developing countries lack a well-managed public transport infrastructure so with the inexorable rise in the number of vehicle journeys, roads are increasingly congested. Congestion is exacerbated by a failing or nonexistence/nonfunctional traffic control infrastructure at crossings and intersections, as well as low level of adherence to traffic rules. Under such conditions, current approaches to traffic flow analysis are much less robust.

Unstructured road lane marking: Existing technologies and applications rely on road infrastructure with proper lane marking. Well-defined lane markings encourage road users to keep their vehicles in a systematic flow pattern. However, in developing countries lane markings might not exist or are

Figure 6.15 Illustration of some challenges in the analysis of traffic flow in developing countries (Pakistan).

not clearly visible on the majority of urban roads. Even if lanes do exist, erratic driver behavior and nonadherence to traffic rules are significant problems.

Highly cluttered and occluded vehicles: Current vehicle detection algorithms are not sufficiently robust to cope with the levels of clutter and occlusion encountered on the roads of developing countries. Clutter arises from both high traffic densities and a lack of pedestrian crossing control that results in pedestrians crossing the road at arbitrary times and locations. Occlusion is also a major factor due to the existence of trees, electricity/telephone poles, and cables over the urban roads. Placing a camera on a pole at a crossing may not provide a clear view of the traffic stream.

Unconventional customized vehicles: A key task of traffic flow analysis is the detection, counting, and classification of vehicles which rely on well-defined knowledge about vehicle size and orientation. In developing countries, key transportation modes for low-income people are motorcycles, bicycles, and unconventional/customized vehicles (rikshaws, motorized rikshaws, trishaws and trolleys, etc.). These vehicles will be a real challenge for existing traffic flow monitoring technologies.

Unawareness of traffic rules: Application and implementation of technologies for traffic flow management and analysis is highly dependent on a driver's awareness of the rules and regulations governing road usage. In many countries, public awareness and adherence of the rules is low and enforcement is patchy or nonexistent. As a consequence, it is very common to see drivers ignoring traffic signals, lane markings, and junction stop signs. This problem is increasing day by day due to increasing population, vehicles on the road, and lack of education.

Extreme weather conditions: Extreme weather conditions can also cause failures in a traffic flow analysis system. Winter fog, heavy rains, and poor lighting are some of the problems that are difficult for a computer vision system to deal with.

Figure 6.15 illustrates some of the challenges that arise on for automating the analysis of traffic flow, in this case in the Indian subcontinent, due to less-structured road environments, unconventional road vehicles, nonexistence of lane markings, and unawareness of traffic rules.

Keeping in view all these highlighted issues, existing technologies have a long way to go to deal with cluttered and occluded environments, multiclass vehicles, less-developed road infrastructure, unstructured traffic patterns, and extreme weather conditions. Therefore, much research is needed to come up with real-time solutions that are robust under such operational conditions.

References

1   R. Fernández, *Elementos de la teoría del tráfico vehicular*. Fondo Editorial, Pontificia Universidad Católica de Valparaíso Universidad de los Andes (Santiago-Chile) Universidad Católica del Perú, Lima, 2011.

2   Transportation Research Board, "*Traffic flow theory—a state-of-the-art report*," Special Report 165, Federal Highway Administration, Washington, DC, 1992.

3   J. G. Wardrop, "Some theoretical aspects of road traffic research," *Proceedings of the Institution of Civil Engineers, Part II*, vol. I, pp. 352–362, 1952.

4   R. Fernández, *Temas de ingeniería y gestión de tránsito*, RIL Editores, Santiago, 2014.

5   IHT, Transport in the Urban Environment, *The Institution of Highways & Transportation*, Mayhew McCrimmon Printers Ltd, Essex, 1997.

6   A. Sobral, A. Vacavant, "A comprehensive review of background subtraction algorithms evaluated with synthetic and real videos," *Computer Vision and Image Understanding*, vol. 122, pp. 4–21, 2014.

7   X. Su, T. M. Khoshgoftaar, X. Zhu, A. Folleco, "Rule-based multiple object tracking for traffic surveillance using collaborative background extraction." In *Advances in Visual Computing*, pp. 469–478, Springer-Verlag, Berlin, 2007.

8   D. Bloisi, L. Iocchi, "Argos—A video surveillance system for boat traffic monitoring in Venice," *International Journal of Pattern Recognition and Artificial Intelligence*, vol. 23, no. 7, pp. 1477–1502, 2009.

9   N. Buch, J. Orwell, S. A. Velastin, "Urban road user detection and classification using 3-D wireframe models," *IET Computer Vision*, vol. 4, no. 2, pp. 105–116, 2010.

10   B. Johansson, J. Wiklund, P. Forssén, G. Granlund, "Combining shadow detection and simulation for estimation of vehicle size and position," *Pattern Recognition Letters*, vol. 30, no. 8, pp. 751–759, 2009.

11   Z. Chen, N. Pears, M. Freeman, J. Austin, "A Gaussian mixture model and support vector machine approach to vehicle type and colour classification," *IET Intelligent Transport Systems*, vol. 8, no. 2, pp. 135–144, 2014.

12   Y. Li, Z. Li, H. Tian, Y. Wang, "Vehicle detecting and shadow removing based on edged mixture Gaussian model," In *Proceedings of the 18th IFAC World Congress*, Università Cattolica del Sacro Cuore, Milano, August 28, 2011–September 2, 2011, pp. 800–805, 2011.

13   Z. Chen, T. Ellis, "A self-adaptive Gaussian mixture model," *Computer Vision and Image Understanding*, vol. 122, pp. 35–46, 2014.

14   N. A. Mandellos, I. Keramitsoglou, C. T. Kiranoudis, "A background subtraction algorithm for detecting and tracking vehicles," *Expert Systems with Applications*, vol. 38, pp. 1619–1631, 2011.

15   Xiaofeng Lu, Takashi Izumi, Tomoaki Takahashi, Lei Wang, "Moving vehicle detection based on fuzzy background subtraction," In *Proceedings of the IEEE International Conference on Fuzzy Systems (FUZZ–IEEE)*, Beijing, July 6–11, 2014, pp. 529–532, 2014.

16   D. Comaniciu, V. Ramesh, P. Meer, "Kernel-based object tracking," *IEEE Transactions on Pattern Analysis and Machine Intelligence*, vol. 25, pp. 564–575, 2003.

17   S. Lefebvre, S. Ambellouis, "Vehicle detection and tracking using Mean Shift segmentation on semi-dense disparity maps," In *Proceedings of the IEEE Intelligent Vehicle Symposium*, University of Alcala, Madrid, Spain, June 3–7, 2012.

18   B. Gorry, Z. Chen, K. Hammond, A. Wallace, G. Michaelson, "Using mean-shift tracking algorithms for real-time tracking of moving images on an autonomous vehicle testbed platform," *International Journal of Computational Science and Engineering*, vol. 1, no. 3, pp. 165–170, 2007.

19   J. Dinger, "An investigation into segmenting traffic images using various types of graph cuts." M.Sc. Engg. thesis, Graduate School of Clemson University, 2011.

20 K. Priyadharshini, S. Vishnupriya, P. Saranya, "Automatic vehicle detection and tracking in aerial surveillance using DBN and graph cut model," In *Proceedings of the IEEE International Conference on Emerging Trends in Computing, Communication and Nanotechnology (ICECCN 2013)*, Tirunelveli, India, March 25–26, 2013.

21 Z. Chen, T. Ellis, S. A. Velastin, "Vehicle detection, tracking and classification in urban traffic," In *Proceedings of the 15th International IEEE Conference on Intelligent Transportation Systems*, Anchorage, AK, September 16–19, 2012.

22 Y. Wang, "Real-time moving vehicle detection with cast shadow removal in video based on conditional random field," *IEEE Transactions on Circuits and Systems for Video Technology*, vol. 19, no. 3, pp. 437–441, 2009.

23 Z. Zheng, X. Wang, G. Zhoub, L. Jiang, "Vehicle detection based on morphology from highway aerial images," In *Proceedings of the 2012 IEEE International Geoscience and Remote Sensing Symposium (IGARSS)*, Munich, July 22–27, 2012, pp. 5997–6000, 2012.

24 B. Sharma, V. K. Katiyar, A. K. Gupta, A. Singh, "The automated vehicle detection of highway traffic images by differential morphological profile," *Journal of Transportation Technologies*, vol. 4, pp. 150–156, 2014.

25 A. Anto, C. Minnu Jayan, "Vehicle detection and classification from high resolution aerial views using morphological operations," *International Journal of Engineering Research and Technology*, vol. 4, no. 6, pp. 853–860, 2015.

26 D. G. Lowe, "Object recognition from local scale-invariant features," In *Proceedings of the International Conference on Computer Vision (ICCV)*, vol. 2, Los Alamitos, CA, September 1999, pp. 1150–1157, 1999.

27 X. Ma, W. E. L. Grimson, "Edge-based rich representation for vehicle classification," In *Proceedings of the 10th IEEE International Conference on Computer Vision*, Beijing, China, vol. 2, October 17–21, 2005, pp. 1185–1192, 2005.

28 T. Gao, Z. G. Liu, W. C. Gao, J. Zhang, "Moving vehicle tracking based on SIFT active particle choosing," In *Advances in Neuro-Information Processing*, vol. 5507, Lecture Notes in Computer Science, pp. 695–702, Springer, Berlin/Heidelberg, 2009. 15th International Conference, ICONIP 2008, Auckland, New Zealand, November 25–28, 2008.

29 H. Bay, T. Tuytelaars, L. Van Gool, "SURF: Speeded-up robust features," In *Computer Vision—ECCV 2006*, vol. 3951, Lecture Notes in Computer Science, pp. 404–417, Springer, Berlin/Heidelberg, 2006. 9th European Conference on Computer Vision, Graz, Austria, May 7–13, 2006.

30 B.-F. Lin, Y.-M. Chan, L.-C. Fu, P.-Y. Hsiao, L.-A. Chuang, S.-S. Huang, M.-F. Lo, "Integrating appearance and edge features for sedan vehicle detection in the blind-spot area," *IEEE Transactions on Intelligent Transportation Systems*, vol. 13, no. 2, pp. 737–747, 2012.

31 J. W. Hsieh, L. C. Chen, D. Y. Chen, "Symmetrical SURF and its applications to vehicle detection and vehicle make and model recognition," *IEEE Transactions on Intelligent Transportation Systems*, vol. 15, no. 1, pp. 6–20, 2014.

32 N. Dalal, B. Triggs, "Histograms of oriented gradients for human detection," In *Proceedings of the IEEE Computer Society Conference on Computer Vision and Pattern Recognition*, CVPR 2005, vol. 1, San Diego, CA, June 20–26, 2005, pp. 886–893, 2005.

33 Q. Yuan, A. Thangali, V. Ablavsky, S. Sclaroff, "Learning a family of detectors via multiplicative kernels," *IEEE Transactions on Pattern Analysis and Machine Intelligence*, vol. 33, no. 3, pp. 514–530, 2011.

34 H. Tehrani Niknejad, A. Takeuchi, S. Mita, D. McAllester, "On-road multivehicle tracking using deformable object model and particle filter with improved likelihood estimation," *IEEE Transactions on Intelligent Transportation Systems*, vol. 13, no. 2, pp. 748–758, 2012.

35 J. Arrospide, L. Salgado, "A study of feature combination for vehicle detection based on image processing," *Scientific World Journal*, vol. 2014, Article ID: 196251, 2014.

36 S. M. Elkerdawi, R. Sayed, M. Elhelw, "Real-time vehicle detection and tracking using Haar-like features and compressive tracking," *Advances in Intelligent Systems and Computing*, vol. 252, pp. 381–390, 2014.

37 H. Wang, H. Zhang "A hybrid method of vehicle detection based on computer vision for intelligent transportation system," *International Journal of Multimedia and Ubiquitous Engineering*, vol. 9, no. 6, pp. 105–118, 2014.

38 J. Wang, X. Sun, J. Guo, "A region tracking-based vehicle detection algorithm in nighttime traffic scenes," *Sensors*, vol. 13, pp. 16474–16493, 2013.

39 W.-C. Chang, C.-W. Cho, "Online boosting for vehicle detection," *IEEE Transactions on Systems, Man, and Cybernetics. Part B, Cybernetics*, vol. 40, no. 3, pp. 892–902, 2010.

40 S. Sivaraman, M. Trivedi, "A general active-learning framework for on-road vehicle recognition and tracking," *IEEE Transactions on Intelligent Transportation Systems*, vol. 11, no. 2, pp. 267–276, 2010.

41 J. W. Hsieh, S. H. Yu, Y. S. Chen, W. F. Hu, "Automatic traffic surveillance system for vehicle tracking and classification," *IEEE Transactions on Intelligent Transportation Systems*, vol. 7, no. 2, pp. 175–187, 2006.

42 B. Morris, M. Trivedi, "Robust classification and tracking of vehicles in traffic video streams," In *Intelligent Transportation Systems Conference (ITSC 2006)*, Toronto, ON, September 2006, pp. 1078–1083, 2006.

43 M. Bernas, B. Płaczek, P. Porwik, T. Pamuła, "Segmentation of vehicle detector data for improved k-nearest neighbours-based traffic flow prediction," *IET Intelligent Transport Systems*, vol. 9, no. 3, pp. 264–274, 2015.

44 S. Teoh, T. Brunl, "Symmetry-based monocular vehicle detection system," *Machine Vision and Applications* (online), vol. 23, no. 5, pp. 831–842, 2012.

45 S. Sivaraman, M. M. Trivedi, "Active learning for on-road vehicle detection: A comparative study," *Machine Vision and Applications*, vol. 25, no. 3, pp. 599–611, 2011.

46 Y. Chen, G. Qin, "Video-based vehicle detection and classification in challenging scenarios," *International Journal of Smart Sensing and Intelligent Systems*, vol. 7, no. 3, pp. 1077–1094, 2014.

47 F. Moutarde, B. Stanciulescu, A. Breheret, "Real-time visual detection of vehicles and pedestrians with new efficient adaBoost features," In *2nd Workshop on Planning, Perception and Navigation for Intelligent Vehicles (PPNIV), at 2008 IEEE International Conference on Intelligent RObots Systems (IROS 2008)*, September 2008, Nice, France, 2008.

48 P. Dalka, A. Czyzewski, "Vehicle classification based on soft computing algorithms," In *Rough Sets and Current Trends in Computing*, vol. 6086, Lecture Notes in Computer Science, pp. 70–79, Springer, Berlin/Heidelberg, 2010.

49 J. Gleason, A. V. Nefian, X. Bouyssounousse, T. Fong, G. Bebis, "Vehicle detection from aerial imagery," In *IEEE International Conference on Robotics and Automation*, Shanghai, May 9–13, 2011.

50 M. Hodlmoser, B. Micusik, M. Pollefeys, M. Liu, M. Kampel, "Model-based vehicle pose estimation and tracking in videos using random forests," In *2013 International Conference on 3D Vision (3DV)*, Seattle, WA, June 29–July 1, 2013, pp. 430–437, 2013.

51 Z. Chen, T. Ellis, "Multi-shape descriptor vehicle classification for urban traffic," In *Proceedings of the International Conference on Digital Imaging Computing Techniques and Applications (DICTA)*, Brisbane, QLD, December 6–8, 2011, pp. 456–461, 2011.

52 B. Tamersoy, J. K. Aggarwal, "Robust vehicle detection for tracking in highway surveillance videos using unsupervised learning," In *Proceedings of the International Conference of Advanced Video and Signal Based Surveillance*, Genoa, September 2–4, 2009.

53 N. Buch, S. A. Velastin, J. Orwell, "A review of computer vision techniques for the analysis of urban traffic," *IEEE Transactions on Intelligent Transportation Systems*, vol. 12, no. 3, pp. 920–939, 2011.

54 J. N. Castaneda, V. Jelaca, A. Frias, A. Pizurica, W. Philips, R. R. Cabrera, T. Tuytelaars, "Non-overlapping multi-camera detection and tracking of vehicles in tunnel surveillance," In *2011 International Conference on Digital Image Computing: Techniques and Applications (DICTA)*, Noosa, QLD, December 6–8, 2011.

55 R. R. Cabrera, T. Tuytelaars, L. V. Gool, "Efficient multi-camera vehicle detection, tracking, and identification in a tunnel surveillance application," *Computer Vision and Image Understanding*, vol. 116, pp. 742–753, 2012.

56 V. Jelaca, J. N. Castaneda, A. Pizurica, W. Philips, "Image projection clues for improved real-time vehicle tracking in tunnels," In *Proceedings of the SPIE 8301, Intelligent Robots and Computer Vision XXIX: Algorithms and Techniques, 83010C*, Burlingame, CA, USA, January 22, 2012.

57 C. Stauffer, W. Grimson, "Adaptive background mixture models for real-time tracking," In *Proceedings/CVPR, IEEE Computer Society Conference on Computer Vision and Pattern Recognition. IEEE Computer Society Conference on Computer Vision and Pattern Recognition*, Ft. Collins, CO, USA, June 23–25, 1999, vol. 2, pp. 246–252, 1999.

58 Z. Zivkovic, F. van der Heijden, "Recursive unsupervised learning of finite mixture models," *IEEE Transactions on Pattern Analysis and Machine Intelligence*, vol. 26, no. 5, pp. 651–656, 2004.

59 Z. Chen, T. Ellis, "Semi-automatic annotation samples for vehicle type classification in urban environments," *IET Intelligent Transport Systems*, vol. 9, no. 3, pp. 240–249, 2014.

60 T. Horprasert, D. Harwood, L.S. Davis, "A statistical approach for real-time robust background subtraction and shadow detection," In *Proceedings of the IEEE ICCV'99 Frame Rate Workshop*, Corfu, Greece, September 21, 1999.

61 J. Heikkila, "Geometric camera calibration using circular control points," *IEEE Transactions on Pattern Analysis and Machine Intelligence*, vol. 22, no. 10, pp. 1066–1077, 2000.

62 A. Bosch, A. Zisserman, X. Munoz, "Representing shape with a spatial pyramid kernel," In *CIVR 2007, Proceedings of the 6th ACM International Conference on Image and Video Retrieval*, Amsterdam, the Netherlands, July 9–11, 2007, pp. 401–408, 2007.

63 Z. Chen, T. Ellis, "Lane detection by trajectory clustering in urban environments," In *IEEE 17th International Conference on Intelligent Transportation Systems*, Qingdao, China, October 8–11, 2014, pp. 3076–3081, 2014.

64 "Design Manual for Roads and Bridges: Volume 6 Road Geometry," Available in http://www.dft.gov.uk/ha/standards/dmrb/vol6/section1/td2705.pdf (accessed July 29, 2015).

7

Intersection Monitoring Using Computer Vision Techniques for Capacity, Delay, and Safety Analysis

Brendan Tran Morris[1] and Mohammad Shokrolah Shirazi[2]

[1] *University of Nevada, Las Vegas, NV, USA*
[2] *Cleveland State University, Cleveland, OH, USA*

Vision-Based Intersection Monitoring

7.1 Vision-Based Intersection Analysis: Capacity, Delay, and Safety

The chapter provides a discussion of traffic analysis at intersections for capacity, delay, and safety analysis. Traffic cameras are utilized for all three goals through vision-based measurement of intersection counts, queue analysis, and finally safety analysis. Sections 7.1.1 and 7.1.2 highlight the importance of intersection monitoring and the underlying advantages of using computer vision techniques. The overview of the intersection analysis system is explained in Section 7.2, and each intersection application is discussed separately in its own section.

7.1.1 Intersection Monitoring

Intersections are planned conflict points on roads which are essential for connecting different geographic areas. They are generally monitored for design and safety goals. The design of intersections includes determining the need and placement of traffic signals, bike paths, and pedestrian walkway features as well as predicting how the specific layout affects existing and future traffic operations. The accuracy of this prediction is challenging since many parameters such as vehicle turning, queuing at intersections, and overall level of service can contribute to this assessment in complex relations. Still, good intersection design has progressed rapidly in recent years. Features like curb extensions, permissive and protected bicycle/pedestrian signal phasing strategies, advanced detection and actuation, signal controller sophistication, and striping and signage have all contributed to intersections that are safer, easier to understand for all users, and more comfortable for bicyclists and pedestrians.

However, a key challenge still remaining at intersections is complete understanding of the interactions between different road users. Motorized and nonmotorized participants come in close proximity when they cross paths, turn, or travel an intersection which highlights a major safety issue. The common way of addressing safety at intersections is to use traffic control devices such as signs, signals, and pavement markings. However, statistical reports stress the need for even more safety techniques and

Computer Vision and Imaging in Intelligent Transportation Systems, First Edition.
Edited by Robert P. Loce, Raja Bala and Mohan Trivedi.
© 2017 John Wiley & Sons Ltd. Published 2017 by John Wiley & Sons Ltd.
Companion website: www.wiley.com/go/loce/ComputerVisionandImaginginITS

(a)   (b)

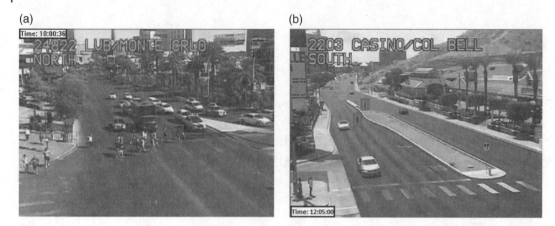

Figure 7.1 Example video frames from intersections which highlight the complexity of mixed participants and important events that occur at intersections: (a) group of pedestrians and (b) two jaywalkers.

systems. According to a 2009 report by Cano et al. [1], 30% of all high severity and 20% of all fatal accidents occur at intersections and junctions. In Europe, the severe and fatal accident rate is in 30–60% and 16–36% range, respectively. In the United States, around two million accidents and 6700 fatalities occur at intersections every year. These constitute 26% of all collisions [2]. As a consequence, improvements in accident avoidance and safety are greatly desired and efforts have begun to address these through more active intersection monitoring systems.

Intersections have complicated interactions between traffic participants such as groups of pedestrians in Figure 7.1a. Figure 7.1b shows two jaywalkers that require real-time detection for safety analysis. Without camera-based monitoring, an accident or a near-accident event would go unnoticed and potential safety hazards left unaddressed.

7.1.2 Computer Vision Application

The traditional method (manual) for intersection monitoring requires a human operator to observe traffic flows and records important traffic parameters such as the crossing count and time. The observation can be performed onsite or through watching video recordings which are obtained by traffic cameras. Traffic monitoring cameras are an innovative and extremely functional use of video surveillance technology. They are usually mounted on top of traffic signals and placed along busy roads and intersections. In addition to recording traffic patterns for the monitoring purpose, they also help cultivate safe driving habits and discourage moving violations [3].

The main difficulty with manual intersection monitoring and data collection is the high cost of labor. Technicians must be trained and paid for their time which limits the scalability of traditional intersection monitoring. In addition, since people are recruited to perform data collection, there is a real possibility of data corruption due to human error. As a result, traffic monitoring cameras at intersection can greatly facilitate the data collection process. Video frames can be manually evaluated offline and important traffic information such as vehicle counts and accidents can be recorded in a more efficient manner [4]. This offline evaluation improves the accuracy of the data collection since complex activities can be viewed at slower speed or repeated. However, this still requires significant human resources (evaluation time) and prohibits real-time usage of traffic data hence is generally used for before–after analysis.

As an alternative, automated vision-based systems can be utilized to detect objects of interest using computer vision techniques to provide long-term trajectories of road users without manual intervention for intersection monitoring [5]. The key advantages of vision-based systems include the following:

- Computer vision systems are relatively inexpensive and they can be easily installed on a vehicle or a road infrastructure element. They can recognize objects without the need for complementary companion equipment.
- Cameras can capture a tremendous wealth of visual information over wide areas. Due to advances in computer vision algorithms, this wealth of visual data can be exploited to identify more subtle changes and distinctions between objects, enabling a wide array of ever more sophisticated applications. For example, vehicle type, count, and speed can be provided concurrently [6].
- Future events such as an imminent conflict or accidents can be predicted in real time. This prediction is crucial for future cooperative advanced driver assistance systems (ADAS) focused on improving safety.
- Finally, infrastructure mounted cameras are able to monitor an intersection without any additional devices or hardware embedded in, physically printed on, or attached to the objects of interest. This gives cameras an operational advantage over radio-frequency identifier (RFID) tags, bar codes, and wireless access points since they require additional installation of complementary readers, scanners, and wireless modems, respectively [7].

7.2 System Overview

A typical framework for automated vision-based intersection analysis is shown in Figure 7.2. Video data (i.e., a sequence of images) are provided as input to the system in either an online or offline manner. The front-end system deals with detecting and classifying objects as road users (i.e., vehicles or pedestrians), while the back-end provides higher level traffic activity analysis.

Feature extraction methods are common for detecting objects in a dense scene. The salient features (e.g., corners) identify parts of an object and can be reliably detected even with occlusion. These features are identified in consecutive video frames and matched through tracking. The features are grouped (e.g., using spatial proximity) to form object proposals for road users. These object proposals are passed to a classification stage which uses predefined criteria or models to

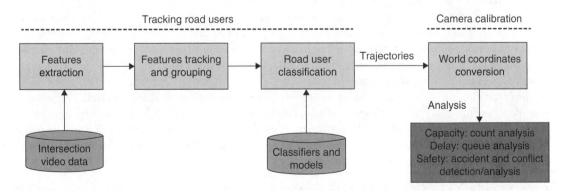

Figure 7.2 Intersection monitoring system overview.

classify the road user proposals as vehicles, pedestrians, or bikes. Classifiers can be simple, such as using a speed measurement, or can be complex appearance models obtained from machine learning techniques.

Intersection activity analysis is conducted on the extracted trajectories of road users for object-specific understanding. Camera calibration is required to convert trajectories in image coordinates to the three-dimensional (3D) world coordinates. The same trajectories can be used for different types of analysis as shown in the bottom right of Figure 7.2. The following sections highlight popular techniques for front-end processing and provide example analysis applications.

7.2.1 Tracking Road Users

The aim of tracking is to generate a consistent trajectory of an object over time. Computer vision has been used to great success for the automatic detection and tracking of vehicles and pedestrians due to many significant advances. However, an appropriate tracking method for intersection analysis must contend with some major challenges as follows:

- Vehicles and pedestrians are not in continuous motion at intersections. Vehicles might stop behind the stop bars and pedestrians remain standing on a curb waiting for a red traffic signal to change. Consequently, detection by using motion (i.e., background subtraction) is not reliable [8].
- Intersections are designed conflict points which put vehicles and pedestrians into close proximity with more complicated motion patterns than on highways. This increases the possibility of interactions and occlusion between participants which is a well-known and challenging problem for vision-based monitoring systems.
- Since intersections are designed for various road users, more complex classifiers must be used to distinguish pedestrians from vehicles. Further, there is a trade-off between wide-area field-of-view (FOV) and object resolution which usually leads to small pedestrians of low quality and small size making them more difficult to detect [8]. Consequently, many vision-based studies only consider vehicle tracking at intersections.

Optical flow is a common tracking method for intersection scenarios [9–13]. Optical flow finds matches between consecutive images to find a displacement vector (pixel translation). Optical flow assumes brightness constancy (consistent appearance) of corresponding pixels over the frames, and can be computed by two well-known methods, Lucas–Kanade [14] and Horn–Schunck [15]. The methods first detect important features (keypoints) on moving objects and matches are found using an optical flow optimization process.

Selecting the right features for object representation is an important step for tracking. Among recent studies, corner features are viewed as the most reliable features for intersection monitoring since they are still visible even when the object is partially occluded [9]. A corner can be defined as a point of intersection between two edges that has a well-defined position and can be robustly detected for a sequence of images. Corner detection for an object varies for different sizes and image quality, which affects the performance of the tracking.

The basic method of feature detection is to search for image patches that are distinct. The popular Harris corner method [16] is a simple and effective keypoint detection method. Consider a grayscale image I and window function w as a rectangular window or Gaussian window which is convolved with the image with displacements u and v in x and y directions. The method finds the difference in intensity for a displacement of (u, v) in all directions. This is expressed in Equation 7.1.

$$E(u,v) = \sum_{x,y} w(x,y) \left[I(x+u, y+v) - I(x,y) \right]^2 .$$

(7.1)

The $E(u,v)$ should be maximized in order to detect corners which means the second term in Equation 7.1 should be maximized. Using Taylor expansion, Equation 7.2 is obtained.

$$E(u,v) \approx \sum_{x,y} \left[I(x,y) + uI_x + vI_y - I(x,y) \right]^2.$$ (7.2)

I_x and I_y are image derivatives in x and y directions, respectively. Expanding the equation and canceling properly yield

$$E(u,v) \approx \sum_{x,y} u^2 I_x^2 + 2uvI_xI_y + v^2 I_y^2$$ (7.3)

$$E(u,v) \approx \begin{bmatrix} u & v \end{bmatrix} \left(\sum_{x,y} w(x,y) \begin{bmatrix} I_x^2 & I_xI_y \\ I_xI_y & I_y^2 \end{bmatrix} \right) \begin{bmatrix} u \\ v \end{bmatrix}.$$ (7.4)

The Harris matrix is defined as

$$M = w(x,y) \begin{bmatrix} I_x^2 & I_xI_y \\ I_xI_y & I_y^2 \end{bmatrix}.$$ (7.5)

Finally, the score $R = \det(M) - k\left(\text{trace}(M)\right)^2$ is computed, where $\det(M) = \lambda_1\lambda_2$ and $\text{trace}(M) = \lambda_1 + \lambda_2$ based on the two eigenvalues λ_1, λ_2 of the M matrix. Any small image window with an R score greater than a sensitivity threshold value is considered a "corner." The method essentially detects a region as a corner, when λ_1 and λ_2 are large and $\lambda_1 \sim \lambda_2$.

Figure 7.3 shows detected corners on vehicles and pedestrians from an intersection. By robustly locating good corner keypoints, they can be matched over a sequence of frames to build object trajectories. After the corner features are detected, they are matched between frames under brightness constancy and small motion constraints using and optical flow tracker such as the Kanade–Lucas–Tomasi (KLT) tracker [17]. Under brightness constancy and small motion (i.e., small u, v), the image of a corner remains the same between frames at time t and $t+1$ as $I(x,y,t) = I(x+u, y+v, t+1)$. A first-order Taylor expansion around points (x, y, t) is taken to linearize the right side as

Figure 7.3 An example of Harris corners detected for an intersection scenario.

$$I(x+u,y+v,t+1) \approx I(x,y,t) + I_x.u + I_y.v + I_t \tag{7.6}$$

$$I(x+u,y+v,t+1) - I(x,y,t) = I_x.u + I_y.v + I_t, \tag{7.7}$$

where I_t is a frame difference, and I_x, and I_y are image derivatives along x and y directions, respectively. Since $I(x+u,y+v,t+1)$ is equal to $I(x,y,t)$ under brightness constancy constraint, Equation 7.8 is obtained.

$$I_x.u + I_y.v + I_t \approx 0. \tag{7.8}$$

The equation needs to be solved to compute u and v displacement values. Since there is one equation with two unknowns, the equation can be extended by assuming all points in a local area move in the same way. Using a 5×5 window results in 25 equations per Harris corner which is compactly defined as follows:

$$\begin{bmatrix} I_x(P_1) & I_y(P_1) \\ I_x(P_2) & I_y(P_2) \\ . & . \\ . & . \\ I_x(P_{25}) & I_y(P_{25}) \end{bmatrix} \begin{bmatrix} u \\ v \end{bmatrix} = - \begin{bmatrix} I_t(P_1) \\ I_t(P_2) \\ . \\ . \\ I_t(P_{25}) \end{bmatrix}. \tag{7.9}$$

Equation 7.9 is in the form of $A_{25 \times 2}\ d_{2 \times 1} = b_{25 \times 1}$ which can be solved using a least-squares solution. The solution for d is $d = (A^T A)^{-1} A^T b$ which is shown in Equation 7.10,

$$\begin{bmatrix} u \\ v \end{bmatrix} = - \begin{bmatrix} \sum I_x I_x & \sum I_x I_y \\ \sum I_x I_y & \sum I_y I_y \end{bmatrix}^{-1} \begin{bmatrix} \sum I_x I_t \\ \sum I_y I_t \end{bmatrix} \tag{7.10}$$

with each of the summations over all pixels in the 5×5 window.

Figure 7.4 shows tracks of corner features between two frames for a typical intersection. The features on moving pedestrians and vehicles are tracked, while the large group of pedestrians on the curb have

Figure 7.4 Tracking features of two frames using the KLT algorithm.

no trajectories since they are not in motion. It is also important to note that in typical low-resolution traffic video, there are few reliably tracked keypoints per intersection participant.

The final step for object tracking is to group features and associate them to vehicles and pedestrians based on some meaningful criteria, such as the amount of movement and their direction. Since the number of vehicles and pedestrians changes over time, a clustering method is required for grouping which does not need to be initialized with number of clusters. As a result, hierarchical clustering algorithms are preferred which utilize criteria such as speed, magnitude, and direction of movement to group tracks. The algorithm steps [9] are as follows:

1) Features detected at frame t are grouped initially only if their overall displacement is large enough to avoid noise. Each feature f_i is connected to other features f_j within a maximum distance threshold $D_{connection}$ to build a correspondence graph.
2) The distance is computed for all pairs of connected features (f_i, f_j) and their minimum and maximum distances are updated. The features are disconnected (i.e., the edge is broken in the correspondence graph) if there is enough relative motion between the features, which indicates inconsistent movement between features.
3) The connected components in the correspondence graph represent vehicles and pedestrians. If all the features that compose a component are no longer tracked, meaning the object has disappeared from the scene, the features are removed from the graph.

The major difficulty in feature-based tracking by optical flow comes from oversegmentation. This problem is most apparent for large vehicles, such as a bus, where multiple feature clusters arise due to its large size. In these scenarios, features at the front of the large vehicle are separated from features at the rear. More domain knowledge, such as vehicle size, can be included to help tackle this problem. Finally, the feature clusters are distinguished as either a vehicle or pedestrian using simple classifiers. A common criterion for the classification is a threshold on average velocity since pedestrians travel more slowly than vehicles.

7.2.2 Camera Calibration

The main purpose of camera calibration is to establish a set of camera parameters in order to find a relationship between the image plane coordinates and world coordinates. This step is essential for translation of an image measurement into meaningful world units. For example, the speed of a vehicle is calculated in an image as pixels per second and should be converted to another metric like miles per hour. The process of converting image coordinates to world coordinates is usually performed for transportation monitoring through ground-plane homography normalization [18].

A 2D point (x, y) in an image can be represented as a homogeneous 3D vector $x_i = (x_1, x_2, x_3)$ where $x = x_1/x_3$ and $y = x_2/x_3$. A homography is an invertible mapping of points and lines on a projective plane P_2 between two images. In order to calculate the homography which maps each x_i to its corresponding x_j in a second image, it is sufficient to compute the 3×3 homography matrix H which only has eight degrees of freedom.

Typically, homographies are estimated between images by finding feature correspondences in those images. The basic method is direct linear transform (DLT) that can be applied to obtain the homography matrix H given a sufficient set of point correspondences. Since we are working in homogeneous coordinates, the relationship between two corresponding points x_i and x_j can be rewritten as

$$cx_j = Hx_i \tag{7.11}$$

$$c\begin{pmatrix} u \\ v \\ 1 \end{pmatrix} = H\begin{pmatrix} x \\ y \\ 1 \end{pmatrix} = \begin{pmatrix} h_1 & h_2 & h_3 \\ h_4 & h_5 & h_6 \\ h_7 & h_8 & h_9 \end{pmatrix}\begin{pmatrix} x \\ y \\ 1 \end{pmatrix},$$ (7.12)

where c is any nonzero-scale constant.

Dividing the first row of Equation 7.12 by the third row and the second row by the third row results in the following two equations:

$$-h_1 x - h_2 y - h_3 + (h_7 x + h_8 y + h_9)u = 0.$$ (7.13)

$$-h_4 x - h_5 y - h_6 + (h_7 x + h_8 y + h_9)v = 0.$$ (7.14)

Equations 7.13 and 7.14 can be written in matrix form as

$$A_i h = 0$$ (7.15)

with

$$A_i = \begin{pmatrix} -x & -y & -1 & 0 & 0 & 0 & ux & uy & u \\ 0 & 0 & 0 & -x & -y & -1 & vx & vy & 1 \end{pmatrix}$$ (7.16)

$$h = \begin{pmatrix} h_1 & h_2 & h_3 & h_4 & h_5 & h_6 & h_7 & h_8 & h_9 \end{pmatrix}^T.$$ (7.17)

Since each point correspondence provides two equations, four correspondences are sufficient to solve for the eight degrees of freedom of H. Figure 7.5 shows an example of defining four-point correspondences for conversion between image and world global positioning system (GPS) coordinates.

Figure 7.5 Four-point correspondence between camera image plane and map-aligned satellite image to estimate the homography (H) matrix and convert image locations to world latitude and longitude.

7.3 Count Analysis

Count analysis provides a basic measure of the usage of an intersection. Vehicle and pedestrian counts can be used to determine where there is demand in a network for planning and operations.

7.3.1 Vehicular Counts

Turning movement (TM) counts are a fundamental measurement for intersection analyses including traffic operations analyses, intersection design, and transportation planning applications. In addition, TM counts are required to develop optimized traffic signal timings leading to various benefits such as fuel consumption reduction, air pollution reduction, and travel time improvement.

Traditionally, TM counts were obtained by field observations where technicians observe an intersection and hand count the number of vehicles. Technicians usually monitor flow of vehicles turning right, left, and straight during the peak hours (i.e., 10 a.m. to noon and 4–6 p.m.). The observations are generally aggregated into a report that provides the counts in 15-min increments for the {north, south, east, west}-bound directions separated into {left, through, right} movements. However, robust automated counting methods are desirable since manual counting has labor limitations with respect to the number of locations, observation time periods and frequency, accuracy, and available budgets [19, 20]. In contrast, automated TM count systems should provide high accuracy as well as long-time operation in order to provide daily traffic patterns.

Most TM count systems use the same basic zone definition framework and count a movement based on the zones traversed [19, 20]. Figure 7.6a shows an example of zone definitions. Each zone is denoted by number and represents the various sides of the intersection and the internal transition area. The zones are used to define a regular sequence (RS) set which provides a list of all acceptable zone traversals. For example, the sequence set {1, 5, 2} indicates a westbound left turn. During tracking, the vehicle location is mapped to a zone and only a record of transitions between zones is recorded to build the tracked zone sequence. If the resulting zone sequence exists in the RS set for the intersection, a counter for the associated TM count is incremented. The zone counting technique is very simple to implement but is not effective when a vehicle is not well tracked, which results in poor zone localization and noise in the tracked zone sequence.

More accurate TM counts can be obtained through better utilization of object trajectories. Sequence comparison techniques can be used to match an observation with the typical paths defined at an intersection. Figure 7.6b shows an example of the known paths for one intersection.

A count system can recognize TMs using temporal alignment techniques which measure the similarity between trajectories and paths. The longest common subsequence (LCSS) distance is a popular technique used to compare unequal length trajectories since it is robust against tracking noise and has been shown to be effective for intersection monitoring and TM counts [6, 21].

In order to determine the TM count, a trajectory is compared with all the typical paths for the intersection to find the best match. The first step for comparison is to compute a warping to best align the trajectory and path using LCSS

$$\mathrm{LCSS}\left(F_i^{T_i}, F_j^{T_j}\right) = \begin{cases} 0 & T_i = 0 \,|\, T_j = 0 \\ 1 + \mathrm{LCSS}\left(F_i^{T_i-1}, F_j^{T_j-1}\right) & d_E\left(f_{T_i}, f_{T_j}\right) < \varepsilon \ \& \ |T_i - T_j| < \delta, \quad (7.18) \\ \max\left[\mathrm{LCSS}\left(F_i^{T_i-1}, F_j^{T_j}\right), \mathrm{LCSS}\left(F_i^{T}, F_j^{T_j-1}\right)\right] & \text{otherwise} \end{cases}$$

(a)

(b)

Figure 7.6 Scene preparation process for TM count: (a) zone definition and (b) models of typical paths.

where T_i is the length of trajectory F_i and F_j is a typical path. The LCSS measure provides the total number of matching points between two ordered sequences. $F^t = \{f_1, ..., f_t\}$ represents the trajectory centroid up to time t, and two important parameters, ε and δ, are empirically chosen based on an intersection configuration. ε is used to find matching points within a small Euclidean distance (i.e., d_E) and δ is a temporal alignment constraint to ensure that lengths are comparable and meaningful. Finally, the matching path is found as the path with the smallest LCSS distance [22].

$$D_{\text{LCSS}}\left(F_i^{T_i}, F_j^{T_j}\right) = 1 - \frac{\text{LCSS}\left(F_i^{T_i}, F_j^{T_j}\right)}{\min\left(T_i, T_j\right)}. \tag{7.19}$$

7.3.2 Nonvehicular Counts

Counts of nonvehicular participants include pedestrians and bicycles are used for design of crosswalks and bicycle routes or trails. Pedestrian counts are also required for designing facilities at crossing points, such as signals and timers. For this purpose, pedestrian counts are carried out over some length on either side of the proposed crossing point (e.g., along a crosswalk). Similar to a TM count, typical trajectories can be modeled along the crosswalk. Unlike for vehicles, nonvehicular traffic is typically bidirectional (e.g., right to left and left to right), which means that typical paths must be defined in both directions of a crosswalk in order to keep track of the direction of travel. The same LCSS trajectory comparison technique used before for vehicles can be implemented for nonvehicle trajectories as well.

In addition to the counts, nonvehicular crossings are analyzed further for crossing characteristics. The trajectories are further analyzed to extract information such as crossing speed since this directly relates to the functionality of an intersection. For instance, a crosswalk countdown timer may be extended if pedestrians are not able to reliably cross during a walk phase.

In the PedTrack [23] system, moving objects were detected using a background subtraction technique and potential pedestrians are determined by their size and width/height characteristics. An inherent cost function was adopted to track subsequent potential objects based on their attributes of size, height, width, and grayscale color distribution, and occlusion reasoning was performed. Waiting zones were defined to register when pedestrians entered the scene and measure their crossing time. The goal of the system was to track pedestrians in complete crossing events (waiting and completing the crossing from one registration line to another). Figure 7.7 shows the waiting zones defined by two trapezoids which register a complete crossing event.

Generally, pedestrian counts are more difficult to obtain than vehicle counts since detection and tracking is a challenging problem in traffic cameras. Pedestrians are small in size, and they tend to walk in group causing heavy occlusion which degrades tracking performance and undercounts the number of people in a grouping. Some efforts have addressed these shortcomings by avoiding tracking all together and instead learning to count individuals using dynamic textures.

Figure 7.8 shows the overview of a crowd counting system. Foreground objects were extracted and segmented into different motion directions using background subtraction and set of features, such as segments, edge and texture, are extracted. The mapping from features to counts is learned using a Gaussian process (GP) regression model [24, 25]. While effective, the nontracking-based counting method has limitations. It is not scalable since new regression models must be learned for every scene which requires an effort to collect ground truth training data. Additionally, the system is limited to providing pedestrian counts in aggregate such that behaviors such as crossing speed are difficult to infer. Finally, tracking methods are able to operate without camera calibration while this system must explicitly deal with perspective distortion.

7.4 Queue Length Estimation

Vehicle queue analysis is used in intersection control models to improve its passing capacity. The estimation of queue and associated delay are useful for devising traffic management strategies that would help to optimize traffic signals and improve the performance of a traffic network. Queue length is usually estimated either using loop detectors [26] or video cameras [27, 28] at junctions. Loop detectors are only able to measure the queue at specific lengths cut into the roadway while cameras can provide more resolution which is desirable.

(a)

(b)

Figure 7.7 Complete crossing event [23] defined by traversal between waiting zones (trapezoids): (a) pedestrian crossing and (b) pedestrian data recorded. *Source*: Cho et al. [23]. Reproduced with permission of National Academy of Sciences.

Vision-based queue analysis can be performed either with or without tracking. Details on nontracking techniques which only rely on detection are presented in Section 7.4.1, while tracking techniques are discussed in Section 7.4.2

7.4.1 Detection-Based Methods

Detection-based queue methods determine the existence of vehicles based on the appearance of features over the road. The general idea is that vehicles can be detected, without tracking, by examining the queue area in a lane (i.e., a predefined zone) based on important features that appear due to presence of vehicles. Local binary pattern (LBP) [27], spatial edges [29], fast Fourier transform (FFT) [29], and image gradients [30] are some features that have been used to detect stopped vehicles.

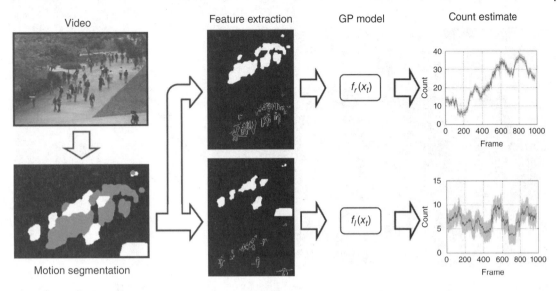

Figure 7.8 Crowd counting system [24]. *Source*: Chan et al. [24]. Reproduced with permission of IEEE.

The same Harris corner features used for tracking are also useful for detecting stopped vehicles when they build a queue [28]. The use of such features is based on the observation that the asphalt is normally uniformly constant while vehicles have a more textured appearance. Therefore, when vehicles are in a queue the zone will be filled with Harris corners, while an empty zone will have none (Figure 7.9a). Note that while corners can also be detected on road markings (e.g., lane lines), with an appropriate lane mask, it is straightforward to reject these road features that separate adjacent lanes. The zone corners are further classified as static or moving features using an optical flow method. Figure 7.9a shows static corners of one lane with dots, while the lane mask is shown by the shaded trapezoid. It demonstrates how the presence of a queue results in a high number of static corners from the queue head (at the bottom of the image) and up to the queue tail.

The queue length is determined by a line fit of the static corner locations. In the work of Albiol et al. [28], only the vertical y coordinate was needed to denote the extent of the queue in the image. The true length of the queue is then the end-to-end length of the line fit after mapping to world coordinates through homography as described in Section 7.2.2. Figure 7.9b shows an example of queue length estimation using the static corners of stopped vehicles. In this example, the length of the queue is provided in meters and an estimate for how long the queue is present in seconds is provided for each lane.

7.4.2 Tracking-Based Methods

Although queue analysis using detection methods is simple, its application is limited to queue length estimation. Tracking-based methods, in contrast, can provide vehicle specific measures of count, delay, and speed. Queue analysis with tracking must contend with the following two challenges [31]:

1) Detectability: Popular motion-based techniques (background subtraction) are not able to detect stopped vehicles in a queue. Further, motion-based detection leads to occlusion for slow moving vehicles due to viewing angle and close proximity [8].
2) Trackability: Optical flow tracking works poorly for stationary objects since it is designed for moving points. Also, with heavy occlusion in queues, there is the possibility of object grouping drift as stopped vehicles are in close proximity and have similar motion.

(a)

(b)

Figure 7.9 Detection-based queue length estimation uses nonmoving keypoints to estimate a queue line [28]. (a) Static corner points and (b) estimated queue length. *Source*: Albiol et al. [28]. Reproduced with permission of IEEE.

Therefore, tracking-based queue analysis utilizes tracking algorithms that are able to reliably segment vehicles even when stopped in a line. One approach to solve this problem is the use of a two-phase tracking algorithm based on the overall vehicle motion state [31]. When a vehicle is in a moving state, an optical flow tracker can successfully track it. When the vehicle is stopped, there is no longer a need to actively track it and a vehicle segmentation (e.g., centroid or bounding box) can be obtained by the last frame of motion. Figure 7.10 shows an example of tracking-based queue length estimation using feature points. Vehicles are detected using optical flow (i.e., points) as they approach the intersection. Once the vehicle stops in the queue, the vehicle position is maintained solely by the feature grouping. Finally, the queue is estimated as before but includes not only the

Figure 7.10 The queue length estimation (lines), detected stopped vehicles (bounding boxes), and feature points (points). Accumulated vehicle count regarding each lane and the number of waiting vehicles in the queue are displayed for each lane.

length but also the count of cars. Note the vehicle at the head of the queue in the left lane was not tracked properly, and hence has no bounding box when stopped.

A time-series analysis of the queue can also be provided. Figure 7.11a shows an example of the queue length estimation, and Figure 7.11b depicts the corresponding number of waiting vehicles in two lanes. These figures depict a similar pattern from which the traffic signal phases can be inferred (no queue during green phase). Note that while the queue length estimate has some variability due to the consistency of features points (e.g., between 24 and 50 s), the number of vehicles is consistent which is an advantage of tracking-based queue estimation.

7.5 Safety Analysis

It is important for intersections to operate safely in addition to efficiently due to the high portion of crashes that occur at these sites. In some cases, a signal is even installed to manage traffic flow and improve safety (e.g., severe angle crashes at a stop-controlled intersection). Due to potential severity, intersections must be systematically and continuously monitored throughout their life.

Historically, safety practitioners have identified intersections with the highest number of crashes in a specified time period and focused their efforts and resources at those intersections. This reactive approach can be effective in addressing a small number of high-crash locations through collection of official accident reports and data mining methods are applied to the crash datasets to find contributing factors and underlying reasons. However, accidents are infrequent and statistical significance may require many years of data and situational changes may have occurred over this time period.

Vision-based systems have emerged as a great tool for safety analysis through automatic behavior observation, accident detection, and conflict analysis. Since they operate in real time, they can generate large volumes of historical data quickly without requiring crash data collection.

As a result, there has been a major trend in video-based traffic analysis to design systems that understand participant behaviors and detect conflicts and accidents [32, 33].

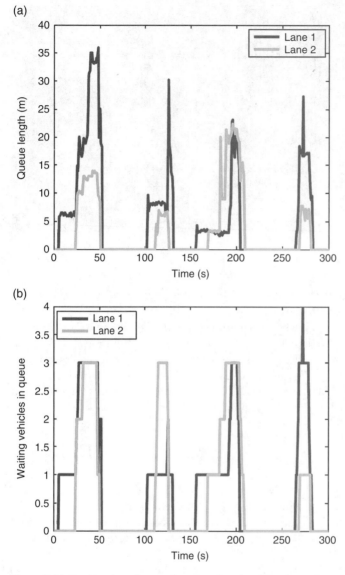

Figure 7.11 Tracking-based queue length estimation: (a) estimated queue length and (b) waiting vehicles in queue.

7.5.1 Behaviors

A behavior exhibited by an object of interest can be a single event (e.g., braking behavior) or a sequence of events indicating an action (e.g., "turning right" behavior includes a declaration and braking behavior). Behavior analysis provides an answer for some challenging questions, such as what an object (vehicle or pedestrian) is doing right now or what is that object going to do in the upcoming seconds. Compared to highways, intersections introduce many more attributes of interest due to more complex layout, structures, and participants. For example, a bus stop may be adjacent to an intersection and behaviors of interest might be the number of pedestrians using the crosswalk to enter the bus or crossing illegally in addition to the wait time for a bus arrival.

In this section, some important behaviors that have been discussed in intersection literature are explained.

7.5.1.1 Turning Prediction

TMs are of particular interest for safety since these paths intersect with other paths which might lead to hazardous situations. Predicting turning behavior involves learning turning patterns, building a model, and finding a match for observed vehicle patterns with the model.

The tracking data up to the current time can be leveraged to infer future behavior and as more information is gathered the prediction is refined. The intent is conditioned on the set of typical paths λ_j allowing for better long-term prediction [34]. The path prediction $\hat{\lambda}$ uses a window of recent data to make the comparison

$$\hat{\lambda} = \mathrm{argmax}\left(\lambda_j \mid w_t \hat{F}_{t+k}\right), \tag{7.20}$$

where w_t is a windowing function and $\hat{F}_{t+k}$ is the trajectory up to the current time t with k predicted future tracking states (obtained by extending the motion model k steps into the future). Different temporal windows will affect the trade-off between the accuracy of the prediction and the delay in recognizing a change in path. When no window function is used, all historical trajectory data up to the current time are used to make predictions. The most likely predicted paths can be seen in Figure 7.12 for a left turn at a simulated intersection at three different times.

7.5.1.2 Abnormality Detection

One of the most important tasks in a surveillance system is the detection of unusual events. Abnormal behaviors can be detected using learned models which account for the typical path locations and travel speeds. When a new trajectory does not fit a model well and there is enough deviation, it can be considered an abnormality. Abnormal patterns can then be detected by intelligent thresholding,

$$p\left(\lambda^* \mid F\right) < L_{\lambda^*}, \tag{7.21}$$

where λ^* is the most likely path and L_{λ^*} a path specific threshold. The threshold can be tuned for each path individually based on its unique characteristics through training. Abnormal speed behavior is usually investigated for contextually defined areas. For instance, Kumar et al. [35] defined special areas before a parking lot to compute speed changes for detection of violation of stop at a checkpost.

7.5.1.3 Pedestrian Crossing Violation

Detection and understanding of nonconforming behavior (violations) can be beneficial in identifying pedestrian movement patterns or design elements that may be causing safety deficiencies and in preparing for a sound safety diagnosis as well as for developing safety countermeasures. The detection process is similar to abnormality detection. Pedestrians are specifically included in crossing violation systems and crossing paths must be considered for pedestrian crossings. As explained in Section 7.2.1, optical flow is generally used for tracking since it is able to better handle occlusion between pedestrians and vehicles in close proximity.

Violation detection starts by identifying a set of movement prototypes that represent what are considered normal crossing prototypes. Crossing violations can then be categorized as either a spatial or temporal type. A comparison is conducted between a pedestrian trajectory and the normal crossing prototypes. Any significant disagreement between the two sequences of positions is interpreted

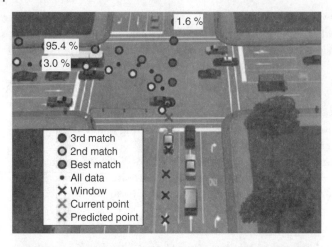

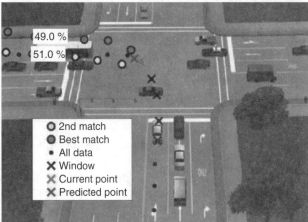

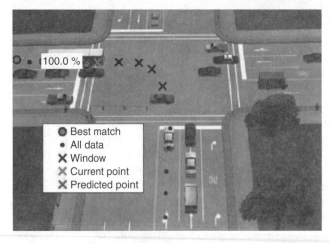

Figure 7.12 Trajectory analysis for left-turn prediction at intersection showing the probability of the top three best paths [34]. *Source*: Morris and Trivedi [34]. Reproduced with permission of IEEE.

(a)

(b)

Figure 7.13 Crossing violation examples [36]. A spatial violation occurs when a pedestrian does not use the crosswalk. A temporal violation occurs when pedestrians cross the intersection during a "red-hand" phase. (a) Spatial violation and (b) temporal violation. *Source*: Zaki and Sayed [36]. Reproduced with permission of National Academy of Sciences.

as evidence a nonconforming pedestrian trajectory. For a temporal violation, traffic signal cycles should be recorded as well to identify pedestrians traversing an intersection segment during an improper signal phase. Figure 7.13 shows examples of temporal and spatial violations.

7.5.1.4 Pedestrian Crossing Speed
Pedestrian walking and crossing speed is another key factor for behavior analysis [37–39]. Contributing factors like age, sex, and external design factors such as signal timers and facilities are also studied to evaluate their influence on pedestrian crossing speed. For instance, Tarawneh [39]

evaluated the effect of age, gender, street width, and number of pedestrians crossing as a group (group size) on their speed. A walking speed and flow rate (ped/min) relationship was found in Ref. [38] for shopping and walking areas. They found that walking speed was fixed at first and then was reduced when the flow rate was less than a 30-ped/min threshold for an indoor walkway. Montufar et al. [37] evaluated normal and crossing speed for different gender and ages, weather conditions, and temperatures. They found that the "normal" walking speed is lower than walking speed while crossing an intersection. In addition, females and older pedestrians walk slower than males and younger pedestrians, respectively.

The aforementioned studies used manual observation, either in the field or with recorded video, which limited the monitoring time period. Vision-based systems can significantly reduce the manual labor required to collect data and provide signal design guidance based on crossing times of elderly pedestrians.

However, there are significant challenges for crossing speed analysis. Pedestrians crossing together in groups result in occlusions which add noise to trajectories. For example, a group bounding box size may vary as the shape changes through bunching and spreading. These variations actually have significant impact on crossing speed estimation since even a small image variation can map to large distances in 3D intersection coordinates. Collection of age and sex is typically not possible with traffic video due to limited resolution which necessitates recordings using more specialized camera configurations. A solution is to build human-in-the-loop assistive systems which are semiautomatic. These systems use automation to reduce the amount of difficult situations that a human must handle to provide a significant analysis speedup. One such system used manual frame detections of pedestrians which were then used for automatic tracking and activity analysis [4].

7.5.1.5 Pedestrian Waiting Time

The waiting time is a fundamental measurement for signal design, and it has been shown as a criterion which affects safety as well. Hamed et al. [40] found in 2001 that long waiting time resulted in impatience to cross which could lead to undesirable behavior.

Given a pedestrian trajectory, the waiting time is calculated as the total time that the pedestrian has a speed value lower than some predefined speed threshold. At each frame, the instantaneous speed is used to characterize the waiting or walking state of a pedestrian. This is important when calculating the average pedestrian walking speed since waiting frames should be excluded from the calculation.

Figure 7.14a shows a heatmap highlighting waiting locations which help transportation engineers understand the locations of high pedestrian probability. Note the high values in red at the designed crosswalk corners. Figure 7.14b shows a snapshot from a vision-based pedestrian tracking system that used optical flow [8]. This system used a fusion of appearance and motion cues for detection which made it possible to track the waiting pedestrian around the store who was talking on his cell phone for a long time.

Fully automated waiting time estimation is difficult since waiting pedestrians lack the necessary motion required for successful detection and tracking in traffic video. With major advances in pedestrian detection [8], the use of both appearance and motion cues has become a promising methodology even in lower resolution video.

7.5.2 Accidents

Traffic accidents are abnormal events in traffic scenes and have a high societal cost. If a real-time system can accurately predict accidents in advance and then generate warnings, many traffic accidents may be avoided. As a result, accident-based safety assessment includes developing systems which

(a)

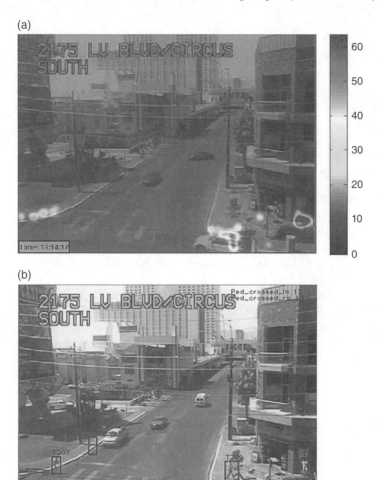

(b)

Figure 7.14 Waiting-time distribution and snapshot of tracking system which indicate a waiting pedestrian who talks with his phone for a long time period: (a) heatmap of waiting pedestrians and (b) snapshot of tracking system.

can track road users at intersections and make accurate predictions to prevent accidents. At the same time, the system can purposefully record the event as it develops. If the accident does indeed occur, the responsibility for the accident can be judged using the captured video sequences.

Vision-based methods address accidents between vehicles by predicting their future state using vehicle dynamics and recent history of movements. Collisions are detected if there is an overlap between the 3D space occupied by different vehicles at the same time. Note that 3D overlap between vehicles is utilized since 2D overlap could result from partial occlusion and not a collision. The complete process includes three steps: 3D model-based vehicle tracking, learning of activity patterns to find movement models, and finally prediction of traffic accidents [41]. Vehicle tracking uses motion-based tracking methods such as background subtraction to find regions of interest (ROIs). The process proceeds with pose prediction and pose refinement by fitting a cube model to foreground regions. A search for the optimal mapping between the projected 3D model and the 2D image data is performed to infer 3D vehicle pose. Figure 7.15a shows a 3D model example of an ROI.

(a)

(b)

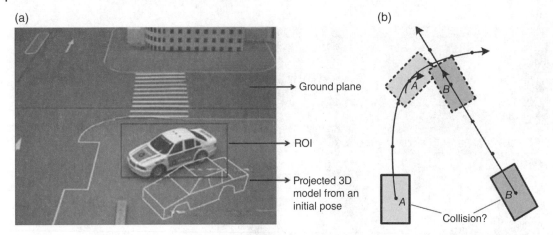

Figure 7.15 The process of collision inference from 3D pose tracking and prediction of impending collision [41]. (a) Pose refinement process and (b) collision judgment. *Source*: Hu et al. [41]. Reproduced with permission of IEEE.

With accurate 3D tracking, collision prediction can then be performed. Suppose that the observed partial trajectory of vehicle A is X and that of vehicle B is Y. The collision probability can be calculated by finding the most probable patterns (i.e., from the learned history of trajectories) for vehicles A and B. The future locations can be predicted using the LCSS method to find the path. A vehicle is represented by its ground plane projection as a rectangular box. Thus, the probability of having a collision can be formulated as whether two rectangular boxes of certain sizes would intersect at a given future time. In Figure 7.15b, the two solid lines represent the trajectory patterns regarding vehicles A and B. The solid points are sample points at equal times in the two corresponding patterns, the arrowheads show the direction of motion, the two solid line rectangles represent the vehicle projections at the current time, and the dashed rectangle lines represent the vehicle projections after some time if the two vehicles move along their associated path.

Suppose that at time t, vehicle A is at position $A(x_1, y_1)$, with direction of motion $(\delta x_1, \delta y_1)$, and the length and width of the rectangle are L_1 and W_1, respectively. The four dashed bounding box lines can be represented [41] as follows:

$$x^* \delta y_1 - y^* \delta x_1 + y_1^* \delta x_1 - x_1^* \delta y_1 \pm \frac{L_1}{2}\sqrt{\delta x_1^2 + \delta y_1^2} = 0 \tag{7.22}$$

$$x^* \delta y_1 - y^* \delta x_1 + y_1^* \delta x_1 - x_1^* \delta y_1 \pm \frac{W_1}{2}\sqrt{\delta x_1^2 + \delta y_1^2} = 0. \tag{7.23}$$

Similarly, suppose that vehicle B is at position $B(x_2, y_2)$ with direction of motion $(\delta x_2, \delta y_2)$, and the length and width of rectangles are L_2 and W_2 at time t. The dashed bounding box can be computed with similar equations.

A process to investigate whether two vehicles will come into collision is described in Algorithm 7.1. Each vehicle position is propagated forward in time along its path and a collision is returned if there is bounding box overlap. Usually, a future overlap is not a good indicator of the likelihood of a collision since it is possible for vehicles A and B to make adjustments to avoid this situation. Instead, a probability of collision is determined based on the time difference $\Delta t = t - t_0$ between the current time t_0 and the future collision time t. A larger time difference produces a lower possibility of

---

Algorithm 7.1   The Collision Judgment [41]

1:       function PREDICTION
2:           t=the current time t_0;
3:           **while** (Sample points in patterns T_i and T_j both exists) **do**
4:               Compute intersection points between two line segments regarding vehicles A, and B at time t;
5:                   **if** there are n points of intersection, **then**
6:                       **for** (each point of intersection) **do**
7:                           compute distance d_1 between the point and $A(x_1, y_1)$;
8:                           compute distance d_2 between the point and $B(x_2, y_2)$;
9:                           **if** $(d_1 < \dfrac{\sqrt{L_1^2 + W_1^2}}{1})$ or $(d_2 < \dfrac{\sqrt{L_2^2 + W_2^2}}{2})$ **then**
10:                              return collision;
11:                          **end if**
12:                      **end for**
13:                  **end if**
14:              t=t+the sampling time;
15:          end while
16:          return no collision;
17:      end function

---

collision, while smaller Δt means a collision is imminent. Therefore, a collision score to rank the danger between A and B can be computed by using Gaussian weighting as follows:

$$S_a = e^{\frac{-(t-t_0)^2}{2\sigma^2}}.$$ (7.24)

Accident detection is a core application of automated recording and reporting systems. In the traffic accident recording and reporting system (ARRS) [33], moving vehicles are detected using frame differencing and different features such as acceleration, and variation rate of area, direction, and position are estimated and used in aggregate to indicate an accident occurrence. The system was evaluated for a 2-week test period and some accidents that were not officially reported were recorded. The major problems of such vision-based systems are a high occurrence of false positives and the inability to operate during the nighttime. The ARRS incorporated complimentary data sources, such as acoustic signals, to improve performance in these situations.

7.5.3   Conflicts

Although safety is emerging as an area of increased attention and awareness, it is difficult to assess due to the lack of strong predictive models of accident potentials and lack of consensus on what constitutes a safe or unsafe facility. As an example, the collision score S_a is simple to compute but does not necessarily reflect a true probability of accident. In general, there is a lack of accident data (i.e., infrequent event) availability to make accurate assessments. Generally, traffic engineers use years of collisions along with exposure data to assess safety and in the meantime, the characteristics of the intersection may have changed. The shortcomings of accident data for pedestrian safety analysis are

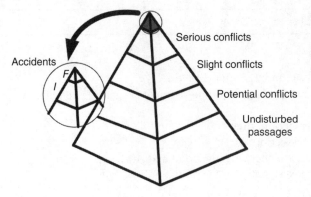

Serious conflicts

Slight conflicts

Accidents

Potential conflicts

Undisturbed
passages

Figure 7.16 Traffic safety pyramid measurement
showing hierarchy of traffic events (F, fatal; I, injury)
[44]. *Source*: Morris and Trivedi [44]. Reproduced
with Permission of SPIE.

Table 7.1 Surrogate safety measurements.

Parameter	Definition
Time to collision (TTC)	The time for two vehicles (or vehicle and pedestrian) to collide if they continue at their present speed on their paths
Distance to intersection (DTI)	The distance until a vehicle reaches a stop bar with current speed. The stop bar is used as reference point
Time to intersection (TTI)	The time remaining until a vehicle reaches a stop bar at its current speed. The stop bar is used as reference point
Time headway	Elapsed time between the front of a lead vehicle passing a reference point on the roadway and the front of the following vehicle passing the same point
Postencroachment time (PET)	Time lapse between the back end of a turning vehicle and the front of an opposing through vehicle (or pedestrian) when it actually arrives at the potential point of collision

even more acute since collisions involving pedestrians are less frequent than other collision types and exposure rates much lower.

As a result of this data deficiency, surrogate safety measures have been introduced and validated to address safety issues [42, 43]. Rather than relying on data from infrequent collisions, other events in the hierarchical traffic safety pyramid (Figure 7.16) are utilized. At the top are the accidents and going down the pyramid are successively greater number of event instances. The most often used safety surrogate is a traffic conflict defined as "an observable situation in which two or more road users approach each other in space and time to such an extent that there is a risk of collision if their movements remained unchanged" [42]. With reliable and consistent definition, conflict-based safety measurements have been proven to be practical for safety assessment [42, 45].

Table 7.1 shows some important metrics which are used for real surrogate safety analysis. Among the common surrogate safety measurements, TTC and PET are used to quantify the severity of a conflict since low values for these are directly related to how likely a collision can be avoided. The severity index (SI) is a criticality measure which can be obtained using the TTC estimate as

$$SI = e^{-\left(\frac{TTC^2}{2PRT^2}\right)},$$ (7.25)

Figure 7.17 Vehicle–vehicle conflict frequency heatmap (conflicts/m^2) providing spatial danger assessment at Burrard and Pacific Intersection [46]. *Source*: Sayed et al. [46]. Reproduced with permission of National Academy of Sciences.

where SI ranges between [0,1] and the perception reaction time, PRT, is generally approximated as 1 s [11]. The minimum TTC over an observation is extracted to denote the maximum severity of the intersection. The aggregated TTC values are then normalized by the number of hours and/or traffic volumes to fully characterize the inherent safety of an intersection. Figure 7.17 shows the vehicle–vehicle conflict heatmap to highlight the most dangerous regions of intersection [46]. The figure shows that even with the best designs, there can be regions of potential danger and these can be identified based on usage patterns and not on design assumptions.

7.6 Challenging Problems and Perspectives

In the chapter, a number of techniques for vision-based intersection monitoring have been presented along with important intersection applications. However, there still remain a number of challenges for widespread implementation and deployment of these systems.

7.6.1 Robust Detection and Tracking

While computer vision techniques have dramatically improved in quality and efficiency over the years, there are still many limitation and open problems that must be addressed before a fully autonomous intersection monitoring system can be operational. These general intersection problems include the following:

- Robustness to occlusion due to viewing angle and scene participants
- Low-resolution pedestrian detection in wide FOVs
- Long-term operation over various environmental conditions
- Trade-off between computational efficiency and accuracy

Generally, there are more vehicles and pedestrians at intersections than at highways which increase the possibility for interactions and occlusion which is a well-known and challenging problem for vision-based monitoring systems. Occlusion is amplified when there is dense traffic or when the viewing angle does not provide an advantageous perspective leading to corrupt trajectories [47]. Even when using feature-based tracking, such as optical flow, with high occurrence of occlusion, it is difficult to disambiguate neighboring objects.

Pedestrian detection and tracking has been a very popular research area since 2005 [48, 49] and detection systems are now in use in various applications such as driver assistance systems. However, most of the research has been for larger pedestrians at ground-level view rather than the small-size and overhead perspective from traffic cameras. In the perspective view, the characteristic arms and legs of pedestrians are difficult to distinguish from complex backgrounds. Motion can be reliable for small objects and appearance can be used to detect when stopped. Contextual fusion of appearance-based classifiers and motion-based detectors improves intersection detection and tracking robustness [8]. Strategies, such as optical flow, can be utilized to handle partial occlusions and further improve performance. Tracking is also challenging due to crowds of people all with very similar appearance and close proximity.

Practical intersection monitoring must be very robust to changing environmental conditions. They must handle direct sunlight, low light, shadows, rain, snow, and even nighttime deployment. Additionally, due to limited funding, inexpensive cameras are used instead of IR or stereo systems. These environmental variations dramatically affect appearance and detection and recognition frameworks.

Although great progress has been made in recent years, the tracking performance of the various systems is difficult to report and compare, especially when many of these systems are not publicly available or their details disclosed, and when benchmarks are infrequent and not systematically used. As a result, computer vision communities have developed centralized benchmarks for the performance evaluation of various tasks including pedestrian detection and tracking. For instance, MotChallenge [50] has presented a unified evaluation platform that allows participants to submit their tracking methods and own data including video and annotations. Since centralized evaluation is performed, the comparison of state-of-the-art methods can be easily inferred using the common criteria and parameters for a typical video. An extension focusing on surveillance and traffic video would be beneficial for the field.

7.6.2 Validity of Prediction Models for Conflict and Collisions

There are few opportunities to assess the quality of conflict assessment systems. Conflicts by definition require participants to be unaware of one another, which is generally not true. The only "real" conflict is a collision which itself is infrequent making it difficult to collect examples for ground truth assessment. Furthermore, the prediction of the future trajectories is unreliable. This is especially true of pedestrians whose movements are not as constrained and restricted as vehicles. Accurate motion predictions utilize more complex motion models that require precise measurements of dynamics (e.g., yaw rate and angle) which is difficult if not impossible with traffic video. Extra constraints on intersection motion, such as predefined paths, help address this but may affect conflict accuracy. The mapping from a conflict to safety is still an open problem.

Table 7.2 Comparison of sensors.

Sensor	Perceived energy	Raw measurement	Hardware cost	Units	FOV
GPS	Microwave radio signal	Satellite distance	Low	–	–
Radar	Millimeter wave radio signal	Distance	Medium	Meters	Small
LiDAR	Nanometer wave laser signal	Distance	High	Meters	Wide
Camera	Visible light	Light intensity	Low	Pixels	Wide

7.6.3 Cooperating Sensing Modalities

Each sensing technology encounters challenges for intersection use. Various sensors have been used for vehicle and pedestrian detection at intersections including GPS, light detection and ranging (LiDAR), radar, and cameras. Table 7.2 compares different sensors for data collection at intersections.

The use of cameras is most common since they are inexpensive and easy to install; in addition, they provide large FOV and camera recordings facilitate the traditional observation process. Radar-based systems trade-off FOV for range and must be tuned for specific object size. Camera-based detection of pedestrians is difficult in highly occluded scenes and during the nighttime. Newer types of nonvisible light sensors, such as thermal infrared and LiDAR, show promising results and their cost is decreasing with mass production. Sensor fusion and integration of sensing modalities with vision can be important for performance enhancement. A video provides a data stream that can be easily understood by humans, while complementary sensors can simplify detection algorithms. Further, infrastructure-based sensors can complement vehicle-based sensing to develop a complete intersection description without blind spots. Practical safety systems, such as the cooperative intersection collision avoidance system (CICAS) or forward warning systems, demonstrate the effectiveness of sensor cooperation.

7.6.4 Networked Traffic Monitoring Systems

The technology of cooperative driving with intervehicle communication [51] is a potential solution to suppress traffic jams and prevent collisions. Driver assistance systems can then be outfit with operational trajectory planning algorithms which consider potential conflicts. However, it is very difficult to infer the intention of another driver using only vehicle-based sensors. Vehicles need to send an intention signal (e.g., turn signal) to the monitoring system. Since traffic cameras have wider FOV, they can, in turn, provide more contextual data for planning used in ADAS. Cooperation between vehicles and infrastructure can be used to augment the effective sensor coverage and provide preferential views for solving problems such as occlusion and situational awareness.

Figure 7.18 shows two common scenarios with occlusion that can be addressed by vehicle-to-infrastructure communication. Both the infrastructure and driver derive mutual benefit through cooperation. The situational awareness of the driver (SV) can be improved since infrastructure-based cameras provide the trajectories of unseen vehicles.

7.7 Conclusion

In this chapter, we presented vision-based infrastructure monitoring techniques for intersection analysis including capacity, delay, and safety. The optical flow tracking method was described as a popular method to provide participant trajectories. These are used as the key input to various

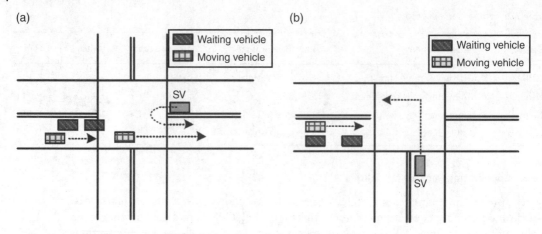

Figure 7.18 Two hazardous gap estimation scenarios. (a) Driver (SV) intends to make U-turn on a flashing yellow, but waiting vehicles mask the driver view of a moving vehicle. (b) Parked cars (i.e., waiting vehicles) occlude the driver view (SV) when attempting to make a left turn at a junction.

application systems which consider TM count, queue length analysis, and safety analysis. Robust detection and tracking is a key element which requires essential improvements such as cooperating sensing modalities and communicating with other vehicles.

References

1 J. A. León Cano, J. Kovaceva, M. Lindman, and M. Brännström, Automatic incident detection and classification at intersections, in *Proceedings of the 21st International Conference on Enhanced Safety of Vehicles (ESV)*, Stuttgart, June 15–18, 2009.

2 Intersection safety. Available: http://safety.fhwa.dot.gov/intersection/ (accessed on October 27, 2015).

3 Video surveillance for traffic. Available: https://www.videosurveillance.com/traffic.asp (accessed on November 20, 2015).

4 M. S. Shirazi and B. Morris, A typical video-based framework for counting, behavior and safety analysis at intersections, in *IEEE Intelligent Vehicles Symposium (IV)*, Seoul, June 28 to July 1, 2015, pp. 1264–1269.

5 M. S. Shirazi and B. Morris, Observing behaviors at intersections: A review of recent studies and developments, in *IEEE Intelligent Vehicles Symposium (IV)*, Seoul, June 28, 2015, pp. 1258–1263.

6 M. S. Shirazi and B. T. Morris, Vision-based turning movement monitoring: Count, speed and waiting time estimation, *IEEE Intelligent Transportation Systems Magazine*, 8(1), 23–34, Spring 2016.

7 S. H. B. Adrian Guan and R. Neelakantan, Connected vehicle insights, trends in computer vision: An overview of vision-based data acquisition and processing technology and its potential for the transportation sector, The Intelligent Transportation Society of America (ITS America), Technical Report. Technology Scan Series, 2011–2012.

8 M. S. Shirazi and B. Morris, Contextual combination of appearance and motion for intersection videos with vehicles and pedestrians, in *Proceedings of the ISVC 2014 10th International Symposium on Visual Computing (ISVC)*, Las Vegas, December 8–10, 2014, pp. 708–717.

9 N. Saunier and T. Sayed, A feature-based tracking algorithm for vehicles in intersections, in *Proceedings of the 3rd Canadian Conference on Computer and Robot Vision*, Quebec, June 7–9, 2006, p. 59.

10 M. H. Zaki, S. Tarek, A. Tageldin, and M. Hussein, Application of computer vision to diagnosis of pedestrian safety issues, *Transportation Research Record: Journal of the Transportation Research Board*, 2393, 75–84, December 2013.

11 T. Sayed, K. Ismail, M. H. Zaki, and J. Autey, Feasibility of computer vision-based safety evaluations: Case study of a signalized right-turn safety treatment, *Transportation Research Record: Journal of the Transportation Research Board*, 2280, 18–27, December 2012.

12 N. Saunier and T. Sayed, Clustering vehicle trajectories with hidden Markov models application to automated traffic safety analysis, in *Proceedings of the IEEE International Joint Conference on Neural Network*, Vancouver, July 16–21, 2006, pp. 4132–4138.

13 K. Ismail, T. Sayed, N. Saunier, and C. Lim, Automated analysis of pedestrian-vehicle conflicts using video data, *Transportation Research Record: Journal of the Transportation Research Board*, 2140, 44–54, December 2009.

14 B. Lucas and T. Kanade, An iterative image registration technique with an application to stereo vision, in *Proceedings of the 7th International Joint Conference on Artificial Intelligence (IJCAI '81)*, Vancouver, April 1981, pp. 674–679.

15 B. Horn and B. Schunk, Determining optical flow, *Artificial Intelligence*, 17, 185–203, 1981.

16 C. Harris and M. Stephens, A combined corner and edge detector, in *Proceedings of the 4th Alvey Vision Conference*, Manchester, August 1988, pp. 147–151.

17 C. Tomasi and T. Kanade, Detection and tracking of point features, International Journal of Computer Vision, Technical Report CMU-CS-91-132, April 1991.

18 Z. Zhang, A flexible new technique for camera calibration, *IEEE Transactions on Pattern Analysis and Machine Intelligence*, 22(11), 1330–1334, November 2000.

19 S. Messelodi, C. M. Modena, and M. Zanin, A computer vision system for the detection and classification of vehicles at urban road intersections, *Pattern Analysis and Applications*, 8(1–2), 17–31, September 2005.

20 J. Gerken and B. Guy, Accuracy comparison of non-intrusive, automated traffic volume counting equipment, Albeck Gerken, Inc., Technical Report, White Paper, Tampa, FL, USA, October 2009.

21 B. Morris and M. Trivedi, Learning trajectory patterns by clustering: Experimental studies and comparative evaluation, in *IEEE Computer Society Conference on Computer Vision and Pattern Recognition Workshops (CVPR)*, Miami, FL, June 20–25, 2009, pp. 312–319.

22 M. Vlachos, G. Kollios, and D. Gunopulos, Discovering similar multidimensional trajectories, in *Proceedings 18th International Conference on Data Engineering*, San Jose, CA, February 26 to March 1, 2002, pp. 673–684.

23 Y. Malinovskiy, Y.-J. Wu, and Y. Wang, Video-based monitoring of pedestrian movements at signalized intersections, *Transportation Research Record: Journal of the Transportation Research Board*, 2073, 11–17, December 2008.

24 A. Chan, Z.-S. Liang, and N. Vasconcelos, Privacy preserving crowd monitoring: Counting people without people models or tracking, in *Proceedings of the IEEE Conference on Computer Vision and Pattern Recognition (CVPR)*, Anchorage, AK, June 23–28, 2008, pp. 1–7.

25 S.-Y. Cho, T. Chow, and C.-T. Leung, A neural-based crowd estimation by hybrid global learning algorithm, *IEEE Transactions on Systems, Man, and Cybernetics, Part B: Cybernetics*, 29(4), 535–541, August 1999.

26 J. J. Bezuidenhout, P. Ranjitkar, and R. Dunn, Estimating queue length at signalized intersections from single loop detector data, in *Proceedings of the Eastern Asia Society for Transportation Studies*, Tokyo, Japan, October 14, 2013.

27 Q. Chai, C. Cheng, C. Liu, and H. Chen, Vehicle queue length measurement based on a modified local variance and LBP, in *Proceedings of the 9th International Conference, ICIC 2013*, Nanning, July 28–31, 2013, pp. 123–128.

28 A. Albiol, A. Albiol, and J. Mossi, Video-based traffic queue length estimation, in *Proceedings of the 2011 IEEE International Conference on Computer Vision Workshops*, Barcelona, November 6–13, 2011, pp. 1928–1932.

29 M. Fathy and M. Siyal, Real-time image processing approach to measure traffic queue parameters, *IEE Vision, Image and Signal Processing*, 142(5), 297–303, October 1995.

30 M. Zanin, S. Messelodi, and C. Modena, An efficient vehicle queue detection system based on image processing, in *Proceedings of the 12th International Conference on Image Analysis and Processing*, Mantua, September 17–19, 2003, pp. 232–237.

31 M. S. Shirazi and B. Morris, Vision-based vehicle queue analysis at junctions, in *Proceedings of the 12th IEEE International Conference on Advanced Video and Signal Based Surveillance (AVSS)*, Karlsruhe, August 25–28, 2015, pp. 1–6.

32 S. Kamijo, Y. Matsushita, K. Ikeuchi, and M. Sakauchi, Traffic monitoring and accident detection at intersections, *IEEE Transactions on Intelligent Transportation Systems*, 1(2), 108–118, June 2000.

33 Y.-K. Ki and D.-Y. Lee, A traffic accident recording and reporting model at intersections, *IEEE Transactions on Intelligent Transportation Systems*, 8(2), 188–194, June 2007.

34 B. Morris and M. Trivedi, A survey of vision-based trajectory learning and analysis for surveillance, *IEEE Transactions on Circuits and Systems for Video Technology*, 18(8), 1114–1127, August 2008.

35 P. Kumar, S. Ranganath, H. Weimin, and K. Sangupta, Framework for real-time behavior interpretation from traffic video, *IEEE Transactions on Intelligent Transportation Systems*, 6(1), 43–53, March 2005.

36 M. Zaki and T. Sayed, Automated analysis of pedestrians' nonconforming behavior and data collection at urban crossings, *Transportation Research Record: Journal of the Transportation Research Board*, 2443, 123–133, 2014.

37 J. Montufar, J. Arango, M. Porter, and S. Nakagawa, Pedestrians' normal walking speed and speed when crossing a street, *Transportation Research Record: Journal of the Transportation Research Board*, 2002, 90–97, December 2007.

38 W. H. Lam and C.-Y. Cheung, Pedestrian speed/flow relationships for walking facilities in Hong Kong, *Journal of Transportation Engineering*, 126(4), 343–349, July 2000.

39 M. S. Tarawneh, Evaluation of pedestrian speed in Jordan with investigation of some contributing factors, *Journal of Safety Research*, 32(2), 229–236, 2001.

40 M. M. Hamed, Analysis of pedestrians' behavior at pedestrian crossings, *Safety Science*, 38(1), 63–82, June 2001.

41 W. Hu, X. Xiao, D. Xie, T. Tan, and S. Maybank, Traffic accident prediction using 3-D model-based vehicle tracking, *IEEE Transactions on Vehicular Technology*, 53(3), 677–694, May 2004.

42 H.-C. Chin and S.-T. Quek, Measurement of traffic conflicts, *Journal of Safety Science*, 26(3), 169–185, 1997.

43 A. Svensson and C. Hyden, Estimating the severity of safety related behaviour, *Accident Analysis and Prevention*, 38(10), 379–385, October 2006.

44 B. T. Morris and M. M. Trivedi, Understanding vehicular traffic behavior from video: A survey of unsupervised approaches, *Journal of Electronic Imaging*, 22(4), 041 113–041 113, 2013.

45 C. Hyden, The development of method for traffic safety evaluation: The Swedish traffic conflict technique, Technical Report. Bulletin 70, ISSN 0346-6256, Lund Institute of Technology, Department of Traffic Planning and Engineering, Lund, Sweden.

46 T. Sayed, M. H. Zaki, and J. Autey, A novel approach for diagnosing cycling safety issues using automated computer vision techniques, *Proceedings of the 92nd Annual Meeting of the Transportation Research Board*, Washington, DC, January 13, 2013.

47 A. J. Lipton, H. Fujiyoshi, and R. S. Patil, Moving target classification and tracking from real-time video, in *IEEE Workshop Applications of Computer Vision*, Princeton, NJ, October 19–21, 1998, pp. 8–14.

48 P. Dollar, C. Wojek, B. Schiele, and P. Perona, Pedestrian detection: An evaluation of the state of the art, *IEEE Transactions on Pattern Analysis and Machine Intelligence*, 34(4), 743–761, April 2012.

49 N. Dalal and B. Triggs, Histograms of oriented gradients for human detection, in *Proceedings of the 2005 IEEE Conference on Computer Vision and Pattern Recognition (CVPR)*, San Diego, CA, June 2005, pp. 886–893, vol. 1.

50 L. Leal-Taixé, A. Milan, I. Reid, S. Roth, and K. Schindler, MOTChallenge 2015: Towards a benchmark for multi-target tracking, *arXiv:1504.01942 [cs]*, April 2015. Available: http://arxiv.org/abs/1504.01942.

51 S. Yamamoto, K. Mizutani, and M. Sato, Aichi DSSS (driving safety support system) field verification test field verification test on vehicle infrastructure cooperative systems, in *Proceedings of the 13th World Congress and Exhibition on Intelligent Transport Systems (ITS) and Services*, London, October 8–12, 2006.

8

Video-Based Parking Management

Oliver Sidla and Yuriy Lipetski

SLR Engineering GmbH, Graz, Austria

8.1 Introduction

The task of the automatic parking management is becoming increasingly essential nowadays. According to a report by Idris et al. [1], the number of annually produced cars has grown by 55% in the past 7 years. In fact, most large cities have a problem of insufficient availability of parking space. Finding a vacant parking space in a busy city business district, commercial area, near touristic sights, stadiums, exhibition halls, etc., is often a challenging task. Having no manner of assistance, a driver can spend a lot of valuable time searching for a parking lot. The result is an unnecessary waste of gas, increased emissions, and traffic [2, 3] with all its negative consequences. Moreover, when looking for a parking spot in vain, people can eventually either double park or block exits, bus stops, etc. In addition, a direct safety issue may also arise—while searching for a parking space, drivers are distracted and so have a higher probability to produce accidents.

Traffic caused by vehicles searching for a parking place can account for up to 35% of all vehicles in central business districts of cities such as New York, London, or San Francisco. According to Ref. [4], half of the fuel consumed in San Francisco is spent by vehicles trying to find a parking space. An automated free parking space detection system that operates in real time could make use of existing parking areas much more efficiently—thus, the parking management process could be greatly optimized. The information about actual parking space availability could be sent to drivers via display panels, smart phone applications, or other mobile devices.

The benefits of driver parking assistance and smart parking management have been proven in the large-scale *SFpark* project [5], which took place as a pilot project in San Francisco. The testing area covered 7,000 of San Francisco's 28,800 metered spaces, and 12,250 spaces in 15 of 20 city-owned parking garages. There were many wireless in-ground parking sensors installed, which provided real-time data about parking space occupancy states. The actual parking availability information and current parking rates could be queried with the help of smart phone applications as well as a Web application, SMS service, signs, and electronic boards.

A core aspect of the project was demand-responsive parking pricing—the rates were increased when parking spots were hard to find, and lowered when demand was low. As result, the amount of

Computer Vision and Imaging in Intelligent Transportation Systems, First Edition.
Edited by Robert P. Loce, Raja Bala and Mohan Trivedi.
© 2017 John Wiley & Sons Ltd. Published 2017 by John Wiley & Sons Ltd.
Companion website: www.wiley.com/go/loce/ComputerVisionandImaginginITS

time that it took for most people to find a space decreased by 43%, traffic volume decreased by 8%, double parking decreased by 22%, and emissions dropped by 30%.

Another project focusing on driver parking assistance, project *iLadezone* [6, 7], took place in the City of Vienna, Austria, in the years 2012–2013. The goal of this pilot project was the monitoring of commercial loading zones, which are designated for delivery trucks (e.g., spots in front of supermarkets). Due to the frequent lack of nearby parking lots and the convenience of the commercial loading zones, private noncommercial vehicles tend to temporarily occupy those areas. As a result, delivery trucks often have to park in a second row (thus blocking the road), or make another lap in the hope that the zone will be free at a later time. In order to monitor the loading areas, video sensors provided by *SLR Engineering* were mounted nearby to observe the spaces and monitor their vacancy state. Information from the camera sensors was sent in real time to a fleet management server, and a smart phone application developed by the company *Fluidtime* [7] provided the truck drivers with the current state of the loading zones. Figure 8.1 gives an example of the user interface presented to the truck driver, and Section 8.4.8 describes the project in more detail.

The Web service *parkon.com* is one more example of parking management via the cloud. It allows users to make a parking lot reservation near the largest US airports for a desired date and time span. The service aggregates information about parking lot availability from different parking area suppliers and shows a list of areas where at least one parking space is vacant—after a reservation, the parking space is marked as "occupied" for the specified time period.

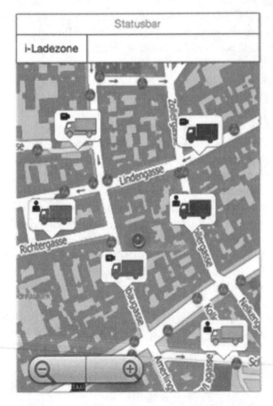

Figure 8.1 Example of a smart phone application developed for use by delivery trucks in the City of Vienna, Austria. Bright-colored truck icons indicate free loading zones, dark icons indicate occupied zones, and mid-gray color means "not in use." Source: Adapted from *Fluidtime*, Vienna.

Parking space usage is optimized with this reservation system. Still, the maintainers of the parking areas have to check frequently and manually if the numbers of reserved and actually occupied parking lots correspond to each other. For example, poorly parked vehicles may occupy two parking spaces at once, or on the contrary, reserved parking spaces may actually not be used and thus wasted for the allotted time span.

8.2 Overview of Parking Sensors

Various commercial sensors with differing sensing technologies can be utilized for automated parking space occupancy detection. For their use in commercial parking lots on the street side, in local communities, campuses, or public places, one has to consider that typically many units have to be installed in order to completely cover an area. Since this normally involves the burying of devices in the asphalt, the cost factor for installation and maintenance needs to be considered as part of the evaluation of a sensor-based system.

The following active/passive detection modalities are commercially available:

1) *Induction loops* measure the presence of the ferrous body of a vehicle over the sensor area when it crosses the loop. Typically, loops are installed at the entrance and the exit of a parking area, thus allowing the estimation of the total occupancy rate of the area. Loop sensors cannot be used in on-street areas where vehicles are to be parked close to the traffic flow.

2) *Magnetic in-ground sensors* allow to determine whether a single parking space is free or occupied. They detect the change of the earth's magnetic field by the vehicle: while a vehicle is parked, it causes a change in the local magnetic field strength, which can be detected. The sensors send the measurement results through a communication channel (e.g., radio) to a collection system, which in turn reports to a management system. Magnetic sensors are buried within the road in each parking place. Thus, this technology requires high installation and maintenance costs, and it is therefore not suitable for large areas were many parking spaces have to be managed. Moreover, electromagnetic noise coming from overhead transit lines may require some parking spaces to have two sensors in order to increase the detection quality [4]. Also, a sensor's battery life is typically limited to 3–5 years, which further may increase the total maintenance costs.

3) *Laser sensors, radar sensors, or ultrasound sensors* allow for parking space monitoring using active sensing. These sensors work on the principle of distance measurement—when a vehicle enters the measurement zone, it will be detected by the sensor, which is positioned above the area to be monitored. One example of a laser-based sensor system can be found in Ref. [8].

4) *Radio frequency identification (RFID)* is already a proven technology for the identification of vehicles at greater distances ("long range"). An RFID sensor consists of three units—transceiver, transponder, and antenna. The transponder unit sends coded information, which is read and processed by the transceiver. Occupancy detection would be basically possible through evaluation of these coded radio signals.

The most relevant sensor technologies for parking space monitoring are summarized in Table 8.1 below.

The development of video-based sensors aimed at the occupancy detection task is still an active research area. Generally, an image- or video-based solution can offer several advantages with regard to other sensor types. One factor is potential cost saving—a single video camera is able to monitor several parking spaces, or even the whole parking area at once. Thus, hardware and installation costs can be much lower in comparison to the aforementioned technologies. Moreover, already existing

Table 8.1 Technologies for the occupancy detection task using nonvisual sensing techniques.

Technology	Application	Disadvantages	Advantages
Metal detectors	Vehicle presence, counting	High installation and maintenance costs, limited battery life, false detections may occur	Robust detection per parking space
Ultrasonic sensors	Vehicle presence, possibly distance for short ranges	Not usable for medium-to-long distance, sensitive to temperature changes and air turbulence	Easy to install
Induction loops	Vehicle presence, double loops for speed measurement	Cost of installation, usage is limited to the counting of entrance/exits of a larger parking area, false triggers may occur	Well-studied technology that provides the best detection accuracy
Passive IR	Vehicle presence, counting, speed measurement, classification	Reduced detection accuracy in the presence of rain, snow, and fog	Easy to install
Radar	Vehicle presence, with Doppler radar also speed measurement	Doppler radar requires auxiliary sensors to be installed to detect stopped vehicles	Insensitive to weather conditions

surveillance cameras can be utilized for the monitoring task, resulting in no additional hardware costs. Since most of the camera-based systems work on static images, a processing time of several seconds would be acceptable for many applications. Therefore, low-power systems or inexpensive smart cameras can be very well suited for the occupancy detection task. Taken together, these properties of vision systems can result in sensors that can cover large areas, are small in size, and are relatively energy efficient.

Another advantage of a video sensor is that it can provide a live view, or at least a recent static image of the observed area. In the best case, the parking sensor can be engaged for surveillance and law enforcement applications as well—additional information like vehicle speed, type, and the number plate can potentially be extracted (albeit at higher cost for hardware, transmission bandwidth, and required power). None of the sensors mentioned so far is perfect—all show a degree of inaccuracy. In the case of a video sensor, one can easily control system performance and fix possible errors, if necessary, by visual inspection.

The design of a computer vision–based occupancy detection system is challenging in many respects. Various sources of occlusions (caused by both parking and moving vehicles, static, or moving vegetation), unfavorable weather conditions (rain, snow, fog), a range of extreme lighting conditions, and the temporary presence of pedestrians within a parking space may disturb the image of a camera sensor.

Another potential issue for visual parking space monitoring is the fact that parking areas naturally show a large variation in size and shape (see Figure 8.2 below). There can be either large or small parking areas, or just a single parking row along the roadside. Depending on the size of the observed area, one or several cameras need to be installed on site. Furthermore, the possible mounting locations for video sensors are usually limited by existing constraints of surrounding infrastructure. This in turn implies that vehicles are viewed under different aspect angles, varying sizes, and perspective distortions in each single installation. All these make it difficult to develop a generic system that is equally well applicable for an arbitrary parking area. Most of the solutions, therefore, need to be adapted (re-trained, or at least fine-tuned) for a specific site to work well.

(a)

(b)

Figure 8.2 Examples for two parking area configurations. Ideally a vision-based system needs only be reconfigured for those two parking area types: (a) a large parking space area (b) parking spaces along the road.

8.3 Introduction to Vehicle Occupancy Detection Methods

As we have outlined, the use of vision-based systems for parking monitoring applications has the advantage that potentially large areas can be covered with only a few well-placed camera sensors. Nevertheless, the number of required cameras can still be considerable for locations of unfavorable geometry. Typical parking monitoring applications are very cost sensitive, so that the hardware infrastructure for a system needs to have the smallest possible cost. The number of imaging sensors must therefore be minimized, as well as the processing load on the server side. Algorithms used in a low-cost vision system therefore need to be simple, but still robust enough to handle low-resolution and low-light situations. They also need to cope with disturbances caused by adverse weather situations, frequent occlusion, and possibly unstable image geometries caused by moving poles on to which the cameras are mounted.

In addition, the deployment and configuration of a new parking monitoring system should not be too complicated for an integrator during the installation process. A detailed configuration of algorithms to tune an installation is not possible for typically available personnel, and even on sites with just a few cameras this would not be economical.

The following Sections 8.4–8.6 describe possible 2D and 3D parking space monitoring algorithms and systems with increasing complexity. We present a concrete example for a 2D parking space management system, as well as an example for a 3D vehicle detection and position measurement application. The two systems have proven their robustness in a real-world extended field test, and they provide a good reference for the benefits which can be achieved by a well-tuned system.

8.4 Monocular Vehicle Detection

8.4.1 Advantages of Simple 2D Vehicle Detection

The potential speed and simplicity of 2D image processing for parking space monitoring makes this class of algorithms a good candidate for inexpensive systems with low processing requirements (and therefore low server loads). Good implementations for many of the algorithms described in this section exist, so that development costs for a new parking space monitoring system can also be quite low.

The disadvantage of 2D algorithms still lies in their sometimes limited robustness and applicability to difficult imaging environments. Typical 2D parking lot monitoring systems can achieve accuracies in the range of 80–95%, which is acceptable for many practical purposes, but may be not sufficient for demanding applications. Optional training and tuning of settings and detectors for every camera of an installation is possible and may help to drastically improve the accuracy of a system, but for an integrator this work is too tedious and most of the time not practical.

8.4.2 Background Model–Based Approaches

This is probably the simplest approach for occupancy detection. Typically, it includes initialization of a background image with an empty parking space, and successive updates of individual pixel values with each new frame according to the chosen background model algorithm. A binary mask is created to differentiate foreground/background pixels by thresholding the difference of the new frame relative to the background model. Morphological operations are used in a postprocessing step to obtain solid foreground blobs, which mask out image regions where vehicles occupy a parking space.

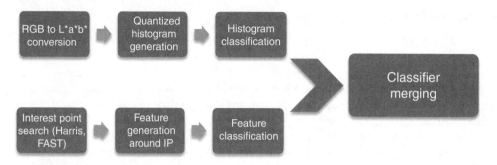

Figure 8.3 The working of a combined color/interest point-based feature detector and classifier. In an off-line processing stage, the parking area is divided into single parking spaces, which are stored in the system as ROIs. Processing as shown in the figure takes place on each of the individual parking spaces.

The main advantage of the background modeling approach compared to other 2D methods is that there is no need to train data for a specific camera view. This is a significant advantage in situations where a large number of cameras need to be installed. Unfortunately, there are drawbacks as well. Most of the algorithms need a "clear" (empty) parking area state without vehicles, which is mostly not possible in real-world scenarios (this is not the case in the system presented by Bulan [9], see later).

There are many challenges to cope with as well—changing weather conditions, rapid illumination changes, and strong reflections from wet road surfaces can easily lead to false detections and disturb the background model. Therefore, additional techniques have to be applied to deal with such situations. This makes the originally simple approach increasingly complicated. Also, scenarios with parking vehicles have low temporal dynamic—a vehicle can stay parked for several hours or even days. In this case, chances can be high that a car becomes part of the background over time.

Modi [10] in his work used foreground detection and a shadow avoidance approach to suppress false detections. A set of video cameras was used to cover a large truck parking area in which the locations of individual parking spaces were specified manually once during system setup. A parking space was recognized as occupied if more than a certain percentage of the background area was occluded by a foreground object. Each pixel was modeled as a Gaussian mixture model (GMM) as described in Ref. [11]. Gray values from a new camera frame updated the mean μ and standard deviation σ of several matched Gaussian distributions for every pixel. A pixel gray value was considered to be part of the distribution, if its value was within 2.5σ of one of the learned distributions per pixel (which were continuously updated). Pixel values, which did not fit any of the Gaussian distributions, were considered to be foreground. The GMM method adapts well to lighting changes and slow moving objects. The main drawback is that shadows are often detected as foreground. To eliminate shadows, the approach described by Joshi et al. [12] for shadow removal is used. The basic idea of Joshi et al. is the observation that the ratio of a pixel RGB values will remain constant despite the change of intensity (the color is constant when the brightness changes).

Modi reports several challenges he faced during design of his system. First, fast illumination changes (e.g., from moving clouds) could trigger false alarms. He tried to overcome this problem by tracking the changes in the background mask over 1–2 s (assuming the ability to process up to 30 frames/s). Another issue was the occupation of a parking space by humans. To resolve this, Modi employed a blob tracking approach with successive trajectory analyses to differentiate such occlusions from a parked vehicle. He reports that all these measures have increased the initial accuracy from 44% up to 100%; however, his testing data set was quite small with only nine vehicles.

8.4.3 Vehicle Detection Using Local Feature Descriptors

The basic idea of a local feature set is the creation of a general image descriptor which makes it easier to distinguish between an empty and an occupied parking space, compared to just looking at raw pixel gray values or comparison with a background model. As described before, the use of simple features for parking space occupancy detection has the advantage of low computational complexity and resulting ease of system implementation. The algorithms employed are usually fast and, therefore, very well suited for real-time operation.

Before any feature generation on an image or region of interest (ROI) of a parking space is performed, an important preprocessing operation should be applied: the parking space segments in the image are to be normalized so that their size and scale become comparable. This includes (i) the removal of lens distortion, (ii) a scale adjustment, and (iii) the correction of perspective distortion. Although in practice normalization cannot be perfectly achieved due to the inherent nonlinearities and perspective changes of viewpoints over the parking area, it is nevertheless an important step in order to maximize detector performance.

Color histograms within defined ROIs (typically manually assigned regions of single parking spaces) are probably the simplest means of feature computation. True et al. [13] have tested color histogram features with reasonably good results (the authors report about 90% of recognition accuracy). The important aspect of the use of color is that a transformation from RGB space, as captured by a camera, into L*a*b or HSV color space can significantly improve the detection performance. This color space transformation decouples the luminance value from the color components and thus makes the color features much more independent from illumination changes.

For classification of a parking space as free/occupied, True et al. quantize the a*b* color values into 32 bin histograms and classify the space using Euclidean distance, or chi-squared distance metric, or with a support vector machine (SVM).

The k-NN classifier computes the Euclidean distance

$$D\left(h_i, h_j\right) = \sum_{k=1}^{n} \sqrt{\left(h_i(k) - h_j(k)\right)^2} \tag{8.1}$$

between a new (unknown) color feature vector h_i and a set of reference vectors h_j of size n, which have been established during the learning phase of the classifier. Similarly, the chi-squared distance uses the following distance metric between a new color feature vector h_i and a set of reference vectors h_j:

$$\chi^2\left(h_i, h_j\right) = \frac{1}{2} \sum_{k=1}^{n} \frac{\left(h_i(k) - h_j(k)\right)^2}{h_i(k) + h_j(k)}. \tag{8.2}$$

If the new histogram is close enough to at least k of the learned reference vectors of a specific class, it is assumed to be member of this class.

SVMs [14, 15] learn a high-dimensional decision surface with maximal margin between two classes. They are clearly superior to k-NN classification and, in the case of a linear kernel, can be computed very efficiently as well.

Simple edge-based methods as slightly more complex methods (at least in computational terms) have been used in recent years as feature descriptors in parking space monitoring systems. Čaklović et al. [16] convert a new RGB image from a camera to grayscale image, then apply a Sobel edge detection operator and convert the result to binary by thresholding the Sobel edge magnitude result. The result is postprocessed using morphological operations in order to obtain the edges of possible cars on the parking spaces. The average count of edge pixels is then used within each parking space polygon to decide if it is occupied or free.

The approach described by Čaklović et al. works well from top view images without occlusions and in favorable environmental conditions, but it is too simple to cope with shadows, changing vegetation, or occlusion.

To take things further, the combination of multiple, simple features seems to be a promising approach because the strengths of each feature type can be combined to form a more robust and stable classifier. This has been examined by True in the previously mentioned work [13]. His implementation can be used as a role model for this general class of occupancy detection algorithms, which is shown in Figure 8.3.

First, L*a*b* color space histograms of the parking space are created and classified. Working in parallel to the color histogram classification, a feature point detection algorithm takes each parking space ROI and finds interest points (IPs) within its polygon using a corner detector like the Harris operator or Rosten's machine learning algorithm [17, 18]. Based on the gray-level appearance of a small window around each IP, it is classified as being part of a vehicle or background. The whole parking space is then classified as being vacant or occupied according to the number of IPs, which fall into the vehicle class. This voting allows the algorithm to better cope with patches of the background (empty parking spaces) that are misclassified because they bear a certain similarity to vehicle structures.

During training, the algorithm creates a vocabulary of positive vehicle features, which are the result of a clustering of similar features. Instead of using unprocessed image patches, alternative features could be the scale-invariant feature transform (SIFT) [19, 20] or speeded up robust feature (SURF) [21] descriptor. They are more complex to compute, but can lead to better detection results because they are illumination-, scale-, and rotation-invariant. True [13] uses the trained appearance–based feature detector for occupancy detection and resorts to the color classifier, when no IPs can be found in the ROI of a parking space polygon.

8.4.4 Appearance-Based Vehicle Detection

By extending the area for feature generation to the whole normalized parking space, occupancy detection is essentially transformed into an appearance-based vehicle detection task. Several methods can be used for this task. Viola and Jones [22] started this class of algorithms with their seminal work in 2001.

A practical implementation for vehicle detection used in parking space management is presented by Fusek et al. [23]. In their work, the authors present a method that employs a boosting algorithm, like Viola et al., for car detection on normalized subimages of parking spaces. They train a cascaded classifier which uses simple Haar features to build a robust and fast vehicle detector. An improvement of the detection rate is achieved by shifting the ROI vehicle detector to several locations around the parking space center position to increase the chance of a positive vehicle detection.

Like Viola et al., Fusek et al. compute their Haar-like features using integral images to significantly speed up the feature generation process (see Section 8.4.5). Haar-like features consist of rectangular white and black regions, and the features are defined as the difference of the sum of pixels between those respective regions and the sign of the difference. Due to their structure, Haar-like features are ideally suited for a computation with integral images.

Fusek et al. report that their approach produces better results than a histogram of oriented gradients (HOGs) detector, as HOG can produce a high number of false-positive detections in noisy images. For the training process, the authors used 4500 positive and 4500 negative manually labeled samples. The detection rate of their classifier was about 94% for images of good quality. On noisy images, Fusek et al. achieved 78% for positive samples and 94% for negative ones, while the HOG detector achieved 94% and 44% correct classification, respectively.

Tschentscher et al. [24] have evaluated different combinations of feature extractors and machine learning algorithms in their work. Their setup consisted of a wide-angle lens camera which observed 15 nonoverlapping manually labeled parking spaces. They evaluated the following features:

- Color histograms
- Gradient histograms
- Difference-of-Gaussian histograms (DoGs)
- Haar-like features

For the color histogram features, RGB-, HSV-, and YUV-color spaces were used. For a binary classification (occupied/vacant), several classifiers were tested and compared to each other on a training set of about 10,000 samples:

- k-NN
- linear discriminant analysis (LDA)
- SVM

As a postprocessing step, temporal integration was applied to increase detection robustness of the classification result $f(t)$,

$$f(t) = \alpha * Y_{t-1} + (1-\alpha) * Y_t. \tag{8.3}$$

Here, alpha is the learning rate, and Y represents the particular class label (vacant/occupied) for each parking space.

In the author's experiments, the color- and DoG features have always shown the best results, the detection rate was up to 97%. Using a combination of those two feature types as well as by applying the temporal integration, the detection accuracy has increased up to 99.8% in their tests.

8.4.5 Histograms of Oriented Gradients

Dalal and Triggs [25] introduced a new feature descriptor method in 2005, which is conceptually somewhat similar to SIFT and SURF, but is easier to compute and therefore requires less computational power. Their HOG descriptor represents a very good baseline (although it is no longer considered to be the state of the art) in terms of feature creation, sample classification, and detection performance.

HOG features of a sample window W are computed and classified using the following algorithm:

---

1. Divide W into nonoverlapping $N \times M$ square cells of size $C \times C$ pixels ($C = 6$, 8 typically).
2. Compute the L2 normalized histogram H_c of edge orientations, for each cell, the orientation bins for the histograms are 22.5° wide, so that each cell histogram has 8 bins.
3. Threshold and saturate H_c, a typical threshold value is 0.2.
4. Combine 2×2 neighboring cell histograms H_c into a block histogram B and normalize B using L2 norm.
5. Combine all block histograms B into a feature vector F_R, and normalize it using L2 norm.
6. Classify the resulting feature vector F_R using a linear SVM.

Figure 8.4 shows the arrangement of cells and blocks for a typical vehicle detection scenario. In order to detect vehicles in parking spaces, either a sliding window of suitable shape can be used to detect vehicles within a larger area of interest, or the HOG features can be computed in a fixed area, which is located directly over the specific parking space to be monitored.

Over time, the classical HOG framework has been improved in several ways to achieve higher processing speed and increased detection robustness and accuracy. By employing a cascade of simple HOG detectors for example, Zhu et al. [26] can significantly speed up the detection process (see Section 8.4.7).

Figure 8.4 The computation of HOG features on a sample location. Cells of size $C \times C$ pixels ($C = 6, 8$, typically) are combined into blocks; overlapping blocks span the whole sample window. The concatenated block histograms, after normalization, form the HOG detector feature vector.

Using a different method for choosing simple HOG features and combining them into cascades, Sidla [27] reports a similar improvement in detection as well as an improvement of the runtime performance.

The concept of integral histograms to speed up the creation of local edge orientation histograms in overlapping windows makes HOG real-time capable. The computation of edge orientation histograms is done using quantized edge orientation channels O_H for each possible orientation bin. The 0–180° resp. 0–360° raw edge orientations from the Sobel operator are for this purpose quantized into the typical eight HOG orientation bins O_H. The computation of an integral histogram I_O from an edge orientation map E is done as follows:

For each orientation channel O,

$$I_O(x,y) = I_O(x-1,y) + I_O(x-1,y) - I_O(x-1,y-1) + M_O(E(x,y)).$$

(8.4)

$M_O(x)$ is a function which maps an edge orientation value E to a discretized edge histogram entry which can be added to the summed edge histogram. This map can be a lookup table for computational efficiency.

The HOGs for a rectangular region at image coordinates from $L(x_l, y_l)$ (left upper point) to $R(x_r, y_r)$ (right lower point) can then be computed from the integral histogram as follows:

For each orientation channel O,

$$H_O = I_O(x_1,y_1) + I_O(x_r,y_r) - I_O(x_r,y_1) - I_O(x_1,y_r).$$

(8.5)

By combining integral histograms with low-level program optimizations to make use of modern CPU instructions (single instruction, multiple data (SIMD) instructions like SSE for x86 architectures,

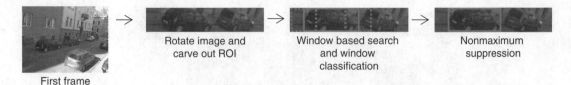

First frame → Rotate image and carve out ROI → Window based search and window classification → Nonmaximum suppression

Figure 8.5 The processing pipeline of the HOG vehicle detector as used by Bulan et al. Raw camera images are rectified first, on an ROI which has been precomputed using background subtraction for change detection. A sliding detector window is then run with the pretrained HOG detector.

and NEON for ARM CPUs), HOG features can be computed an order of a magnitude faster compared to conventional implementations.

Scale invariance of the HOG feature descriptor is achieved by either resizing the original input image, and running the detector again with identical cell/block size. Alternatively, the HOG detection window and its corresponding cells/blocks can be scaled and tested on the same input image size. The integral histogram makes the latter approach especially easy to apply, and it allows the computation of HOG features in real time. As can be seen from Equation 8.5, a local histogram can be computed using $4 \times$ (histogram size) memory accesses only and the remaining normalizing and scaling operation of the feature vectors can be efficiently parallelized using SIMD vector instruction techniques.

The resulting high-dimensional normalized feature vector is classified using a linear SVM, which is again very well suited for optimization using SIMD programming techniques.

A practical implementation of a vehicle detector using HOG for a parking monitoring system in the city of Boston is presented by Bulan et al. [9]. Their system can detect the occupancy of parking spaces from distorted camera images using a multistage algorithm.

Candidate ROIs are first defined by foreground difference blobs, which are separated from blobs created by motion or occlusion with the help of (i) a simple background model, and (ii) double image differencing for motion detection. The background model B accumulates all new camera frames using

$$B_{t+1} = f * F_{t+1} - \left(1 - f\right) * B_t,$$ (8.6)

where F_{t+1} is the next frame from the camera and f is an update factor (typically around 0.05). Note that single pixels will be updated only in areas where (i) no motion, (ii) no occlusion, and (iii) no parked vehicle have been detected.

The location of candidate regions in a frame is stored as a binary mask. Since these regions may be caused not only by a parked vehicle but also may occur due to changes from varying illumination, shadows, glare, reflections, and other objects, a verification stage is needed. To this end, a HOG classifier, which has been trained in an off-line phase, tests the candidate ROIs in sliding window search mode. By computing HOG only on the masked image pixels, the runtime performance of detection is significantly improved so that Bulan's implementation becomes real-time capable.

The HOG detector is furthermore used on the camera images after applying a perspective correction, so that the approximate scale of vehicles is similar for all parking spaces under consideration. Figure 8.5 shows the HOG detection pipeline of the system. Bulan could achieve greater than 91% average detection accuracy over a test period of 5 days. This would be higher than the required accuracy of in-ground sensors in San Francisco. The processing time was about 5 frames/s. In addition to the occupancy-state detection, Bulan proposed methods for parking angle violation detection and parking boundary violation detection.

8.4.6 LBP Features and LBP Histograms

The appearance–based object detection principle relies on the fact that object classes exhibit certain visual properties that can be represented by a compact statistical description. These properties, or features, which are computed using SIFT, SURF, or HOG are very often based on some form of edge information. But many other descriptors are possible.

The local binary pattern (LBP) descriptor [28] is a fast and robust measure for texture description, which uses an efficient strategy to describe local image intensity differences. For some time, LBP features have been used very successfully in machine vision applications for texture analysis, for example, in wood sorting machines. In recent years, they have found their way into systems for face detection, face recognition, and pedestrian detection.

The advantage of the LBP approach is that it is easily and efficiently to compute and that it is relatively insensitive to illumination changes. Variations of LBP—rotation invariant, and uniform binary patterns (see below)—are even more robust to small perturbations and illumination changes of local image patches.

Local binary patterns are computed from binarized intensity differences in a small neighborhood around a central pixel G_c. Typical distances R for the neighborhood are 1 or 3 pixels. Every neighborhood around a pixel G_c can be sampled by an arbitrary number of P locations, but for the convenience of storage and representation, P is usually chosen to be 8 or 16.

At every location P_i around the center pixel G_c, we have to sample a gray value g_i from the image as follows:

$$g_i = \left(R\sin\left(\frac{2\pi i}{P}\right), R\cos\frac{2\pi i}{P} \right), i = 0\ldots7. \tag{8.7}$$

Here, R is the radius chosen (1 or 3) and P is the number of LBP features. The gray values at the resulting positions g_i are calculated by applying a bilinear interpolation using four neighboring pixel values. Then each local binary value B_i for the center pixel G_c is computed as follows:

$$B_i(g_c) = \begin{cases} 1, g_c - g_i > T \\ 0, g_c - g_i \leq T \end{cases}. \tag{8.8}$$

Here, g_c is the gray value of the center pixel, g_i is the gray value at the neighborhood P_i, and T is an empiric gray value difference threshold to reduce noise influence, and it often ranges from $T = 2$ to 10 in practical implementations.

The pixel gray value information from Equation 8.8 can be transformed into a binary value number representation by applying the binomial factor 2^p.

$$\text{LBP}_{P,R} = \sum_{p=0}^{P-1} \left(B_p(g_c) \right) 2^p \tag{8.9}$$

For $P = 8$, using input from all eight neighboring pixels, a total amount of 256 combinations is possible. Thus, our feature vector as descriptor for the single pixel location g_c has a size of 8 bits and can be stored as byte value (Figure 8.6).

A collection of LBPs within a region of interest as a histogram, similar to edge histograms in the HOG descriptor, then becomes a powerful texture descriptor, which can be directly used for object classification. The unprocessed LBP histogram has a size of 256 entries.

The raw local binary patterns, or their histograms, provide the most detailed representation of a pixel neighborhood at an image location. Each neighbor pixel P_i contributes 1 bit of information. By applying a selection and transformation of the raw LBPs, one can achieve rotation invariance as well. In order to become rotation invariant, the patterns $\text{LBP}_{P,R}$ need to be rotated bitwise to the right so that the least significant bit becomes a "1".

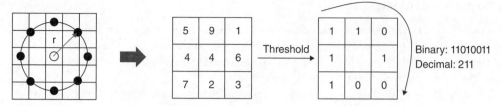

Figure 8.6 Creation of the decimal encoding of an LBP. Source: Delibaltov [29]. Reproduced with permission of IEEE.

$$\text{LBP}_{P,R}^{ri} = \min\left\{\text{ROR}\left(\text{LBP}_{P,R},i\right),|\,i=0\ldots P-1\right\}. \tag{8.10}$$

$\text{ROR}(x, i)$ is a function which rotates x bitwise, circular, to the right. All in all, 36 rotation-invariant patterns $\text{LBP}_{R,8}^{ri}$ exist.

A further improvement and invariance of the LBP descriptor can be achieved by choosing "uniform" patterns—these are all bit patterns that have at most 2 bitwise transitions from 0–1 or 1–0. For example, 00000000_2 has 0 bitwise transition and 00001111_2 has 1 bitwise transition. The set of rotation-invariant patterns, $\text{LBP}_{R,8}^{ri2}$, has a total of nine uniform patterns, which are:

$$00000000_2,\ 00000001_2,\ 00000011_2,\ 00000111_2,\ 00001111_2,$$
$$00011111_2,\ 00111111_2,\ 01111111_2,\ 11111111_2$$

A histogram of this type of LBP would therefore contain only 9 bins and leads to a very compact and efficient texture representation of an ROI.

8.4.7 Combining Detectors into Cascades and Complex Descriptors

Different types of descriptors have their own unique properties of computational complexity, locality, invariance to geometric transformations, etc. By combining descriptors into cascades, a balance between detection accuracy and computational resources can be achieved.

Beginning with Viola et al. [22], the concept of detector cascades has proven to be extremely worthwhile. By selecting HOG features in a shallow cascade of detectors, Lipetski et al. [30] have shown that a very robust and real-time capable object detector can be trained. This is achieved by using simpler, but less discriminative, HOG features in the first cascades stages, and testing the more complex and computationally demanding HOG descriptors only in the last stages of the cascade.

LBPs have been combined by Wang et al. [31] with HOG features in a similar manner in order to create a strong object detector. The advantage of the LBP approach is that it is computationally very efficient, and so does not significantly increase computing times. Armanfad et al. [32] use LBP in their texture-edge descriptor in blocks of pixels to segment background patches from object candidates with promising results.

The addition of new feature dimensions by adding LBP also improves the recognition performance of the vehicle detector occupancy detector, which is outlined in Section 8.4.8. Lipetski [30] demonstrates that the combination of HOG and LBP features can achieve almost 100% accuracy even in adverse and varying outdoor conditions.

8.4.8 Case Study: Parking Space Monitoring Using a Combined Feature Detector

This section describes a real-world pilot project for a parking space monitoring application (Lipetski [30]), which had to operate under demanding outdoor conditions. The detection approach is based on appearance based modeling—it combines the HOG feature descriptor with LBPs in order to

Figure 8.7 The loading zones which had to be monitored for the pilot project in the City of Vienna. This is a nighttime image of the two empty parking spaces. Harsh lighting situations, reflections, and adjacent heavy traffic complicate the detection task.

create a strong vehicle occupancy detector. Figure 8.7 shows a typical image of two parking spaces, which the system had to monitor. The parking spaces are reserved as loading zones for delivery trucks, and their occupation status (empty/free) is reported to a fleet management server.

Due to the camera viewpoint, the wide-angle view, and the environmental conditions on site, the vision system had to deal with strong perspective distortion, extreme lighting conditions, and disturbance by adjacent vehicle and pedestrian traffic.

The occupancy detection algorithm implemented for this application has been optimized for a smart camera. It could execute the vehicle detection task using its built-in CCD sensor and processing unit, as well as handle communication with an external fleet management system over a UMTS connection.

The visible parts of the observed parking area were divided into two individual ROIs, which were independently monitored. Their shapes are marked as polygons in Figure 8.7. It is worthwhile noting that the classification has been computed on these image patches directly, without geometric normalization. For feature computation and occupancy state detection only pixels, which lie inside these predefined areas, are considered.

The first part of the classifier relies on HOG feature descriptors, which are trained using manually annotated samples and a linear SVM as classifier. The vehicle detection areas in the camera image were of several meters length—every parking space was split into 4×4 blocks, each of which was divided into 4×4 nonoverlapping cells. A gradient histogram with a resolution of 8 bins over a 180° orientation range was constructed for each cell, and all cell histograms were normalized and merged. Thus, the whole HOG feature vector had a length of $4 \times 4 \times 4 \times 4 \times 8 = 2048$ elements.

The second part of the classifier computed raw LBP histograms within the detection ROIs, which were also classified using linear SVMs (see Table 8.2 for an evaluation of the performance of different LBP variants).

For the final decision, the results of the HOG detector (Figure 8.8) and LBP classifier were fused and smoothed in time over several camera frames. Due to their different nature, length, and descriptive power, the LPB and HOG feature vectors were not merged into a large vector, which was then being classified. Instead, the results of both detectors were combined using some simple rules. When both detectors recognized a parking space as *occupied*, the output was flagged as

Table 8.2 Detection scores for different LBP configurations, and a comparison with the HOG detector results.

LBP configuration	Correct detection rate (%)
LBP, $R = 1$, uniform 2, bilinear interpolation	67.2
LBP, $R = 1$, uniform 2, nearest-neighbor interpolation	33.4
LBP, $R = 3$, raw, bilinear interpolation	78.1
LBP, $R = 3$, uniform 2, bilinear interpolation	74.4
LBP, $R = 3$, rotation invariant, bilinear interpolation	74.3
HOG descriptor	94.4

Interpolation relates to the method which has been used to sample the gray values g_i (see Eq. 8.7) from the input image.

(a)   (b)

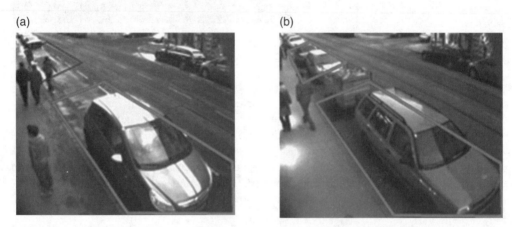

Figure 8.8 Typical detection results using the combined HOG–LBP detector. The disadvantage of site- and camera-specific training is offset by the excellent detection rate and very low false-positive rate of this approach. (a) Presence of pedestrians does not produce false alarms. (b) Cars' positive detections.

occupied. When both marked it as *empty*, the output was flagged as *empty*. In the other two contradicting situations, the state of the parking space was not changed and the previous occupancy state was kept.

The complete workflow of the detector was as follows:

1) The camera triggered image acquisition in 2-s intervals.
2) Features were extracted from predefined regions (polygons representing parking spaces):
   2.1 HOG features, *FH*
   2.2 LBP features, *FL*.
3) Feature vectors *FH* and *FL* were normalized to unit vectors.
4) Classification of *FH* and *FL* using the pretrained SVM for both feature types:
   4.1. From a generic database of training images
   4.2. From a database of images taken at the site.
5) Fusion of HOG and LBP detector results.
6) The last *N* detection results were kept and the occupancy state only changed if all of them were identical.

The additional time series filtering of Step 6 made the occupancy sensor much more stable in that it removed spurious false positives, which could be created by passing vehicles, or pedestrians which came into the field of view, or strong reflections of headlights. The last N occupancy states (typically $N = 7$) were taken, and only if all of them were equal to each other, the sensor accepted this value as the final decision.

Two smart cameras have been installed at test sites in the City of Vienna and operated constantly day and night during a period of several months. The results of an extensive evaluation were encouraging: 97.4% of all vehicles were detected, at a false positive rate of 0%—despite the fact that the parking spaces were constantly perturbed by pedestrians, shadows, and reflections. The robustness against pedestrians can be explained by the fact that people have mostly vertical edges, while vehicles also contain horizontal edges. All of the 2.6% of the vehicles that have not been detected were parked in such a way that they occupied only a fraction (mostly <50%) of the parking space area.

Summarizing the approach by Lipetski et al., the combination of HOG and LBP achieved a very good detection rate with negligible false positives for several reasons as follows:

- The system was made more robust by the combination of two different detectors. Only when both detectors recognized the presence of a vehicle, the state of a parking space could change to *occupied*.
- A history of N successive detector answers over several images was considered in order to suppress single false detections.
- The site-specific training optimized the detector for the geometry of the site. This is important for both HOG and LBP features. *Note:* this effort would need to be eliminated in a larger production system to avoid the overhead of manual training for every camera view.

For the evaluation of the performance of different LBP configurations, and a comparison with HOG, Lipetski et al. have taken video data from one of the test sites for a detailed performance analysis. The authors used two arbitrary days for training the classifiers and two additional days for evaluation. From the two training days, they labeled images taken in 10 s intervals. By choosing only images which had been taken between 6 a.m. and 7 p.m., Lipetski gathered 4680 training samples per parking space (since the parking spaces were part of a delivery zone, typical business times were sufficient for evaluation).

After training, Lipetski et al. tested the occupancy detector on the evaluation set. For comparison, they applied different neighborhood radii and LBP pattern configurations. The results are shown in Table 8.2. For comparison, the HOG detector results for the same set of evaluation images are listed in the last row of Table 8.2.

Considering only the LBP features, it can be seen from Table 8.2 that detectors with $R = 3$ clearly provide better results than detectors with $R = 1$. One can also see that the HOG detector significantly outperforms any of the LBP feature combinations. The best LBP result of 78.1% is much worse compared to the classification rate of 94.4% obtained by the HOG descriptor alone. One of the main reasons is the fact that spatial information from the scene is neglected by the LBP descriptor—the LBP patterns serve as input for only a few histograms, each of which cover a larger image area. The HOG approach in contrast creates a much more detailed scene representation.

8.4.9 Detection Using Artificial Neural Networks

Artificial neural networks (ANNs) represent a learning approach, which is inspired by biological NNs. They are widely used in computer vision for object classification tasks, with a resurgence of interest over the recent years.

Typically, an NN consists of the input layer (data to classify), and the output (result) layer and one or several hidden layers of interconnected neurons. Information is forward propagated over the layers from the input to the output. The connections between neurons are linearly weighted; the weights are to be learned in the training stage. A neuron output signal is calculated as the weighted sum of connected neurons from the previous layer, and a nonlinear activation function is typically applied afterward.

$$y_j = K\left(\sum_i w_i x_{ij} + b_i\right). \tag{8.11}$$

Here, x_i is a value of an input neuron i, w_{ij} is a connection weight from the input neuron x_i to the output neuron y_i, b_i is a bias value, and $K()$ is an activation function—for example, a sigmoid S function can be used as activation,

$$S(t) = \frac{1}{1+e^{-t}}. \tag{8.12}$$

The neuron output response is passed to the neurons of the next layer, and the calculation is repeated for all layers, until values of the output layer are finally calculated (Figure 8.9).

NN-based approaches have become more and more popular and powerful over the recent years, as new techniques have been developed. One of the essential contributions is the work of LeCun [33], where he describes his convolutional NN (CNN) approach. Since they preserve spatial information and the mutual context of features, CNNs are very well suited for object classifications tasks in computer vision (Figure 8.10).

In recent years, effective algorithms for the training of large networks with many hidden layers (deep networks) have become available, making it possible to perform object classification tasks with unprecedented accuracy. In fact, approaches that use deep networks show the best results in practically all actual benchmarks and competitions related to the object classification benchmarks. Deep networks are even beginning to outperform humans [34] in certain specific recognition tasks. With today's available data sets, which sometimes range into 10 million sample images, deep networks have practically unlimited training potential.

The main drawback of the NN approach is the necessity of a tedious and time-consuming training process. To perform well, a NN needs a huge set of manually annotated data. Concerning the parking space occupancy detection task, it is difficult to build a generic NN for arbitrary input data sets because in different installations, vehicles appear under very different viewing angles. To achieve the best detection results, the training process should be performed individually for each camera installation,

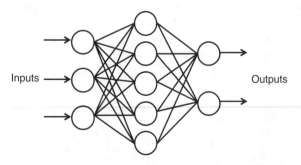

Inputs → Outputs

Figure 8.9 Data processing in a fully connected NN with input, intermediate, and output layer.

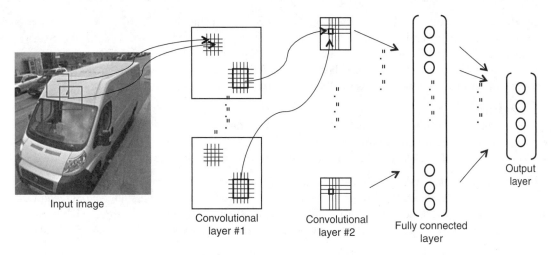

Input image   Convolutional layer #1   Convolutional layer #2   Fully connected layer   Output layer

Figure 8.10 CNN configuration with input layer, two convolutional layers, fully connected layer, and output layer. The input layer is a normalized grayscale input image. Each of the convolutional layers consists of several feature maps; the output layer represents the classification results.

and the training data should include all possible weather and lighting conditions. Ideally, data from all seasons should be included to cover dry and wet streets, snow, and rain.

There are not many NN-related approaches used for parking space detection available in the literature to date. Jermsurawong et al. [35] train two separate networks for day- and nighttime conditions. A single camera was used to simultaneously observe 126 nonoverlapping parking spaces (9 lanes with 14 parking spaces in each lane). The resolution of a single parking space was only 30 × 14 pixels in this scenario. Such a setup was possible due to the installation of the camera on the 29th floor of a building. The number of manually annotated training data was about 40,000 and 14,000 samples for the daytime and the nighttime network. For the network input layer, different features were extracted—brightness-related, pixel-related, color-related, time-related, and edge features.

A test over 24 h has shown 99.9% accuracy for occupied spots and 97.9% for empty parking spaces. The lower second value can be explained with much fewer training samples of empty parking spaces compared to the occupied ones (about 3% of the whole training set), as the parking area was occupied most of the time.

8.5 Introduction to Vehicle Detection with 3D Methods

Even the most sophisticated 2D vision systems have to deal with situations were visual analysis is very difficult or practically impossible. Strong occlusions by vegetation, extreme illumination conditions, perspective distortions, large depth of fields, or even very high required accuracies can limit the usability of pure 2D image analysis for parking space monitoring.

A 3D monitoring approach can provide the means to cope with some of the aforementioned problems, and it can offer additional benefits because more usable information can be extracted with this approach. For example in heavy traffic, occlusions from vehicles on the street can reliably be separated from parking vehicles on the street side, even when the available 3D depth resolution is only in the range of several centimeters.

Active methods for 3D sensing use laser scanners or light sources to build up 3D depth maps one point at a time. Quite a few commercial laser scanners are available (e.g., devices from company *Sick*), but they all suffer from disadvantages by being

- too expensive
- not suitable for continuous use over a long period of time
- often not robust enough for outdoor operation
- too slow for a parking space monitoring application with traffic flowing next to observed areas

CMOS chips, which work as time-of-flight (TOF) sensors, are gradually becoming practicable; however, for the intended large distances and areas which need to be covered, their spatial resolution and range is not yet sufficient. In addition, TOF sensors still have difficulty operating reliably in outdoor environments. Nevertheless, they could be an interesting alternative for 3D sensing in the coming years.

Section 8.6 describes algorithms for parking space monitoring using passive 3D sensing with stereo vision. Sections 8.6.5 and 8.6.6 give an example for a practical 3D parking space monitoring system, which cover an area of approximately 25 m in length with a medium-resolution camera pair.

The reconstruction of 3D information from a single monocular image is generally not possible without additional means of information gathering (e.g., light stripe projectors). Delibaltov et al. [29] nevertheless try to approximate a 3D volume measurement for parking space occupancy detection. They model the volume of parking spaces using geometric information, which is manually configured during system setup. They infer occupancy of even partially occluded spaces using a vehicle detector and the inferred volume of each of the parking spaces. For each pixel, the probability of it belonging to a specific parking space is modeled so that occlusions can be handled and considered for a final occupancy decision. The probability density function for each pixel depends on (i) the distance to the center of the parking space and (ii) the number of possible occlusions from other parking spaces. Figure 8.11 shows the camera views and the approach for volume approximation, which Delibaltov used in her work.

The vehicle detector of Delibaltov et al. is based on LBP with an SVM as classifier, very similar to the approach described in Section 8.4.6. In addition, Delibaltov uses TextonBoost features as described in Ref. [36]. The aim of the TextonBoost algorithm is that of automatic image segmentation. This is a trained approach that builds a learning model, based on the extraction and combination of different kind of visual information. In Delibaltov's work the following three types of features are utilized: (i) textural appearance, (ii) layout, and (iii) content. First, a so-called texton map based on textons [37] is created. For this purpose, training images are convolved with a 17D filter bank, which consists of Gaussians, their derivatives, and Laplacians of Gaussians. Thus, each image pixel is represented by a 17D feature vector. The resulting vectors are clustered using K-means clustering algorithm. Finally, image pixels are assigned to the nearest cluster center of a corresponding feature vector, filling thus a texton map.

On the next stage, texture-layout filters are applied. One filter consists of a pair (r, t) of a rectangular image region r, and a texton t. A random set of candidate rectangles is generated and the feature response at location i is the proportion of pixels having texton index t,

$$v_{\text{r,t}}(i) = \frac{1}{\text{area}(r)} \sum_{j \in (r+i)} [T_j = t]. \tag{8.13}$$

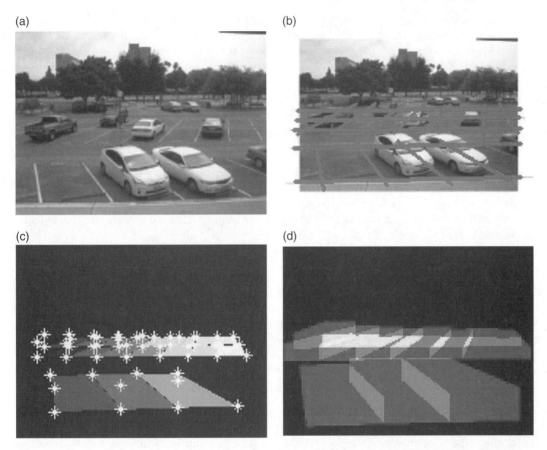

Figure 8.11 The volume modeling method from Delibaltov: (a) an example image of a parking space, (b) the parking layout from manual setup, (c) region of 2D parking spaces, and (d) regions of 3D parking space volume. Source: Delibaltov [29]. Reproduced with permission of IEEE.

To combine different kinds of information (texture, layout, color, location, and edge) into a single unified framework, Delibaltov uses a conditional random field (CRF) model [38]. Finally, an adapted version of the Joint Boosting Algorithm [39] is employed. It iteratively selects discriminative texture-layout filters as "weak learners," and combines them into a strong classifier.

Delibaltov used three test sets to evaluate the performance of her detector with 62, 15, and 10 images, respectively. She used 30 images for training of her vehicle detector, and the rest for evaluation. On average Delibaltov could achieve 76.8% accuracy with a pure TextonBoost classifier, 71.3% with a pure LBP classifier, and 78.1% with a combined TextonBoost/LBP classifier.

8.6 Stereo Vision Methods

8.6.1 Introduction to Stereo Methods

The method of 3D image analysis with the stereo correspondence method has been in practical use for more than 25 years and can be considered a proven technology. Depth estimation from two or more cameras relies on the triangulation principle from computed point correspondences and a known camera calibration. For details to the basic geometric setup and distance calculation, refer to Section 5.2.2.1.

Once a precise calibration of the camera pair is established, the generation of a 3D point cloud respective depth image can be achieved in, or close to, real time. At practical resolutions of 1–4 megapixels, current stereo algorithms can achieve a 0.25–3 Hz measurement rate on a CPU, and up to 25 Hz with GPU support. This beats other sensors in terms of price/performance ratio.

The main properties of stereoscopic imaging for parking space analysis can be summarized in the following text.

Advantages of Stereo Measurement Methods
- Similar to 2D vision systems, a relatively large spatial area can be covered with low-cost hardware, compared to magnetic sensors or other nonoptical methods.
- Lower installation costs than alternative methods (pole mount of the stereo rig instead of subsurface installation of sensors).
- Flexible configuration of parking space delimiters using graphical user interfaces.
- Additional benefits for the operator: visual control and inspection are possible as by-product.

Disadvantages of Stereo Measurement Methods
- Image analysis requires complex algorithms.
- Relatively high computational burden with resulting higher power requirements, larger housings, and generally higher system complexity.
- The semantic analysis of the measured 3D data (interpretation of the captured scene) can become complex.
- Possible reduced range and accuracy in adverse weather conditions.

3D stereo image analysis has the following benefit: in addition to the processing of depth images, standard visual 2D image analysis can also be performed. This additional information can be used to estimate the velocity of objects, or support 3D analysis by generating hypothesis for the semantic interpretation of a scene (e.g., it might be easier to do pedestrian detection in the 2D data than in the depth image).

8.6.2 Limits on the Accuracy of Stereo Reconstruction

The stereoscopic imaging geometry and achievable sub-pixel resolution of the pixel matching algorithm set the limit for the stereo camera depth accuracy, which can be achieved. Typical sub-pixel matching accuracies of state-of-the-art algorithms lie in the range of 0.1–0.5 pixels. The possible theoretical depth resolution dz at a measurement distance z_0 (for focal length $f \ll z_0$) depends on the camera baseline B, the focal length f, and the sensor element (pixel) distance D [40],

$$dz = \frac{z_0^2 D}{Bf},$$

(8.14)

where B, z_0 is given in millimeter, f is defined in millimeter, and D is the effective matching accuracy on the sensor plane, which considers the pixel distance on the sensor plane weighed by the possible sub-pixel matching accuracy. Figure 8.12 plots the achievable distance measurement resolutions dz for some practical camera/lens configurations. For example, with system parameters $D = 0.25 \times px$ (1/4 pixel matching resolution), working distance $z_0 = 40{,}000$ mm, baseline

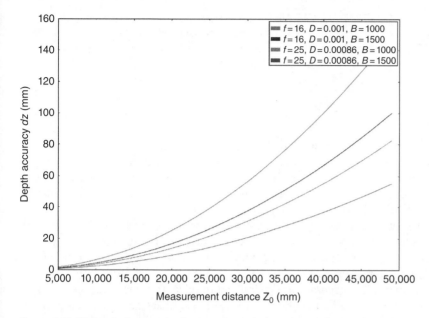

Figure 8.12 Theoretical accuracies for several practical camera configurations.

$B = 1{,}500\,$mm, and a focal length $f = 16\,$mm, an approximate theoretical accuracy of $67\,$mm in depth resolution can be achieved.

8.6.3 Computing the Stereo Correspondence

3D measurements can be extracted from a stereo image pair by means of geometric triangulation. This process requires the computation of corresponding image locations (ideally single pixels), which are then, given intrinsic end extrinsic camera calibration, converted into 3D distance measurements (see Section 5.2.2 for details).

The solution to this so-called correspondence problem is the most important task of a stereo matching algorithm, and depending on the captured scene, it is not always trivial. Areas of homogeneous texture, overexposed image regions, or repeating patterns may lead to errors in matching, which can severely affect the 3D reconstruction process. Over the past 25 years, various matching algorithms have been developed and used in practical systems. For the application in parking space monitoring, stereo matching need not be extremely precise, since only the relatively large volume of a car needs to be considered. More important are (i) robustness of the 3D data points (not too many outliers) and (ii) the computational performance of the algorithm.

The minimum sum of absolute difference (SAD) [41] algorithm calculates gray value differences within local areas of the image and identifies pairs of points with the minimal sum of gray value differences. The advantage of this approach is the high attainable speed, with the disadvantage being that the quality of the resulting 3D data for many applications and scenarios is not sufficient.

The semi-global matching (SGM) method by Hirschmüller [42] computes pixel-wise matching costs and refines them with dynamic programming to effectively optimize disparities globally over an image. The idea of Hirschmüller to regularize the costs from several different directions along paths around a matched pixel avoids the problem of streaking, which affects other local optimization algorithms.

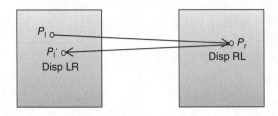

Figure 8.13 Computation of the back-matching distance using the disparity images disp LR and disp RL. The back-matching distance is the distance between points P_l and P_l'.

Many SGM implementations use the census transform operator as descriptor for the local patch around a pixel. Interestingly, the census transform is very similar to the concept of LBP, and it can be implemented efficiently on a CPU as well as on a GPU. It allows sub-pixel matching and works well on the typical textures found in street scenes. SGM is considered to be among the best available stereo algorithms in terms of balance between accuracy and runtime performance.

The pyramidal feature vector matching (PFVM) method (originally developed by Paar [43]) describes each pixel as a vector of local filter responses (gradient, brightness distribution, etc.), which are called "features." Matching is performed by searching for the local feature distance minimum between the feature vector of a reference point in one image and the feature vectors within a search area in the corresponding image. To speed up processing and to make the algorithm more robust, stereo matching is performed on a resolution pyramid of the original image pair. PFVM is accurate and robust, but it is computationally intensive. Its advantage is that it can be parallelized well and that it can deliver dense point clouds by design.

A practical method for checking disparities for consistency (and outlier rejection) is the computation of the back-matching distance [44] as shown in Figure 8.13. Every valid matched point P_l is projected into the right image using its disparity, giving point P_r. P_r is similarly projected back into the left image using its local disparity, resulting in point P_l'. In the ideal case P_l and P_l' are identical, their residual distance $P_l - P_l'$ is called back-matching distance. By accepting only matches of a certain maximal back-matching distance, consistently matched points can be chosen for 3D reconstruction.

8.6.4 Simple Stereo for Volume Occupation Measurement

Wah tried calibrated and uncalibrated stereo reconstruction in her study [45]. First, the Harris and Stephens [17] corner detection method was used to extract representative features from single images. The Harris corner detector searches for points with high variation of gradients in both vertical and horizontal directions in their neighborhood. After that, feature correspondences between the point sets in both images were established using cross-correlation. Finally, Wah then fitted a model using RANSAC [46] in order to remove outliers.

Once feature matching across images was achieved, the 3D reconstruction was performed using uncalibrated and calibrated camera setups [47, 48]. Wah used the already present landmarks in the image pairs to calibrate the camera rigs, so that no external calibration target was necessary.

A decision about parking space occupation was made by counting the number of 3D points within each parking space volume. If the number of points was greater than a predefined threshold, a parking space was considered to be occupied.

8.6.5 A Practical System for Parking Space Monitoring Using a Stereo System

For the purpose of monitoring parking spaces, which are located alongside tracks of trams, the prototype of a stereo measurement system has been deployed by Sidla [27] in a pilot project for the

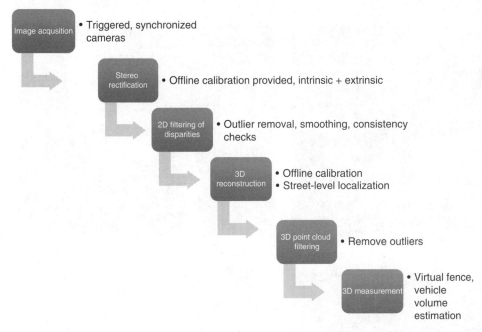

Image acqusition
- Triggered, synchronized cameras

Stereo rectification
- Offline calibration provided, intrinsic + extrinsic

2D filtering of disparities
- Outlier removal, smoothing, consistency checks

3D reconstruction
- Offline calibration
- Street-level localization

3D point cloud filtering
- Remove outliers

3D measurement
- Virtual fence, vehicle volume estimation

Figure 8.14 The data flow of the 3D vehicle detection pipeline.

City of Vienna. The purpose of the system was to measure as precisely as possible the location and position of cars relative to the tram tracks. The aim was to detect situations when a car would potentially block a passing tram and trigger a warning, thus avoiding the costly disruptions of the tram service.

A vertical imaging sensor arrangement with a baseline of 1500 mm has been chosen for the imaging setup because it can be mounted easily on a supporting mast. The stereo camera rig observed a stretch of parking spaces at a distance between 33 and 55 m. Experimental results had previously shown that the computed stereo disparities from a vertical configuration seem to be slightly more robust than those from a horizontal sensor configuration. For the system parameters of $D = 0.0010$ mm (stereo matching accuracy), a medium working distance of $z_0 = 40,000$ m, a camera baseline $B = 1500$ mm, and focal length $f = 16$ mm, an approximate theoretical depth resolution of 67 mm could be achieved.

In practice, the authors of this work observed a localization accuracy of the side of the cars of around 25–60 mm (at 33,000–55,000 mm working distance), which roughly corresponds to the theoretical result—probably a better disparity estimation accuracy than the assumed 1/4 pixel resolution could be achieved.

The processing steps for vehicle detection from the stereo measurement are depicted in Figure 8.14. Intermediate results from a test setup are shown in Figure 8.15. By using the camera calibration data and an automatic alignment of the 3D point cloud with the street level, the location of vehicles relative to the tracks can be computed with acceptable accuracy and robustness.

The system has been successfully tested under real-world conditions for a period of several months. It is currently (Summer 2015) being rebuilt with better resolution and more computing power for long-term tests at several locations at problem hot spots in the City of Vienna.

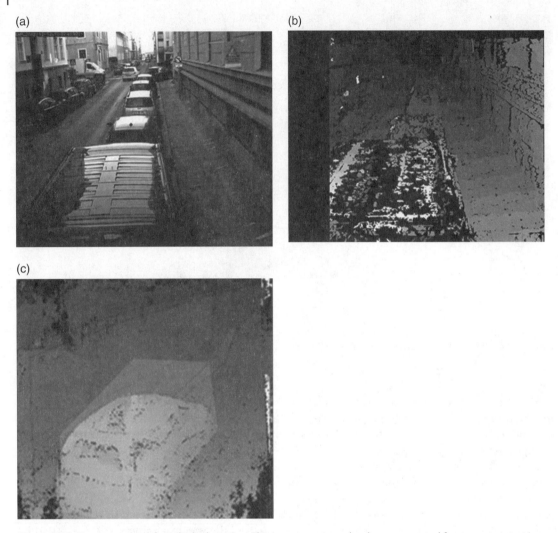

Figure 8.15 Stereo matching for vehicle detection. This image sequence has been generated from a system test in the City of Graz prior to installation. (a) One camera image of the stereo rig, resulting stereo disparities are depicted in (b), and the resulting interpolated 3D point cloud is shown in (c). For vehicle detection, the volume of the vehicle (shown as cube in (c)) is approximated and its boundaries are tested against a virtual fence, which is shown as a vertical plane left of the cube in (c).

8.6.6 Detection Methods Using Sparse 3D Reconstruction

Cook [49] proposes an interesting and extensive vehicle detection approach via sparse 3D scene reconstruction. A parking area is equipped with several video sensors, each of which observes the area from different angles (the author has used 12 sensors), so that depth information can be extracted. The proposed algorithm uses dual 3D sparse reconstruction—one for an empty parking space, and another one for an occupied parking space. First, an identification of parking places is performed in a semiautomated mode. An empty parking space is manually labeled with four-colored fiduciary markers (the marker positions are found in images automatically by the algorithm). The goal of the empty parking space reconstruction is to locate the plane of unoccupied pavement. Once this is done, only a box normal to this plane is analyzed in operation mode—if a sufficient number of reconstructed points of vehicles are found, a parking space is marked as occupied.

To generate sparse 3D points, Cook utilizes the *Bundler* tool chain [50], which in turn uses SIFT [19] features and bundle adjustment. SIFT as descriptor for point matching has the following advantages: it is invariant to scale, translation, rotation, and local affine distortions. SIFT descriptors are also partially invariant to illumination and 3D camera viewpoint, and they are therefore well suited for reidentification across a collection of images. The drawback of SIFT lies in its complexity and slow calculation time.

To speed-up processing, a staged filtering approach is used, which allows to run computationally intensive algorithms only when needed. First, positions of SIFT keypoints are extracted.

A DoG function is computed in scale space (an image pyramid is built), and the keypoints are selected at local maxima or minima of the function. The convolution with the 2D Gaussian function is computed by applying the 1D Gaussian in the horizontal and vertical direction [19],

$$g(x) = \frac{1}{\sqrt{2\pi\sigma}} e^{-x^2/2\sigma^2}. \tag{8.15}$$

For the keypoint localization algorithm, the input image is first convolved with the Gaussian function using $\sigma = \sqrt{2}$. The resulting image A is convolved with $\sigma = \sqrt{2}$ again, giving image B. The difference of Gaussian function is the pixel-wise difference $B - A$.

Maxima and minima are defined at each pyramid level. Every pixel value is compared with its eight neighbors—if it is one of the extrema, its closest pixel location is calculated at the next lowest level of the pyramid and the process is repeated.

Finally, the local image regions around the keypoints are described with a 1D feature vector in such a way that the resulting descriptor is invariant to the location, scale, and orientation. For a particular keypoint, four small (4×4 size) regions are created. Within each region, a HOG in eight directions is created, and the orientations are measured relative to the keypoint orientation.

The resulting high-dimensional feature vector is distinctive and useful to find point correspondences between images. Matching of SIFT features across images is realized with a nearest-neighbor search using Euclidean distance metric. To increase the matching robustness, the second closest neighbor in feature space is considered as well. The ratio of distances between the best and the second best match has to be above a certain threshold; otherwise, the match is considered ambiguous.

After SIFT correspondences are found, the reconstruction with bundle adjustment begins. This is an iterative optimization process, performed with a sparse implementation of nonlinear least squares. The goal is to minimize location error of the reprojection of the predicted locations of 3D points back into the original images. The reprojection error that has to be minimized is defined as follows [49, 51]:

$$\sum_{i,j} d\left(P_i X_j, x_{ij}\right)^2. \tag{8.16}$$

Here, X_j is the 3D location of a feature j, and x_{ij} is the estimated 2D position of that point in an image i. P_i represents the camera matrix of image I, and d is the image distance.

For the nonlinear least-square minimization, Cook uses a Levenberg–Marquard optimization procedure. As feature locations in the images contain noise, a maximum-likelihood estimation is provided by bundle adjustment as well. Figure 8.16 shows a reconstruction result obtained by Cook.

(a)

(b)

Figure 8.16 The result of 3D reconstruction by Cook [49]. (a) Empty parking space with colored markers. (b) Reconstruction example—scenario with four vehicles. Source: Cook [49]. Reproduced with permission of University of Minnesota.

The Bundler output is often too sparse and does not produce a sufficient number of features needed for a confident decision about vehicle presence. That is why an additional reconstruction is performed with PMVS approach [52]. PMVS runs parallel to SIFT and bundle adjustment. It generates its own features based on Harris and DoG detectors. Each image is divided into a grid with patches size $\beta \times \beta$ (with $\beta = 32$ pixels). Each patch is processed with both detectors and the η strongest responses are taken with $\eta = 4$ (see Refs. [52] and [49] for details to PVMS).

Tests on different parking spaces show very good and robust results—also in rain and after snowfall. All vehicles were detected with high confidence and without false alarms. The system can cope with occlusions very well and is, therefore, applicable for larger parking areas. The main drawback of the proposed approach is that it is slow—depending on the number of sensors and their resolution, calculation time for one set of images can take about 6–74 min. Besides that, the approach requires a parking area to be empty during the initialization stage. Finally, a large number of sensors is needed, which increases hardware and installation costs.

Acknowledgment

The work described was partially funded by the Austrian Federal Ministry for Traffic, Innovation and Technology (BMVIT): FFG project *iLadezone*, project number 831775, and FFG project *TRAM-in-TAKT*, Project number 835756.

References

1 M. Y. I. Idris, Y. Y. Leng, E. M. Tamil, N. M. Noor, and Z. Razak, Car park system: A review of smart parking system and its technology, *Information Technology Journal* 8, 101–113 (2009).
2 D. C. Shoup, *The High Cost of Free Parking*, Planners Press, American Planning Association, Chicago (2005).
3 D. C. Shoup, Cruising for parking, *Transport Policy* 13(6), 479–486 (2006).
4 R. P. Loce and E. Saber, Fuel Consumption in San Francisco, http://spie.org/x1816.xml (accessed September 16, 2013).
5 sfpark.com (accessed September 16, 2013).
6 https://www2.ffg.at/verkehr/projektpdf.php?id=805&lang=en (accessed September 16, 2013).
7 http://www.fluidtime.com (accessed September 16, 2013).
8 Garage Laser Park by MAXSA Innovations, LCC. http://www.maxsainnovations.com/PARK-RIGHT-Dual-Garage-Laser-Park-with-LED-Light/ (accessed September 16, 2013).
9 O. Bulan, R. P. Loce, W. Wu, Y. Wang, E. Bernal, and Z. Fan, Video-based real-time on-street parking occupancy detection system, *Journal of Electronic Imaging* 22(4), 041109 (2013).
10 P. Modi, Counting Empty Parking Spots at Truck Stops Using Computer Vision, Artificial Intelligence, Robotics and Vision Laboratory, Department of Computer Science and Engineering, University of Minnesota, Twin Cities, March 2011.
11 C. Stauffer and W. E. L. Grimson, Adaptive Background Mixture Models for Real-Time Tracking, *Proceedings of the International Conference on Computer Vision and Pattern Recognition*, Ft. Collins, CO, June 23–25, 1999, Vol. 2.
12 A. J. Joshi, S. Atev, O. Masoud, and N. Papanikolopoulos, Moving Shadow Detection with Low and Mid-Level Reasoning, *Proceedings of the IEEE International Conference on Robotics and Automation*, Rome, April 10–14, 2007.
13 N. True, Vacant parking space detection in static images, University of California, San Diego, Technical Report, 2007.
14 V. Vapnik, *Statistical Learning Theory*, John Wiley & Sons, Inc., New York (1998).
15 N. Cristianini and J. Shawe-Taylor, *An Introduction to Support Vector Machines and Other Kernelbased Learning Methods*, Cambridge University Press, New York (2000).
16 T. Čaklović and I. Aleksi, Managing and monitoring of parking lot by a video camera, Željko Hocenski University of Osijek, Faculty of Electrical Engineering Kneza Trpimira Croatia.
17 C. Harris and M. Stephens, A Combined Corner and Edge Detector, *Proceedings of the Fourth Alvey Vision Conference*, Manchester, August 31 to September 2, 1988, pp. 147–151.

18 E. Rosten and T. Drummond, Machine Learning for High-Speed Corner Detection, *Proceedings of the 9th European Conference on Computer Vision*, Graz, May 7–13, 2006, pp. 430–443.

19 D. G. Lowe, Object Recognition from Local Scale-Invariant Features. *Proceedings of the 7th IEEE International Conference on Computer Vision*, Corfu, September 20–27, 1999, Vol. 2, pp. 1150–1157.

20 D. G. Lowe, Distinctive image features from scale-invariant keypoints, *International Journal of Computer Vision* 60(2), 91–110 (2004).

21 H. Bay, T. Tuytelaars, and L. Van Gool, Surf: Speeded up Robust Features, *Proceedings of the 9th European Conference on Computer Vision*, Graz, May 7–13, 2006, pp. 404–417.

22 P. Viola and M. Jones, Rapid object detection using a boosted cascade of simple features, *Proceedings of the IEEE Conference on Computer Vision and Pattern Recognition* 1, 511–518 (2001).

23 R. Fusek, K. Mozdren, M. Surkala, and E. Sojka, AdaBoost for parking lot occupation detection, *Advances in Intelligent Systems and Computing* 226, 681–690 (2013).

24 M. Tschentscher and M. Neuhausen, Video-Based Parking-Space Detection, *Proceedings of the Forum Bauinformatik*, Bochum, September 26–28, 2012, pp. 159–166.

25 N. Dalal and B. Triggs, Histograms of Oriented Gradients for Human Detection, *Proceedings of the IEEE Computer Society Conference on Computer Vision and Pattern Recognition, (CVPR 2005)*, San Diego, CA, June 20–25, 2005, pp. 886–893.

26 Q. Zhu, M.-C. Yeh, K.-T. Cheng, and S. Avidan, Fast Human Detection Using a Cascade of Histograms of Oriented Gradients, *Proceedings of the IEEE Computer Society Conference on Computer Vision and Pattern Recognition*, New York, NY, June 17–22, 2006, IEEE Computer Society, Vol. 2.

27 O. Sidla and Y. Lipetski, Methods for Vehicle Detection and Vehicle Presence Analysis for Traffic Applications, *Proceedings of the SPIE Conference on Video Surveillance and Transportation Imaging Applications*, San Francisco, CA, USA, February 2, 2014, Vol. 9026.

28 T. Ojala, M. Pietikäinen, and T. Mäenpää, Multiresolution gray-scale and rotation invariant texture classification with local binary patterns, *Proceedings of the IEEE Transactions on Pattern Analysis and Machine Intelligence* 24(7), 971–987 (2002).

29. D. L. Delibaltov, W. Wu, R. Loce, and E. Bernal, Parking Lot Occupancy Determination from Lamp Post Camera Images, *Proceedings of the 16th International IEEE Conference on Intelligent Transportation Systems: Intelligent Transportation Systems for All Modes, ITSC 2013*, The Hague, October 6–9, 2013.

30 Y. Lipetski and O. Sidla, Vehicle Presence Analysis for Law Enforcement Applications and Parking Lot Management, *Proceedings of the SPIE Conference on Video Surveillance and Transportation Imaging Applications*, Burlingame, CA, February 4–6, 2013, Vol. 8663.

31 X. Wang, T. Han, and S. Yan, An HOG-LBP Human Detector with Partial Occlusion Handling, *Proceedings of the IEEE 12th International Conference on Computer Vision, ICCV 09*, Kyoto, September 29 to October 2, 2009.

32 N. Armanfard, M. Komeili, and E. Kabir, TED: A texture-edge descriptor for pedestrian detection in video sequences, *Pattern Recognition* 45, 983–992 (2012).

33 Y. LeCun, L. Bottou, Y. Bengio, and P. Haffner, Gradient-based learning applied to document recognition, *Proceedings of the IEEE* 86, 2278–2324 (1998).

34 D. Ciresan, U. Meier, and J. Schmidhuber, Multi-column deep neural networks for image classification, Arxiv preprint arXiv:1202.2745 (2012).

35 J. Jermsurawong, M. Umair Ahsan, A. Haidar, H. Dong, and N. Mavridis, Car Parking Vacancy Detection and Its Application in 24-Hour Statistical Analysis, *Proceedings of the 10th International Conference on Frontiers of Information Technology (FIT 2012)*, Islamabad, December 17–19, 2012, pp. 84–90.

36 J. Shotton, J. Winn, C. Rother, and A. Criminisi, Textonboost for image understanding: Multi-class object recognition and segmentation by jointly modeling texture, layout, and context, *International Journal of Computer Vision* 81(1), 2–23 (2009).

37 J. Malik, S. Belongie, T. Leung, and J. Shi, Contour and texture analysis for image segmentation, *International Journal of Computer Vision* 43(1), 7–27 (2001).

38 J. Lafferty, A. McCallum, and F. Pereira, Conditional Random Fields: Probabilistic Models for Segmenting and Labeling Sequence Data, *Proceedings of the 18th International Conference on Machine Learning*, Williamstown, MA, June 28–July 1, 2001, pp. 282–289.

39 A. Torralba, K. P. Murphy, and W. T. Freeman, Sharing visual features for multiclass and multiview object detection, *IEEE Transactions on Pattern Analysis and Machine Intelligence* 19(5), 854–869 (2007).

40 D. Gallup, J.-M. Frahm, P. Mordohai, and M. Pollefeys, Variable Baseline or Resolution Stereo, *Proceedings of the IEEE Conference on Computer Vision and Pattern Recognition (CVPR 2008)*, Anchorage, AK, June 23–28, 2008.

41 F. Tombari, S. Mattoccia, L. Di Stefano, E. Addimanda, Classification and Evaluation of Cost Aggregation Methods for Stereo Correspondence, *Proceedings of the IEEE International Conference on Computer Vision and Pattern Recognition (CVPR 2008)*, Anchorage, AK, June 23–28, 2008.

42 H. Hirschmüller, Stereo processing by semiglobal matching and mutual information, *IEEE Transactions on Pattern Analysis and Machine Intelligence* 30(2), 328–341 (2008).

43 G. Paar, O. Sidla, and W. Pölzleitner, Genetic Feature Selection for Highly Accurate Stereo Reconstruction of Natural Surfaces, *Proceedings of SPIE Conference on Intelligent Robots and Computer Vision XVII: Algorithms, Techniques, and Active Vision*, Boston, MA, November 1, 1998, Vol. 3522.

44 W. Pölzleitner and O. Sidla, Accuracy optimization in hierarchical stereo disparity computation using the back-matching distance, *Proceedings SPIE the International Society for Optical Engineering* 3164, 482–493 (1997).

45 C. Wah, Parking Space Vacancy Monitoring. Projects in Vision and Learning (2009).

46 M. A. Fischler and R. C. Bollee, Random sample consensus: A paradigm for model fitting with applications to image analysis and automated cartography, *Communications of ACM* 24(6), 381–395 (1981).

47 Y. Ma, S. Soatto, J. Kosecka, and S. Sastry, *An Invitation to 3D Vision: From Images to Geometric Models*, Springer Verlag, Heidelberg (2003).

48 J. Y. Bouguet, Camera Calibration Toolbox for MATLAB, California Institute of Technology, Pasadena, CA, (2001).

49 D. Cook, Parking Space Occupancy Detection Utilizing 3D Reconstruction Techniques, M.S. thesis, University of Minnesota, March 2012.

50 N. Snavely, S. M. Seitz, and R. Szeliski, Photo Tourism: Exploring Image Collections in 3D, *Proceedings of ACM SIGGRAPH 2006 ACM Transactions on Graphics*, Boston, MA, July 30–August 3, 2006.

51 R. I. Hartley and A. Zisserman, *Multiple View Geometry*, Cambridge University Press, Cambridge (2004).

52 Y. Furukawa and J. Ponce, Accurate, Dense, and Robust Multi-View Stereopsis, *Proceedings of the IEEE Computer Society Conference on Computer Vision and Pattern Recognition*, Chicago, June 17–22, 2007.

9

Video Anomaly Detection

Raja Bala[1] and Vishal Monga[2]

[1] *Samsung Research America, Richardson, TX, USA*
[2] *Pennsylvania State University, University Park, PA, USA*

9.1   Introduction

The ability to detect anomalous or unusual events has numerous applications in the transportation domain, including identifying traffic violations, accidents, unsafe driver behavior, street crime, and other dangerous and suspicious activities [1]. In many surveillance settings, anomaly detection requires significant human intervention, and is hence not scalable to high volumes of video footage. Thus a large fraction of transportation video is simply stored without review. Automatic and reliable detection of anomalies from natural traffic scenarios is clearly of benefit. This problem is challenging for several reasons. First, video captured in natural urban settings often contains a large amount of clutter amid a complex dynamically varying scene. Second, since anomalies are by definition rare events, it is difficult to obtain a sufficiently large number of samples to accurately characterize anomalous events and behavior. In this chapter, we present a framework for video anomaly detection, provide a brief survey of state-of-the-art methods, and elaborate on a few selected techniques that attempt to address the aforementioned challenges. Last, directions for future investigations in this topic are pondered. Note that anomaly detection can be thought of as a general framework that includes as special cases applications like traffic violation detection discussed in previous chapters.

Figure 9.1 shows a simple yet general framework for anomaly detection. At the high level, there are two stages: event encoding and an anomaly detection model.

In the event encoding stage, raw video is converted into a feature representation suitably chosen to describe events of interest. Common feature descriptors in transportation applications include trajectories and spatiotemporal volumes (STVs). Recently, deep convolutional neural networks (CNNs) have been built that learn relevant motion-related representations from video. These event encoding schemes are described in further detail in Section 9.2.

Given an event encoding, the anomaly detection model makes a binary decision as to whether or not an event (as represented by the feature descriptor) is an anomaly. It is useful to categorize the scenarios for anomaly detection based on how much information and structure is known about the nominal and anomalous samples, as in the following text.

Supervised anomaly detection refers to the setting where anomalous samples are labeled into known classes. Conversely, unsupervised anomaly detection refers to the scenario wherein there are

Computer Vision and Imaging in Intelligent Transportation Systems, First Edition.
Edited by Robert P. Loce, Raja Bala and Mohan Trivedi.
© 2017 John Wiley & Sons Ltd. Published 2017 by John Wiley & Sons Ltd.
Companion website: www.wiley.com/go/loce/ComputerVisionandImaginginITS

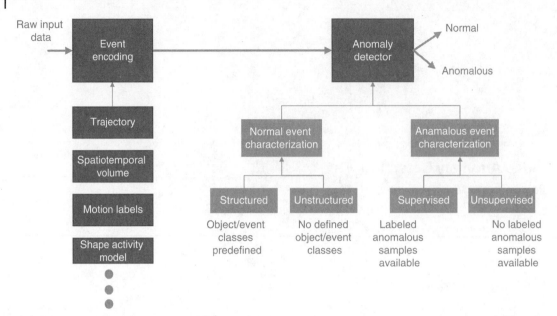

Figure 9.1 Flow diagram of video anomaly detection.

not a sufficiently large number of labeled anomalous samples to warrant characterization of an anomalous event class. The latter is the more realistic and frequently encountered setting in real-world applications. Another dimension to consider is how much structure is known of the nominal classes. Structured anomaly detection models assume that labels exist for objects, motion, or actions corresponding to nominal events. Conversely, unstructured models assume that no information is known about nominal events beyond the fact that they are nominal. The former category allows for more informative inference of anomaly versus nominal, but it requires more effort to create structured models. The latter offers the benefit of requiring less knowledge of nominal samples while possibly compromising accuracy of anomaly detection. The proposed taxonomy of settings is depicted in Figure 9.2. Selected state-of-the-art techniques are characterized on this plot in terms of how much structure or supervision is assumed. Several of these techniques will be elaborated in Section 9.3. In particular, one family of techniques based on sparse representations has shown recent promise, and will be discussed in detail in Section 9.4. Note that the two stages of Figure 9.1 (i.e., event encoding and detection model) are often closely coupled, with the encoding lending itself to a particular type of model, or vice versa.

9.2 Event Encoding

This stage extracts features from raw video that provide compact descriptive representations of appearance and/or motion in the scene. Here, we will review two common types of event encoding methods—trajectories and spatiotemporal descriptors—keeping in mind that the appropriate choice depends on the specific application setting, performance requirements, and computational considerations. Additional encoding methods that are intimately tied to anomaly detection models will be mentioned in Section 9.3.

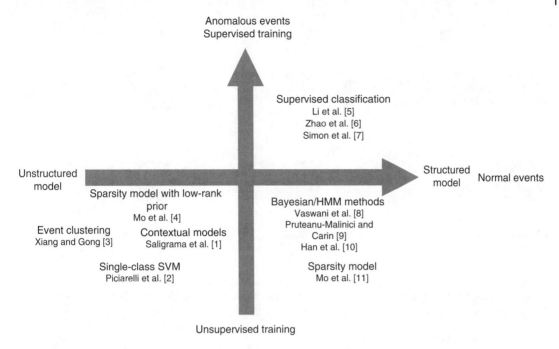

Figure 9.2 Taxonomy of anomaly detection scenarios.

9.2.1 Trajectory Descriptors

A common event encoding for characterizing moving vehicles and pedestrians is the trajectory generated by a tracking algorithm. Fortunately, this problem is very well studied, and many state-of-the-art tracking techniques can be readily leveraged such as background subtraction, mean shift tracking, Kalman filtering, Kanade–Lucas–Tomasi tracking, and particle Filtering. An excellent survey of object tracking can be found in Yilmaz et al. [12]. In its simplest form, a trajectory is defined as a vector $[x_i, y_i]$ of spatial coordinates of a path traveled by the object. Additional motion-related variables such as velocity and acceleration, as well as appearance-related variables such as size, moments, shape, and color at each location along the path, can also be incorporated into a trajectory vector.

Object tracking finds application in many problems across transportation imaging. (e.g., tracking is described in Chapter 5 in the context of traffic violation detection.) We now present in detail an exemplary tracking approach used in Mo et al. [11] that is particularly well suited for rigid objects with certain constraints on size and shape, such as vehicles. A flow diagram is shown in Figure 9.3.

First background subtraction is accomplished via the use of a Gaussian mixture model (GMM) as described in Stauffer and Grimson [13] and also in the Chapter 6 on traffic flow. Specifically, denote the sequence of pixel values from the last t frames as $X_0, \ldots, X_{t-1}$, where $X_i \in \mathbb{R}^N$, and commonly $N = 1$ or 3 for grayscale or color video input, respectively. This sequence is modeled as a mixture of K Gaussian distributions, and the probability of observing a current pixel value X_t is given by

$$P(X_t) = \sum_{k=1}^{K} w_{k,t} * \eta(X_t, \mu_{k,t}, \pounds_{k,t}),\tag{9.1}$$

where $w_{k,t}$ is the weight or relative importance of the kth Gaussian distribution with mean $\mu_{k,t}$ and covariance $\Sigma_{k,t}$. For simplicity, we assume for the case of color video that the R, G, and B channels are

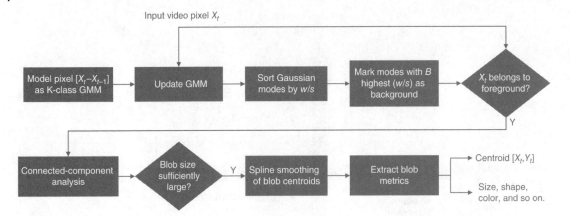

Figure 9.3 Object trajectory generation for anomaly detection.

independent with equal variance so that $\Sigma_{k,t} = \sigma_{k,t}^2 I$. When a new pixel X_t arrives, it is checked against the existing K distributions. A match is found if the pixel is within 2.5 standard deviations of a distribution. The matching Gaussian mode (denoted by index m) is then updated as follows:

$$\mu_{m,t} = (1-\rho)\mu_{m,t-1} + \rho X_t$$
$$\sigma_{m,t}^2 = (1-\rho)\sigma_{m,t-1}^2 + \rho (X_t - \mu_{m,t})^T (X_t - \mu_{m,t}). \tag{9.2}$$

Here, $\rho = \alpha\eta(X_t, \mu_m, \sigma_m)$ and α is the learning rate. If none of the distributions match, the least probable distribution is replaced with a distribution with mean X_t, a high initial variance, and low prior weight. Additionally, prior weights for the remaining distributions are updated as follows:

$$w_{k,t} = (1-\alpha)w_{k,t-1} + \alpha (M_{k,t}). \tag{9.3}$$

Here, $M_{k,t}$ is 1 only for the matching distribution and 0 otherwise.

Next, the Gaussian modes are sorted by the quantity w/σ. The larger the value of this ratio, the more stable the mode, and thus indicative of background pixels. The top B modes are marked as background modes, where B is given by

$$B = \operatorname{argmin}_b \left(\sum_{k=1}^{b} w_k > T \right). \tag{9.4}$$

Pixels that belong to the remaining modes are denoted as foreground. Standard connected-component analysis can be used to create foreground "blobs" representing objects of interest. The centroid of each blob denotes its location and the temporal sequence of centroids forms the trajectory. In Mo et al. [11], only those blobs whose sizes exceed a predefined threshold are retrained for trajectory extraction. The size threshold is chosen to track vehicles and pedestrians and eliminates other noisy moving entities in the scene.

Two issues now remain. First trajectory locations are likely to be noisy; hence, some form of smoothing is desired. Second, many anomaly detection models necessitate that all object trajectories are of the same length. To solve both issues, a functional approximation such as a spline curve can be used to represent trajectories. Specifically, in Mo et al. [11] a B-spline function with 50 knots (i.e., 50 x- and y-coordinates) is fit to the raw trajectory data.

The final outcome is a fixed-length vector of $[x_i, y_i]$ marking the path of an object of interest. As stated earlier, other parameters such as object size, shape, color, moments, and texture can also be computed from the blobs and tracked at each index i along the trajectory. For anomaly detection, it is logical to track those quantities which are likely to provide strong indicators of anomalous events.

9.2.2 Spatiotemporal Descriptors

Trajectories track a given spatial point or appearance attribute over time, and thus pay separate attention to the spatial and temporal dimensions. In contrast, spatiotemporal descriptors encode video information simultaneously in space and time. Many techniques view a video signal as a three-dimensional (3D) function of two spatial and one temporal dimension, and extend 2D interest points and feature descriptors used for analyzing still images to the 3D video counterpart. For example, Laptev [14] extends Harris interest points for still images to spatiotemporal interest points for video. Other techniques include analyzing the determinant of the Hessian matrix [15] and spatiotemporal extensions of image descriptors, such as 3D-SIFT [16], HOG3D [17], extended SURF [15], spatiotemporal Gabor and Gaussian filtering [18], local trinary patterns [19], and combination of histogram of oriented gradient (HoG) and histogram of optical flow (HoF) [6].

We describe in detail two flavors of spatiotemporal descriptors: one based on HoG/HoF features [6] and the other based on STV [7]. In the method presented by Zhao et al. [6], interest points are first identified within the video. These points may be dense (e.g., a regular 3D lattice) or sparse (e.g., points exhibiting only a certain type of spatiotemporal behavior). The approach by Zhao et al. [6] employs the interest point detector proposed by Dollar et al. [18], which applies to each pixel a series of tuned Gaussian and Gabor filter cascades in the spatial and temporal dimensions, respectively, and returns only those points with significant filter response. Next, a spatiotemporal cuboid is extracted around each given interest point. The dimensions of the cuboid may be fixed or adapted to the scale of the detected interest point. The simplest feature descriptor is a flattened list of pixel values within the cuboid. However, more sophistication can be brought to bear to offer the desired tradeoff between compactness, discriminability, and robustness to variations in appearance and motion induced by camera jitter, lighting effects, and so on. Popular approaches include normalization, local gradients, and histogramming. The method in Zhao et al. [6] applies all of these principles. First, HoG is performed [20] to capture spatial appearance information. The cuboid is divided into (n_x, n_y, n_t) cells, and a histogram of gradient orientations is accumulated within the cell. Specifically, gradient magnitude and angle are computed at each pixel; the angle is used as bin index into the histogram, and the vote in that bin is incremented with a weighting proportional to gradient magnitude. Local photometric normalization is applied in order to make the HoG descriptor more resilient to lighting and shadow effects. (See Dalal and Triggs [20] for details.) Cell histograms are then concatenated to form a composite HoG vector for the cuboid. In a similar fashion, HoF descriptors are computed with motion vectors in the x and y directions replacing the spatial gradients computed for HoG. Optical flow can be computed using any standard method such as the Lucas–Kanade algorithm [21]. HoG and HoF vectors are concatenated to form the final descriptor X_i for the ith spatiotemporal cuboid. Finally, a given video clip is encoded as a set of spatiotemporal descriptors $(X_1, ..., X_N)$ and forms the input for subsequent anomaly detection modeling.

Simon et al. [7] proposes a different type of spatiotemporal descriptor based on STV. First, background subtraction is performed, as described earlier in this section. This produces for each video frame a binary mask separating foreground from background regions. An STV is defined as a collection of all foreground pixels that are contiguously connected in space and time. At a particular time instance t, define a window of size $(2k + 1)$, ranging from $t - k$ to $t + k$. Let (x_i, y_i) and (w_i, h_i), respectively, denote the centroid and bounding box dimensions of the STV slice at time i, where $tk \leq i \leq t + k$.

Three groups of features are computed from these quantities. The first set of features describes the average location and dimensions of the STV as follows:

$$f_x = \frac{1}{2k+1} \sum_{i=t-k}^{t+k} x_i$$

$$f_y = \frac{1}{2k+1} \sum_{i=t-k}^{t+k} y_i$$

$$f_w = \frac{1}{2k+1} \sum_{i=t-k}^{t+k} w_i \qquad (9.5)$$

$$f_h = \frac{1}{2k+1} \sum_{i=t-k}^{t+k} h_i$$

$$f_a = f_w \dot{f}_h$$

The second set of features describes temporally localized motion within the STV via changes in centroid location and slice dimensions across adjacent slices as follows:

$$f_{\text{speed}} = \frac{1}{2k+1} \sum_{i=t-k}^{t+k} \sqrt{(x_i - x_{i-1})^2 + (y_i - y_{i-1})^2}$$

$$f_{\text{velocity}} = \sqrt{\frac{x_{t+k} - x_{t-k}}{2k+1} + \frac{y_{t+k} - y_{t-k}}{2k+1}}$$

$$f_{\text{area}} = \frac{1}{2k+1} \sum_{i=t-k}^{t+k} \frac{w_i h_i - w_{i-1} h_{i-1}}{w_{i-1} h_{i-1}} \qquad (9.6)$$

$$f_{\text{height}} = \frac{1}{2k+1} \sum_{i=t-k}^{t+k} \frac{h_i - h_{i-1}}{h_{i-1}}$$

$$f_{\text{width}} = \frac{1}{2k+1} \sum_{i=t-k}^{t+k} \frac{w_i - w_{i-1}}{w_{i-1}}$$

The third and final set of features describes pair-wise interactions between the *m*th and *n*th STV as follows:

$$f_D = \frac{1}{2k+1} \sum_{i=t-k}^{t+k} \sqrt{(x_i^m - x_i^n)^2 + (y_i^m - y_i^n)^2} \; \forall m \neq n$$

$$f_{\text{DS}} = \left| f_{\text{speed}}^m - f_{\text{speed}}^n \right| \forall m \neq n$$

$$f_{\text{DV}} = \left| f_{\text{velocity}}^m - f_{\text{velocity}}^n \right| \forall m \neq n \qquad (9.7)$$

$$f_R = \frac{\left\| (x_{t+k}^n - x_{t-k}^n, y_{t+k}^n - y_{t-k}^n) + (x_{t+k}^m - x_{t-k}^m, y_{t+k}^m - y_{t-k}^m) \right\|}{\left\| (x_{t+k}^n - x_{t-k}^n, y_{t+k}^n - y_{t-k}^n) \right\| + \left\| (x_{t+k}^m - x_{t-k}^m, y_{t+k}^m - y_{t-k}^m) \right\|}$$

Here, the last feature f_R is the ratio between the norm of sums and sum of norms of the velocity vectors of the *m*th and *n*th STVs, and tends to 0 and 1 when the two STVs are moving, respectively, in the opposite or same direction.

The choice of trajectory versus spatiotemporal descriptor depends on the application scenario. The first approach relies upon robust trackers, and is best suited for tracking objects with constrained or well-defined paths such as vehicles and a small number of pedestrians at a traffic intersection, captured from a stationary camera. The second approach is more general in its joint treatment of spatial and temporal dimensions and may be applicable in scenarios involving complex and unstructured motion such as at a crowded intersection or parking lot.

Recently, dense trajectories have been proposed for video action recognition [22] and can be thought of as a means to combine the benefits of both trajectory and spatiotemporal approaches. Here, scene motion is described in terms of a dense set of trajectories computed from a dense optical flow algorithm. Descriptors including HOGs, HOF, and motion boundary histograms (MBHs) are computed within small STVs in steps centered along a given trajectory, and together encode local appearance and motion information along that trajectory. MBH is particularly effective for canceling out camera motion. Finally, the standard bag-of-visual-words (BOV) approach is used to create an aggregated feature representation for a given video clip. This approach performs well, particularly when there is significant camera motion, and object motion is dense and unstructured (as in crowded scenes).

9.3 Anomaly Detection Models

In this section, we discuss three basic flavors of anomaly detection models. The first one (classification methods) is a supervised form of anomaly detection, wherein labeled anomalous samples are available during training. The remaining two approaches handle the unsupervised case where no labeled anomalous samples are available. In this scenario, the problem is one of estimating the probability of a test event and deeming the event to be anomalous if the probability is below a threshold. Alternatively, the problem can be posed as computing the distance between the test event and classes or clusters of normal events according to a chosen feature space and distance metric, and casting distant outliers as anomalies.

9.3.1 Classification Methods

In the structured scenario where both normal and anomalous events are precategorized, anomaly detection boils down to a classification problem. If the final class is an anomalous one, the event is labeled as an anomaly. We thus have at our disposal a rich suite of classification techniques that can be brought to bear. In the work by Simon et al. [7], for example, spatiotemporal features described in Section 9.2.2 are extracted as the event encoding, and decision trees are trained to *categorize* input events as normal or anomalous classes. Decision trees are employed in two stages. First, STV features are clustered in an unsupervised fashion into local patterns using binary trees. At each node of the tree, a binary test on a given STV feature such as size or velocity is used to split the STVs represented by that node into two child nodes. When the tree is completely constructed, the leaves represent local clusters of spatiotemporal patterns $T_1, ..., T_N$. For example, one leaf node may represent small sedans moving within a certain velocity range in the lower left region of the field of view. In the second stage, temporal and causal relationships between these patterns are derived using a second binary decision tree. Again, nodes represent binary tests to assert whether one pattern occurs before, after, or simultaneously with another pattern. The leaves denote global events comprising certain sequences of local patterns, and are each associated with a class label during training. Simon et al. [7] demonstrate their technique on the *CAVIAR* dataset in order to identify group interactions such as *fighting, meeting, and pocket-picking*. Some of these events are prelabeled as anomalies due to a low probability of occurrence. Note that many combinations of event encodings and classification techniques are conceivable, and are generally selected based on the constraints and performance requirements for a given application.

9.3.2 Hidden Markov Models

The approach described before are either frame based, or treat an entire video sequence as a spatiotemporal block and perform anomaly detection holistically on the entire block. Another class of techniques treats the video as a temporally evolving signal, and thus applies sequence modeling techniques to characterize video events. One popular choice is the hidden Markov model (HMM). Briefly, an HMM is a special form of a Bayesian network, and is characterized by a sequence of observed event encodings $\mathbf{O}_1, ..., \mathbf{O}_T$ that are governed probabilistically by a set of latent state variables $s_1, ..., s_T$ each taking on one of N possible states. A schematic is shown in Figure 9.4.

Fundamentally, an HMM assumes that observation $\mathbf{O}_t$ is governed only by hidden state s_t, and that states satisfy the Markov property, so that given the value of state s_{t-1}, the value of the current state s_t is conditionally independent of all states prior to $t-1$. HMM model parameters are given by $(\pi, \mathbf{A}, \mathbf{B})$. π is the N-dimensional probability vector for the initial state s_1; $\mathbf{A}$ is the $N \times N$ matrix of state transition probabilities; and $\mathbf{B}$ is the $N \times T$ matrix of state output probabilities. That is, $b_{si,Oj}$ is the probability of observing output $\mathbf{O}_j$ given state s_i. As is commonly done, we assume the case of a *homogeneous* HMM where $\mathbf{A}$ is time invariant. During inference either the likelihood, $p(\mathbf{O}_t \mid s_t)$ or the posterior distribution $p(s_t \mid \mathbf{O}_t)$ may be estimated with an iterative multipass approach [23]. Probability distributions for π and $\mathbf{B}$ usually take on a parametric form, such as a GMM, to enable tractable computation.

Note that any of the previously described event encodings can be used as observation vectors $\mathbf{O}_t$ for the HMM. In the work of Pruteanu-Malinici and Carin [9], the following three sets of linear transformations on the input video are compared as feature descriptors: shift-invariant wavelet transform (SIWT), independent component analysis (ICA), and independent subspace analysis (ISA). An infinite-state HMM is trained for anomaly detection, and conditional observation probabilities $\mathbf{B}$ are defined by GMMs. An event is determined to be an anomaly when its likelihood for the corresponding hidden state is below a threshold. Good results are shown on events around a traffic intersection involving pedestrians, bikes, and different types of vehicles. Among feature descriptors, ISA performs the best.

9.3.3 Contextual Methods

A third class of techniques models events in terms of the context and behavior of objects in the scene. Adopting the definition in Saligrama et al. [1], context in a video refers to the spatiotemporal coordinates (i.e., location and time) of an object passing through the camera's field of view. Behavioral attributes include the size, speed, direction, and color of the object passing by a specific location at a specific time instance. Note the stark contrast between the trajectory descriptor that follows the path

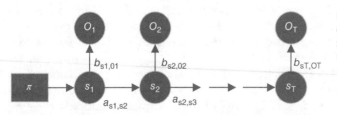

Figure 9.4 Hidden Markov model characterized by initial state probability π, state transition matrix **A**, and state output matrix **B**.

of a single object over space and time, and the contextual/behavioral description that observes the movement of all objects through a single space–time instance.

We now present briefly the contextual model proposed by Saligrama et al. [1]. First, a binary motion label is computed at each pixel with the two states corresponding to "static" or "moving." Such labeling can be accomplished with known techniques such as background subtraction described earlier. At a fixed 2D pixel location $\mathbf{x}$, label $L_t(\mathbf{x})$ denotes a temporal binary sequence, with alternative busy (moving) and idle (static) period. Additional features such as size, shape, color, and speed of an object passing through pixel $\mathbf{x}$ during a busy period are collected into a feature vector $\mathbf{F}_t(\mathbf{x})$. Saligrama et al. [1] show that the busy and idle periods are respectively independent random variables, and that the features $\mathbf{F}_t(\mathbf{x})$ across different busy periods are also independent. This forms the basis for a statistical model, namely a Markov chain with two states corresponding to "busy" and "idle," a 2×2 state transition matrix for the chain, and a probability distribution for features $\mathbf{F}_t(\mathbf{x})$ conditioned upon the underlying state. An exemplary choice for the feature distribution is a Gibbs–Markov model. The overall statistical model assumes both independence among the busy and idle periods of $L_t(\mathbf{x})$ and conditional independence of feature vectors $\mathbf{F}_t(\mathbf{x})$ when conditioned on the state. The latter statement intuitively makes sense, since the size of one vehicle passing through $\mathbf{x}$ is independent of the sizes of preceding or subsequent vehicles.

Based on the aforementioned premise, Saligrama et al. [1] propose an event encoding and probabilistic model for that encoding as follows. Define a time window $W = [t - w + 1, t]$, and define an event encoding that combines the motion labels and appearance features within this window: $\mathbf{l}_W = \{L_{t-w+1}(\mathbf{x}), \ldots, L_t(\mathbf{x}), \mathbf{F}_{t-w+1}(\mathbf{x}), \ldots, \mathbf{F}_t(\mathbf{x})\}$. For convenience, $\mathbf{x}$ is herein omitted from the following formulation but implicitly assumed. The aforementioned Markov chain and Gibbs–Markov assumptions define a probability distribution $g(\mathbf{l}_W)$ for the event encoding. Saligrama et al. [1] assert that the log likelihood of events is given by

$$
\begin{aligned}
\Lambda_{\mathbf{x}}(\mathbf{l}_W) &= -\log\big(g(\mathbf{l}_W)\big) \\
&= \sum_{k=t-W+1}^{t} \Big(L_k(\mathbf{x})\big[A_1 + A_2 V_k(\mathbf{x})\big] + A_3 \kappa_W(\mathbf{x})\Big),
\end{aligned}
\tag{9.8}
$$

where A_1, A_2, and A_3 are constants; V_k is the potential function from the Gibbs–Markov model for feature vector $\mathbf{F}_t(\mathbf{x})$; and $\kappa_W(\mathbf{x})$ is a term that is proportional to the total number of busy–idle transitions within time window W at location $\mathbf{x}$. While the derivation of Equation 9.8 is outside the scope of this chapter (see Saligrama et al. [1] for details), we note that likelihood Λ is a function of both space and time, and thus provides the ability to generate spatiotemporally localized maps of nominal events from training videos. Specifically in Saligrama et al. [1], the maximum value of $\Lambda_{\mathbf{x}}$ over a time window at each pixel location forms a background behavior image $B_{\max}(\mathbf{x})$ representing nominal (background) events during that time period. Test events are processed through Equation 9.8 and compared against $B_{\max}(\mathbf{x})$ in a form of "background behavior subtraction" to determine if an anomaly is present at a given location and time window. A visualization of the model applied on urban transportation video is reproduced from Saligrama et al. [1] in Figure 9.5.

Note again that this contextual model is operable in the unsupervised setting in that there is no explicit characterization or labeling required of anomalies. It also supports the unstructured scenario in the sense that there is no need to generate distinct categories of anomalous events; rather, it is sufficient to collectively characterize all nominal events by a tractable statistical model.

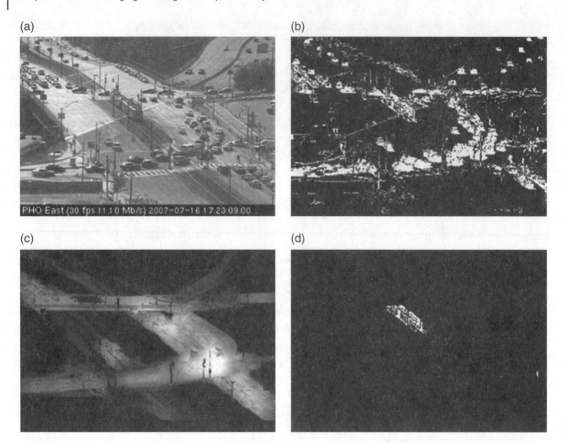

Figure 9.5 Contextual anomaly detection model: (a) input video captured at busy traffic scene, (b) motion label L_t, (c) background behavior image $B_max(\mathbf{x})$, and (d) anomaly detected via behavior subtraction. *Source*: Saligrama et al. [1]. Reproduced with permission of IEEE.

9.4 Sparse Representation Methods for Robust Video Anomaly Detection

This section describes in detail a fourth flavor of anomaly detection models that can be termed "analysis by synthesis." Essentially, the idea is to reconstruct a test event in terms of known normal events, and quantitatively assess some aspect of the reconstruction to determine if the event is normal or anomalous. Specifically, we investigate sparse reconstruction methods as a recent novel and promising idea in the field of video anomaly detection, and associate anomaly detection with a carefully derived measure of sparsity. To place this class of approaches in context of our anomaly detection framework of Figure 9.2, sparsity-based methods can support both structured and unstructured scenarios, as well as both supervised and unsupervised settings. The various flavors will be explained later in this section.

In a typical approach [5], trajectories of objects corresponding to normal events are extracted from video by traditional object tracking algorithms. Each trajectory is then represented as a feature vector by a polynomial spline curves approximation representation [5] and collected into a dictionary of normal events. The fundamental underlying assumption of these methods is that any new normal trajectory can approximately be modeled as a (sparse) linear combination of training trajectories (or equivalent features). Conversely, adequate reconstruction of anomalous events will require a dense (i.e., nonsparse) combination of normal events. The advocacy of sparsity over other competing anomaly detection methods, as surveyed in Section 9.2, is based on two arguments: (i) recent work in face recognition [24] has shown that sparsity-based classification can be powerful even as feature descriptions are missing, for example, occlusion of objects leading to missing trajectory information; (ii) the sparse coefficients can additionally withstand noise and quality distortions to the video. We next discuss sparsity-based video anomaly detection methods in both structured and unstructured scenarios.

9.4.1 Structured Anomaly Detection

Let each trajectory representation lie in $\mathbb{R}^n$, and let T denote the number of training samples (i.e., example trajectory representations) from each of K different classes, that is, behavior patterns in a video which may be normal or anomalous. The T training samples from the ith class are arranged as the columns of a matrix $\mathbf{A}_i \in \mathbb{R}^{n \times T}$. The dictionary $\mathbf{A} \in \mathbb{R}^{n \times KT}$ of training samples from all classes is formed as follows: $\mathbf{A} = [\mathbf{A}_1 \, \mathbf{A}_2 \ldots \mathbf{A}_K]$.

Given a sufficient number of training samples from the mth trajectory class, a test trajectory $\mathbf{y} \in \mathbb{R}^n$ from the same class is conjectured to approximately lie in the linear span of those training samples. Any trajectory feature vector is synthesized by a linear combination of the set of all training trajectory samples as follows:

$$\mathbf{y} \approx \mathbf{A}\alpha = \left[\mathbf{A}_1 \, \mathbf{A}_2 \ldots \mathbf{A}_K \right] \begin{bmatrix} \alpha_1 \\ \alpha_2 \\ \vdots \\ \alpha_K \end{bmatrix}. \tag{9.9}$$

Here, each $\alpha_i \in \mathbb{R}^T$. Typically for an example trajectory $\mathbf{y}$, only one of the α_i's will be active (corresponding to the class/event from which $\mathbf{y}$ is generated). Thus, the coefficient vector $\alpha \in \mathbb{R}^{KT}$ is *sparse* and can be recovered by solving the following equation:

$$\check{\alpha} = \arg\min_{\alpha} \| \alpha \|_1 \text{ subject to } \| \mathbf{y} - \mathbf{A}\alpha \|_2 < \varepsilon . s \tag{9.10}$$

Here, the objective is to minimize the number of nonzero elements in α. It is well known from the compressed sensing literature that minimizing the l_0 norm leads to an NP-hard problem [25]. Thus, the l_1 norm is used as an effective substitute. The residual error between the test trajectory and each class behavior pattern is computed to find the class to which the test trajectory belongs.

$$r_i(\mathbf{y}) = \| \mathbf{y} - \mathbf{A}_i \hat{\alpha}_i \|_2 \quad i = 1, 2, \ldots, K. \tag{9.11}$$

Figure 9.6 shows an example of classification using sparsity model. The training dictionary consists of two classes, and each class contains four different trajectories. The test trajectory can be well represented by the linear combination of trajectory no. 1 and trajectory no. 3 from class 1 (see Figure 9.6). This is in fact tantamount to saying that the coefficient vector α is indeed sparse—in this example, two of eight entries being active.

Extensions: Several extensions of the aforementioned sparsity model have been developed. To overcome the limitations of linear models, kernel sparse representations have been proposed by

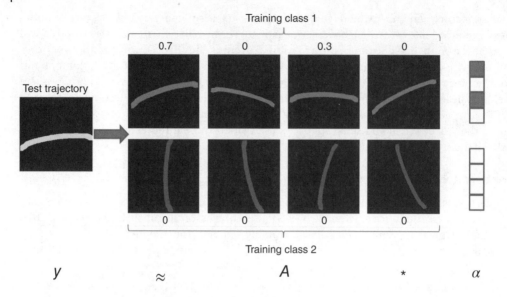

Figure 9.6 An example illustration of sparsity-based trajectory classification.

Mo et al. [11]. Note also that in Equation 9.9, *fixed* dictionaries comprising training trajectories per class are used. There are two practical issues with this approach as follows: (i) raw dictionaries of training samples are uneconomical and often large dictionaries are required for adequately good performance; (ii) as new data/training is accumulated, it is desirable to adapt the dictionaries. Computationally efficient learning of dictionaries for practical video anomaly detection has been pursued in Zhao et al. [6] and Mo and Monga [26].

9.4.1.1 A Joint Sparsity Model for Anomaly Detection

Various anomaly detection algorithms have been designed for video surveillance. However, only few of them have considered the interaction between multiple objects [8, 10, 27]. While it is true that anomalies are generated by atypical trajectory/behavior of a single object, "collective anomalies" that are caused by the joint observation of objects are also significant. For example, in the area of transportation, some events, for example, accidents and dangerous driver–pedestrian behavior, are indeed based on joint and not just individual object behavior. It is possible in fact that the individual events corresponding to each object's behavior are not necessarily anomalies by themselves. Take the example of a vehicle accidentally changing lanes due to an inattentive driver. Another vehicle in close proximity may have to also suddenly change lanes in order to avoid colliding with the first vehicle. Both lane changes, as isolated events are not necessarily anomalous, but when viewed in conjunction should logically be flagged as a joint anomaly.

Previous methods have employed probabilistic models to learn the relationship between different individual events. Han et al. [10] and Vaswani et al. [8] use an HMM-based method to track multiple trajectories followed by defining a set of rules to distinguish between normal and anomalous events. Wang et al. [27] present an unsupervised framework using hierarchical Bayesian models to model individual events and interactions between them.

The sparsity-based approach reviewed thus far does not capture interactions to detect two or more object anomalies. We describe next a new "joint sparsity model" developed recently by Mo et al. [11]

for video anomaly detection which incorporates multiple object trajectories and their interactions. Hence, even if individually the trajectories may be considered normal, "collective anomalies" could occur and can be successfully detected in this framework.

In detail, we are interested in detection of anomalies involving $P \geq 1$ objects. Their corresponding P trajectories can be represented as a matrix: $\mathbf{Y} = [\mathbf{y}_1\ \mathbf{y}_2 \dots \mathbf{y}_P] \in \mathbb{R}^{n \times P}$, where $\mathbf{y}_i$ corresponds to ith trajectory. The training dictionary can be defined as $\mathbf{A} = [\mathbf{A}_1\ \mathbf{A}_2 \dots \mathbf{A}_P] \in \mathbb{R}^{n \times PKT}$, where each dictionary $\mathbf{A}_i = [\mathbf{A}_{i,1}\ \mathbf{A}_{i,2} \dots \mathbf{A}_{i,K}] \in \mathbb{R}^{n \times KT}$, $i = 1,2,\dots,P$, is formed by the concatenation of the subdictionaries from all classes belonging to the ith trajectory. The crucial aspect of this formulation is that the training trajectories for any class j, that is, $\mathbf{A}_{i,j}$, $i = 1,2,\dots,P$ are observed "jointly" from example videos. This generalizes the setup of Li et al. [5] and Zhao et al. [6].

The P test trajectories can now be represented as a linear combination of training samples as follows:

$$\mathbf{Y} \approx \mathbf{AS} = \left[\mathbf{A}_{1,1}\ \mathbf{A}_{1,2} \dots \mathbf{A}_{1,K} \dots \mathbf{A}_{P,1}\ \mathbf{A}_{P,2} \dots \mathbf{A}_{P,K} \right] \left[\alpha_1 \dots \alpha_P \right]. \tag{9.12}$$

Here, the coefficient vectors α_i lie in $\mathbb{R}^{PKT}$ and $\mathbf{S} = [\alpha_1 \dots \alpha_i \dots \alpha_P]$.

It is important to note that the ith object trajectory of any observed set of test trajectories should *only lie* in the span of training trajectories corresponding to the ith object. Therefore, the columns of $\mathbf{S}$ should have the following structure:

$$\alpha_1 = \begin{bmatrix} \alpha_{1,1} \\ \alpha_{1,2} \\ \vdots \\ \alpha_{1,K} \\ \mathbf{0} \\ \mathbf{0} \end{bmatrix}, \alpha_i = \begin{bmatrix} \mathbf{0} \\ \alpha_{i,1} \\ \alpha_{i,2} \\ \vdots \\ \alpha_{i,K} \\ \mathbf{0} \end{bmatrix}, \alpha_P = \begin{bmatrix} \mathbf{0} \\ \mathbf{0} \\ \alpha_{P,1} \\ \alpha_{P,2} \\ \vdots \\ \alpha_{P,K} \end{bmatrix}. \tag{9.13}$$

Here, each of the subvectors $\{\alpha_{i,j}\}_{j=1}^{K}$, $i = 1,2,\dots,P$ lies in $\mathbb{R}^T$, while $\mathbf{0}$ denotes a vector of all zeros in $\mathbb{R}^{KT}$. As a result, $\mathbf{S}$ exhibits a block-diagonal structure.

From Li et al. [5], we know that for a single object, its trajectory can be represented by a sparse linear combination of all the training samples. For the multiple trajectories scenario, we assume that training samples with nonzero weights (in the sparse linear combination) exhibit one–one correspondence across different trajectories. In other words, if the ith trajectory training sample from the jth class is chosen for the ith test trajectory, then it is necessarily that other $P-1$ trajectories choose from jth class with very high probability, albeit with possibly different weights.

We take a simple scenario which only has two objects and two training classes (normal and anomalous class) as an example to explain the structure of Equation 9.12. In this situation, $P = 2$, $K = 2$, Equation 9.12 becomes

$$\mathbf{Y} \approx \mathbf{AS} = \left[\mathbf{A}_{1,1}\ \mathbf{A}_{1,2}\ \mathbf{A}_{2,1}\ \mathbf{A}_{2,2} \right] \begin{bmatrix} \alpha_{1,1} & \mathbf{0} \\ \alpha_{1,2} & \mathbf{0} \\ \mathbf{0} & \alpha_{2,1} \\ \mathbf{0} & \alpha_{2,2} \end{bmatrix}. \tag{9.14}$$

The test trajectory sample is thought of as a collective event. Therefore, all trajectories of the sample should be classified into one class. If the first trajectory is classified into jth class, the second trajectory should also be classified into jth class, which means $\alpha_{1,j}$ and $\alpha_{2,j}$ should be activated simultaneously. This characteristic that some coefficients should be activated jointly captures the interaction between objects.

Further, define a new matrix $\mathbf{S}'$ as follows:

$$\mathbf{S}' = \begin{bmatrix} \alpha_{1,1} & \alpha_{2,1} \\ \alpha_{1,2} & \alpha_{2,2} \end{bmatrix}. \tag{9.15}$$

In $\mathbf{S}'$, "joint coefficients" are moved into the same row. The joint information can be captured by enforcing selected rows of $\mathbf{S}'$ to be activated simultaneously.

In general, when there are K classes and P objects, the structure of $\mathbf{S}'$ is

$$\mathbf{S}' = \begin{bmatrix} \alpha_{1,1} & \cdots & \alpha_{i,1} & \cdots & \alpha_{P,1} \\ \alpha_{1,2} & \cdots & \alpha_{i,2} & \cdots & \alpha_{P,2} \\ \vdots & \vdots & \vdots & \vdots & \vdots \\ \alpha_{1,K} & \cdots & \alpha_{i,K} & \cdots & \alpha_{P,K} \end{bmatrix} \in \mathbb{R}^{KT \times P}. \tag{9.16}$$

The question that remains to be addressed is the particular way of transforming $\mathbf{S}$. Such a transformation is realized by defining matrices $\mathbf{H} \in \mathbb{R}^{PKT \times P}$ and $\mathbf{J} \in \mathbb{R}^{KT \times PKT}$:

$$\mathbf{H} = \begin{bmatrix} 1 & 0 & \cdots & 0 \\ 0 & 1 & \cdots & 0 \\ \vdots & \vdots & \vdots & \vdots \\ 0 & 0 & \cdots & 1 \end{bmatrix}, \ \mathbf{J} = \begin{bmatrix} \mathbf{I}_{KT} & \mathbf{I}_{KT} & \cdots & \mathbf{I}_{KT} \end{bmatrix}. \tag{9.17}$$

The vectors $\mathbf{1}$ and $\mathbf{0}$ are in $\mathbb{R}^{KT}$ and contain all 1s and 0s respectively, and $\mathbf{I}_{KT}$ is the KT-dimensional identity matrix.

$$\mathbf{H} \circ \mathbf{S} = \begin{bmatrix} \alpha_{1,1} & \cdots & \mathbf{0} & \cdots & \mathbf{0} \\ \alpha_{1,2} & \cdots & \alpha_{i,1} & \cdots & \mathbf{0} \\ \vdots & \cdots & \alpha_{i,2} & \cdots & \alpha_{P,1} \\ \alpha_{1,K} & \cdots & \vdots & \cdots & \alpha_{P,2} \\ \mathbf{0} & \cdots & \alpha_{i,K} & \cdots & \vdots \\ \mathbf{0} & \cdots & \mathbf{0} & \cdots & \alpha_{P,K} \end{bmatrix}, \tag{9.18}$$

then we have

$$\mathbf{J}(\mathbf{H} \circ \mathbf{S}) = \mathbf{S}', \tag{9.19}$$

where the $\circ$ indicates matrix Hadamard (entry-wise) product.

Therefore, we can now solve for the sparse coefficients via the following optimization problem:

$$\begin{aligned} &\text{minimize} \quad \|\mathbf{J}(\mathbf{H} \circ \mathbf{S})\|_{\text{row},0} \\ &\text{subject to} \quad \|\mathbf{Y} - \mathbf{AS}\|_{\text{F}} \leq \varepsilon. \end{aligned} \tag{9.20}$$

Here, $\|\cdot\|_{\text{row},0}$ refers to the number of nonzero rows in the matrix and the cost function minimization seeks the $J(\mathbf{H} \circ \mathbf{S})$ with the minimum number of nonzero rows, while the constraint ensures good approximation ($\|\cdot\|_F$ denotes the Frobenius norm).

Algorithm 9.1

Simultaneous Orthogonal Matching Pursuit (SOMP)

Input: Dictionary $\mathbf{A} = [\mathbf{a}_1 \; \mathbf{a}_2 \ldots \mathbf{a}_{PKT}]$, data matrix $\mathbf{Y} = [\mathbf{y}_1 \; \mathbf{y}_2 \ldots \mathbf{y}_P]$, a stopping criterion {Make sure all columns in $\mathbf{A}$ and $\mathbf{Y}$ have unit norm}
Initialization: residual $\mathbf{R}_0 = \mathbf{Y}$, index set Λ_0: empty set, iteration counter $k = 1$

1: **while** stopping criterion has not been met
   1. Find the index of the atom that best approximates all residuals: $\lambda_k = \arg\max_i \| \mathbf{R}_{k-1}^T \mathbf{a}_i \|_2$
   2. Update the index set $\Lambda_k = \Lambda_{k-1} \cup \{\lambda_k\}$
   3. Compute $\mathbf{G}_k = (\mathbf{A}_{\Lambda_k}^T \mathbf{A}_{\Lambda_k})^{-1} \mathbf{A}_{\Lambda_k}^T \mathbf{Y}$, $\mathbf{A}_{\Lambda_k}^T$ consists of the k atoms in $\mathbf{A}$ indexed in Λ_k
   4. Determine the residual $\mathbf{R}_k = \mathbf{Y} - \mathbf{A}_{\Lambda_k}^T \mathbf{G}_k$
   5. $k \leftarrow k+1$
2: end while

Output: Index set $\Lambda = \Lambda_{k-1}$, the sparse representation $\mathbf{S}$ whose nonzero rows indexed by Λ are k rows of the matrix $(\mathbf{A}_\Lambda^T \mathbf{A}_\Lambda)^{-1} \mathbf{A}_\Lambda^T \mathbf{Y}$

The well-known row sparsity problem

$$\begin{aligned} \text{minimize} \quad & \|\mathbf{S}\|_{\text{row},0} \\ \text{subject to} \quad & \|\mathbf{Y} - \mathbf{AS}\|_F \leq \varepsilon, \end{aligned} \tag{9.21}$$

is nonconvex but can be solved using greedy pursuit algorithms widely used in the literature. SOMP [28]—enumerated in Algorithm 9.1—is among the most popular algorithms used. In SOMP, the support of the solution is sequentially updated (i.e., the atoms in the dictionary $\mathbf{A}$ are sequentially selected). At each iteration, the atom that simultaneously yields the best approximation to all of the residual vectors is selected.

$$\lambda_k = \arg\max_i \| \mathbf{R}_{k-1}^T \mathbf{a}_i \|_2 . \tag{9.22}$$

The joint sparsity model for representing multiple-object trajectories involves solving Equation 9.20, which looks quite similar to Equation 9.21, but the Hadamard operator from $\mathbf{S}$ to $\mathbf{S}'$ makes the problem much more involved. We can observe from Algorithm 9.1 that the original SOMP algorithm effectively gives k_0 distinct atoms from a dictionary $\mathbf{A}$ that best approximates the data matrix $\mathbf{Y}$ for k_0 iterations; we apply the general formulation even when the Hadamard operator is present. At every iteration k, SOMP measures the residual for each atom in $\mathbf{A}$ and creates an orthogonal projection with the highest correlation.

This idea can also be extended to solve the problem in Equation 9.20. If the atom of jth trajectory that we selected comes from ith training, the other $P-1$ atoms of trajectories should also be chosen from ith training. Then, Equation 9.22 in SOMP can be modified as follows:

$$\lambda_k = \arg\max_i \sum_j \| \mathbf{R}_{j,k-1}^T \mathbf{a}_{j,i} \|_2 \,. \tag{9.23}$$

Here, $\mathbf{R}_{j,k-1}^T$ refers to the residual of jth trajectory in iteration $k-1$, and $\mathbf{a}_{j,i}$ represents the ith training of jth trajectory. After employing this special rule of choice for atom selection, each row of parameter matrix $\mathbf{S}'$ will be activated simultaneously or inactivated simultaneously; thus, the row sparsity requirements will inherently hold. The implementation details of this algorithm can be found in a technical report [29].

Auxiliary Optimization Problem: The problem in Equation 9.20 is nonconvex, and the solution obtained using any known optimization solvers is suboptimal. To improve the solution, Mo et al. [11] propose an auxiliary optimization problem which is in fact convex, with guarantee of global minima. From the optimization problem (Equation 9.20), we can first get a suboptimal result: $\hat{\mathbf{S}}$. A *membership matrix* $\mathbf{E} \in \mathbb{R}^{PKT \times P}$ is defined which has zeros at locations of nonzero entries in $\hat{\mathbf{S}}$ and ones elsewhere. With this enforcement of locations of nonzero entries, the following optimization problem results:

$$
\begin{aligned}
&\text{minimize} \quad \| \mathbf{Y} - \mathbf{AS} \|_F \\
&\text{subject to} \quad \mathbf{s}_i^T \mathbf{e}_i = 0, \; i = 1, 2, \ldots, P
\end{aligned} \tag{9.24}
$$

Here, $\mathbf{s}_i$ and $\mathbf{e}_i$ refer to the ith column of $\mathbf{S}$ and $\mathbf{E}$, respectively. The initial choice of $\mathbf{S}$ is the $\hat{\mathbf{S}}$ described previously.

This problem can be further simplified to mitigate computational complexity. Each column of $\mathbf{S}$ can be optimized in parallel, since the constraints are separable. So we have

$$
\begin{aligned}
&\text{minimize} \quad \| \mathbf{y}_i - \mathbf{As}_i \|_2 \\
&\text{subject to} \quad \mathbf{s}_i^T \mathbf{e}_i = 0.
\end{aligned} \tag{9.25}
$$

9.4.1.2 Supervised Anomaly Detection as Event Classification

Having obtained the sparse coefficient matrix $\mathbf{S}$, we compute class-specific residual errors and identify the class of the test event $\mathbf{Y}$ as that which gives the minimum residual,

$$\text{identity}(\mathbf{Y}) = \arg\min_i \| \mathbf{Y} - \mathbf{A}\delta_i(\mathbf{S}) \|_F, \tag{9.26}$$

where $\delta_i(\mathbf{S})$ is the matrix whose only nonzero entries are the same as those in $\mathbf{S}$ associated with class i (in all P trajectories). When sufficient representation (example training trajectories) for anomalous events is available, then anomalous classes are simply one or more of the K classes in this joint sparsity based classification framework.

9.4.1.3 Unsupervised Anomaly Detection via Outlier Rejection

If training for anomalies is missing/statistically insignificant, we cannot use Equation 9.26 to identify anomalies. Inspired by the outlier rejection measure in Wright et al. [24],

Figure 9.7 An example illustration of anomalous event versus normal event.

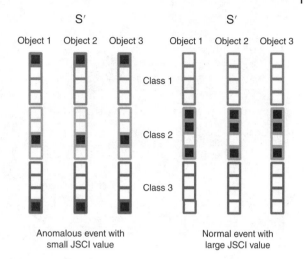

SCI(α) = (K · maxᵢ ‖ρᵢ(α)‖₁ / ‖α‖₁ − 1) / (K − 1)

$$\text{SCI}(\alpha) = \frac{K \cdot \max_i \|\rho_i(\alpha)\|_1 \, / \, \|\alpha\|_1 - 1}{K - 1}$$
$$\text{SCI}(\alpha) < \tau_1 \rightarrow \mathbf{y} : \textbf{outlier},$$
(9.27)

and $\rho_i(\alpha)$ is the new vector whose only nonzero entries are the entries in α that are associated with class i. We model anomalies as outliers, given training from expected normal event classes that form the dictionary $\mathbf{A}$. Equation 9.27 can be used to detect single object anomalies. This may be extended to the multiple-object case as follows:

$$\text{JSCI}(\mathbf{S}) = \frac{K \cdot \max_i \|\delta_i(\mathbf{S}')\|_{\text{row},0} \, / \, \|\mathbf{S}'\|_{\text{row},0} - 1}{K - 1}.$$
(9.28)

Here, JSCI is the joint SCI. If $\text{JSCI}(\mathbf{S}) < \tau_2$, a multiple-object anomaly is identified.

Figure 9.7 shows an example of anomalous event versus normal event. The left figure shows the sparse coefficients of an anomalous event. The activated coefficients are scattered all over the normal classes; therefore, the corresponding event cannot be classified into either class. In this case, it can be regard as an outlier (anomalous event). The JSCI value in Equation 9.28 will be small. On the other hand, the activated coefficients in the right figure are gathered in normal class 2. Note further that $0 \leq \text{JSCI}(\mathbf{S}') \leq 1$, if $\text{JSCI}(\mathbf{S}')$ is close to 0, the event is normal, and if $\text{JSCI}(\mathbf{S}')$ is close to 1, the event is anomalous. A nominal choice of $\tau_2 = 0.5$ can be made, but this can further be optimized experimentally based on the underlying video dataset by observing the range of the measure $\text{JSCI}(\mathbf{S}')$ for normal events.

9.4.2 Unstructured Video Anomaly Detection

The setup in Equation 9.9 and the subsequent joint sparsity model (JSM) assumes that training is available from both normal and anomalous events, and hence anomaly detection reduces to a classification problem. In the absence of training from anomalous events (the more practical scenario), outlier rejection measures as in Equations 9.27 and 9.28 may be used to detect anomalies. Both cases assume a structured sparsity model (SSP), and thus a careful preparation of the dictionary $\mathbf{A}$ is needed often with training examples that are manually labeled to belong to particular event classes.

(a)　　　　　　　　　　　　　　　　　(b)

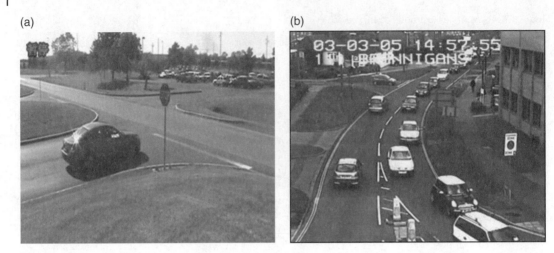

Figure 9.8 (a) Structured scenario and (b) unstructured scenario.

Such elaborate preparation of the dictionary is sometimes unrealistic and invariably burdensome requiring a preanalysis of video footage prior to anomaly detection. Figure 9.8a (the video is available at http://youtu.be/M6_PJigg5CY) shows an example video frame of a structured scenario (detection of stop sign violations) where preparation of a dictionary clearly separated into class-specific subdictionaries is possible. In many other settings, however, multiple objects and features are simultaneously extracted and a clear separation into normal event classes is difficult. An example of a video frame from such a scenario is shown in Figure 9.8b (the video is available at http://youtu.be/jEzLkWF65Io).

We therefore seek a more general and practical sparsity prior which can deal with unstructured scenarios wherein we do not have a well-classified training dictionary. Note that coefficient matrices S and S' in JSM usually contain a small number of active rows per class; hence in those cases, the sparse coefficient matrix is also of low rank. Inspired by this observation and known connections between low-rank and sparse matrices [30], Mo et al. [4] propose using a low-rank sparsity prior (LRSP). In addition, if rank is substituted by its convex nuclear norm alternative, then significant computational benefits can be obtained over existing methods in sparsity-based video anomaly detection. We next describe the LRSP model in detail.

LRSP for Anomaly Detection: In this framework, there is no need to group training trajectories into different normal event classes. All observed training trajectories corresponding to normal events are collected together as a big dictionary: $A \in \mathbb{R}^{N \times T}$. We also collect M test trajectories extracted from the video into a matrix $Y = \{y_i\} \in \mathbb{R}^{N \times M}$, $i = 1, \ldots, M$.

Under a linear model $Y \approx AS$, and given sufficient training, the coefficient matrix $S \in \mathbb{R}^{T \times M}$ is expected to be sparse. Making a departure from the typical $\|\|_{row,0}$ norm, we propose to use a low-rank structure to measure the sparsity of S. Then, we propose to replace Equation 9.10 by

$$\begin{aligned} \text{minimize} \quad & \text{rank}(S) \\ \text{subject to} \quad & \|Y - AS\|_F \le \varepsilon, \end{aligned} \tag{9.29}$$

A convex relaxation of Equation 9.29 can be obtained via substituting rank $(\mathbf{X})$ by $\|\mathbf{X}\|_* = \sum_i \sigma_i(\mathbf{X})$ (where $\|\|_*$ denotes nuclear norm and $\sigma_i(\mathbf{X})$ is the ith singular value of $\mathbf{X}$) [31]. This results in a convex optimization problem,

$$
\begin{aligned}
\text{minimize} \quad & \|\mathbf{S}\|_* \\
\text{subject to} \quad & \|\mathbf{Y} - \mathbf{A}\mathbf{S}\|_F \leq \varepsilon.
\end{aligned}
\tag{9.30}
$$

While low-rank and sparse matrix structures often simultaneously exist (as is expected here as well), in general the two are not the same, and low rank does not imply sparsity. To encourage sparse matrices which are simultaneously low rank, we further add a l_1 regularization term to the cost function and convexity still holds,

$$
\begin{aligned}
\text{minimize} \quad & \|\mathbf{S}\|_* + \lambda \|\mathbf{S}\|_1 \\
\text{subject to} \quad & \|\mathbf{Y} - \mathbf{A}\mathbf{S}\|_F \leq \varepsilon.
\end{aligned}
\tag{9.31}
$$

Anomaly Detection: Once we get the optimal coefficient matrix $\hat{\mathbf{S}}$, the recovered trajectory can be computed using columns of $\hat{\mathbf{S}} = \{\hat{\mathbf{s}}_1, ..., \hat{\mathbf{s}}_M\}$.

$$
\hat{\mathbf{y}}_i = \mathbf{A}\hat{\mathbf{s}}_i.
\tag{9.32}
$$

Here, those test trajectories which are very similar to the recovered trajectories can be regarded as normal trajectories.

$$
\frac{\|\mathbf{y}_i - \hat{\mathbf{y}}_i\|_2}{\|\mathbf{y}_i\|_2} < \tau \rightarrow \mathbf{y}_i \text{ is \textbf{normal}.}
\tag{9.33}
$$

Computational Complexity: The problem in Equation 9.31 can in fact be cast as a semidefinite program (SDP) [32]. This SDP can then be solved using a "custom" interior point method [33, 34] and has an *average* complexity of $O(N^2 TM)$ where N, T, M are as stated before.

On the other hand, Equation 9.10 (minimizing the l_0 norm) is well known to be an NP-hard problem. Thus, the l_1 norm is often used as an effective approximation to l_0. Several fast l_1-minimization algorithms have been published [35]. The homotopy method is among the most popular algorithms and has a computational complexity at the jth iteration as $O(jN^2 + jNT)$ [36]. Let J denote the number of iterations, the total complexity becomes $O\left(\sum_{j=1}^{J} j(N^2 + jNT)\right) = O(J^2 N^2 + J^2 NT)$. Here, J depends on the number of nonzero elements in $\boldsymbol{\alpha}$, so $O(J) = O(N)$. Therefore, the computational complexity of evaluating one test trajectory (event representation) using Equation 9.10 is $O(N^4 + N^3 T)$. Since there are M trajectories, the total computational complexity is $O(N^4 M + N^3 TM)$.

9.4.3 Experimental Setup and Results

In all subsequent experiments, object trajectories are used to represent events. The background subtraction model described in Section 9.2 is used to extract trajectories as collections of coordinate pairs $[x(t), y(t)]$. We approximate a raw trajectory using a basic B-spline function [37] with 50 knots (50 x- and y-coordinates), and these knots are finally used to form the trajectory feature vector.

9.4.3.1 Anomaly Detection in Structured Scenarios

For single-object anomalies, we test on CAVIAR data set [38] in Figure 9.9a and the Xerox Stop Sign video database—represented in Figure 9.9b. For multiple-object anomalies, we test on AVSS data set [39] —see Figure 9.10a, the Xerox Stop Sign data set—multi-object example in Figure 9.10b, and Xerox Intersection data set—see Figure 9.10c. We also compare our experimental results against three well-recognized techniques in trajectory-based anomalous event detection: (i) the recent sparsity-based technique of Li et al. [5], (ii) the approach of Piciarelli et al. [2] using one-class SVMs, (iii) the sparsity model with online dictionary learning of Zhao et al. [6], and (iv) the multiple-object tracking and rule-based anomaly detection technique of Han et al. [10].

9.4.3.2 Detection Rates for Single-Object Anomaly Detection

For the CAVIAR data set, we test on 27 video clips from which 170 trajectories are extracted. Our training dictionary consists of 10 normal trajectory classes and 3 anomalous trajectory classes; each class contains 10 different training trajectories. A total of 21 normal trajectories and 19 anomalous trajectories are used as independent test data.[1] Because training for anomalous events is well represented in this database, Equation 9.11 is used to classify the test trajectories as normal or anomalous. We note the benefits of sparsity-based anomaly detection in the form of improved detection rates of Li et al. and Zhao et al. over Piciarelli's approach. Second, Zhao et al. perform better than Li et al. because dictionaries are adapted in Zhao et al. [6] (Table 9.1).

For the Xerox Stop Sign data set, 118 trajectories from 39 video clips are extracted. The training dictionary comprises nine normal trajectory classes (containing eight trajectories each) and one anomalous trajectory class (containing four trajectories). Again, we have training for anomalies. So we use Equation 9.11 to classify a given test trajectory. An independent set of 34 normal trajectories and 8 anomalous trajectories are used to test our approach. Table 9.2 shows the confusion matrices of four approaches. Again, the benefits of classification using sparsity model versus SVM-based classifier are readily apparent.

9.4.3.3 Detection Rates for Multiple-Object Anomaly Detection

For multiple- (here two) object anomaly detection, we will compare against Han et al. [10],[2] Piciarelli et al. [2], and Zhao et al. [6]. Since the one-class SVM method in Piciarelli et al. and the approach of Zhao et al. are really proposed for single-object anomaly detection, we build two intuitively motivated extensions to evaluate its performance in the multiple- (here two) object setting:

1) If either of the trajectories corresponding to the two objects is an anomaly, the joint event is called anomalous. We denote these methods as Piciarelli et al._1 and Zhao et al._1, respectively.
2) If each of the two trajectories corresponding to the two objects is individually found to be anomalous, only then the joint event is anomalous. We denote these methods as Piciarelli et al._2 and Zhao et al._2, respectively.

To ensure the application of Picarelli et al. and Zhao et al., we separate every two-object event into two individual events with known individual class labels (normal or anomalous) so that we can obtain results for the aforementioned extensions, that is, Piciarelli et al._1, Piciarelli et al._2, Zhao et al._1, and Zhao et al._2.

---

1 All training and test trajectories both normal and anomalous are manually hand labeled. The training and test sets are complete nonoverlapping.
2 The predefined anomaly detection rules in Han et al. are based on underlying scenario.

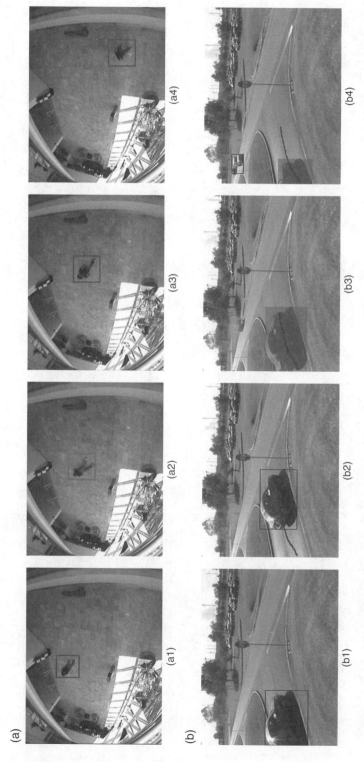

Figure 9.9 Example frames of single-object anomalies: (a) a man suddenly falls on floor—from the CAVIAR data set and (b) a driver backs his car in front of stop sign—from the Xerox Stop Sign data set.

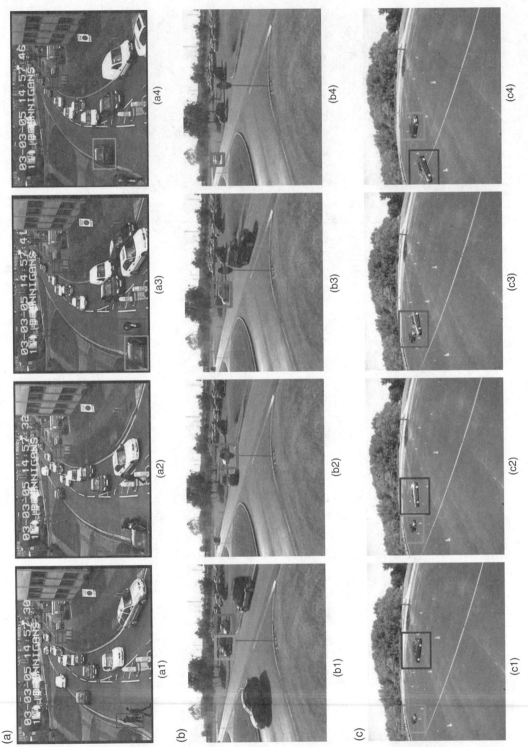

Figure 9.10 Example frames of multiple-object anomalies: (a) a vehicle almost hits a pedestrian—from the AVSS data set, (b) a car violates the stop sign rule—from the Xerox Stop Sign data set, and (c) a car fails to yield to oncoming car while turning left—from the Xerox Intersection data set.

Table 9.1 Confusion matrices of proposed and state-of-the -art trajectory-based methods on CAVIAR data set—single-object anomaly detection.

	Piciarelli et al. [2]		Li et al. [5]		Zhao et al. [6]	
	Normal (%)	Anomaly (%)	Normal (%)	Anomaly (%)	Normal (%)	Anomaly (%)
Normal	85.7	26.3	90.5	15.8	95.2	15.8
Anomaly	14.3	73.7	9.5	84.2	4.8	84.2

Table 9.2 Confusion matrices of proposed and state-of-the-art trajectory-based methods on the Xerox Stop Sign data set—single-object anomaly detection.

	Piciarelli et al. [2]		Li et al. [5]		Zhao et al. [6]	
	Normal (%)	Anomaly (%)	Normal (%)	Anomaly (%)	Normal (%)	Anomaly (%)
Normal	85.3	37.5	91.2	25.0	94.1	25.0
Anomaly	14.7	62.5	8.8	75.0	5.9	75.0

Table 9.3 Confusion matrices of proposed and state-of-the-art trajectory-based methods on the Xerox Stop Sign data set—multiple-object anomaly detection.

	Piciarelli et al._1 and Piciarelli et al._2	Zhao et al._1 and Zhao et al._2	Han et al. [10]	JSM
Detection rates	3/3, 1/3	2/3, 2/3	2/3	3/3

For the Xerox Stop Sign data set, we identify four different two-object normal event classes. Each class contains 15 training trajectory pairs. This database inherently contains three multiple-object anomalies, one of which is illustrated in Figure 9.10b with the actual video at http://youtu.be/tPR-LI3NmiMhttp://youtu.be/tPR-LI3NmiM. Using the outlier rejection in Equation 9.28, all three multiple-object anomalies are successfully detected by our proposed JSM. The detection rates of all methods are shown in Table 9.3.

For the AVSS data set, three different two-object normal event classes (containing 24 training trajectory pairs each) are chosen. The database was experimentally found to contain two different anomalies—corresponding videos can be seen at http://youtu.be/mU5R056zInc and http://youtu.be/jEzLkWF65Io. Our outlier rejection measure in Equation 9.28 was again successfully able to detect both these anomalies. Table 9.4 shows the detection rates of six methods.

In Xerox Intersection data, there are 91 trajectory pairs extracted from 13 video clips. We manually build our training dictionary into six different two-object normal event classes (containing six trajectory pairs each) and six different anomalous classes (containing four trajectory pairs each). A total of 17 normal trajectory pairs and 14 anomalous trajectory pairs are used for testing. The confusion matrices of our method against the three competing trajectory-based techniques are shown in Table 9.5.

It can be easily seen from Table 9.5 that the proposed JSM method leads to the best detection rates. The improvement over the techniques of Piciarelli et al. [2] and Zhao et al. [6] is expected since these

Table 9.4 Detection rates of proposed and state-of-the-art trajectory-based methods on AVSS data set—multiple-object anomaly detection.

	Piciarelli et al._1 and Piciarelli et al._2	Zhao et al._1 and Zhao et al._2	Han et al. [10]	JSM
Detection rates	2/2, 0/2	2/2, 1/2	2/2	2/2

Table 9.5 Confusion matrices of proposed and state-of-the-art trajectory-based methods on the Xerox Intersection data set—multiple-object anomaly detection.

	Normal (%)	Anomaly (%)	Normal (%)	Anomaly (%)
	Piciarelli et al._1		Piciarelli et al._2	
Normal	58.8	7.1	94.1	64.3
Anomaly	41.2	92.9	5.9	35.7
	Zhao et al._1		Zhao et al._2	
Normal	64.7	14.3	94.1	57.1
Anomaly	35.3	85.7	5.9	42.9
	Han et al. [10]		JSM	
Normal	82.4	35.7	88.2	14.3
Anomaly	17.6	64.3	11.8	85.7

techniques is really for single-object anomaly detection, and the extensions Piciarelli et al._1, Piciarelli et al._2, Zhao et al._1, and Zhao et al._2 will either strongly compromise detection or lead to high false alarm. In Han et al. [10], anomalies are detected using context-based rules on the result of multiple-object tracking. This puts an unreasonable burden on defining these rules and is often restrictive in practice, that is, not all anomalies can be anticipated. In the proposed joint sparsity model, interactions between distinct object trajectories are better captured and departures from expected "joint behavior" (particularly in the case training for anomalies is absent/limited) is employed which enables the improvement in detection rates.

9.4.3.4 Anomaly Detection in Unstructured Scenarios

For anomaly detection in unstructured scenarios, we work with the *Public Dataset of Traffic Video (PDTV)* [40] video data set. Figure 9.11a and b shows an example anomaly in the PDTV data, where a car fails to yield to an oncoming vehicle while turning left.

Comparison against State-of-the-Art Trajectory-Based Video Anomaly Detection: The afore-mentioned method for anomaly detection in unstructured scenarios will henceforth be referred to as LRSP. A comparison of LRSP is reported next against a widely cited method by Piciarelli et al. [2] which is based on trajectory extraction and one-class SVMs. For the experiment involving the PDTV data set, a training dictionary consisting of 319 normal event trajectories is obtained (no training corresponding to anomalous events was used). A total of 117 normal trajectories and 24 anomalous trajectories are used as independent test data. The confusion matrices of LRSP are compared with Piciarelli et al. [2] in Table 9.6. The benefits of LRSP are readily apparent.

Figure 9.11 (a and b) Example anomaly in PDTV data, (c and d) example anomaly in Xerox Stop Sign data set, and (e and f) example frames that show object occlusion.

Table 9.6 Confusion matrices of PDTV data set.

	LRSP		Piciarelli et al. [2]	
	Normal (%)	Anomaly (%)	Normal (%)	Anomaly (%)
Normal	78.6	37.5	70.9	41.7
Anomaly	21.4	62.5	29.1	58.3

Table 9.7 Confusion matrices of Stop Sign occluded data set.

	LRSP		Piciarelli et al. [2]	
	Normal (%)	Anomaly (%)	Normal (%)	Anomaly (%)
Normal	76.9	33.3	61.5	50.0
Anomaly	23.1	66.7	38.5	50.0

For the Xerox Stop Sign data set, the training dictionary comprises 72 normal trajectories. This experiment was conducted to specifically compare the performance of the two methods when object occlusion is involved, that is, *occluded trajectories* are used as our test data. Figure 9.11e and f shows an example where a car is occluded by another car (the video is available at http://youtu. be/4Azh2yZjA4o). An independent set of 13 *normal but occluded* trajectories and 6 *anomalous but occluded trajectories* are used to test the automated anomaly detection approaches. The confusion matrices of both methods—LRSP and the method in Picarelli et al. are reported in Table 9.7. In this case, LRSP is vastly better. This can be reasoned as follows: The optimization problem in Equation 9.31 is well conditioned. A set of occluded trajectories $\mathbf{Y}_o$ can be thought of as a perturbation on nonoccluded trajectories $\mathbf{Y}$; if $\| \mathbf{Y}_o - \mathbf{Y} \|_2$ is small enough, then by perturbation theory the solution $\hat{\mathbf{S}}$ under occlusion should only change slightly. This robustness of LRSP is a major practical benefit in real-world surveillance videos for example where noise and occlusion are typical.

Performance Variation with Regular Parameter λ: In our optimization problem in Equation 9.31, there is parameter λ which controls the relative importance of $\| \cdot \|_*$ and $\| \cdot \|_1$ terms. In Figure 9.12, we plot the detection rate curves against the value of λ for PDTV data. Figure 9.12 reveals that $\lambda \in [0.25, 0.75]$ leads to good performance. Both excessively low and high values of λ lead to a loss in performance. In particular, when λ is large, the cost function reduces largely to the $\| \cdot \|_1$ matrix norm and the performance drop is very significant. This emphasizes the value of the low-rank term which allows greater generality over row sparsity and can capture sparse matrix structures arising in real-world scenarios. Note the results of LRSP in Tables 9.6, 9.7, and 9.8 are reported using the "best" λ.

Computational Benefits and Trade-Off: We now compare LRSP against SSPs[3] of Li et al. [5], Mo et al. [11], and Zhao et al. [6]. Since SSP can only detect video anomaly in structured scenarios, the Xerox Stop Sign data set is used to test both LRSP and SSP.

---

3 Although methods in Li et al. [5], Zhao et al. [6], and Mo et al. [11] use varying event representations, their underlying sparsity model is the same. We use the abbreviation "ESP" to represent these three techniques.

Figure 9.12 Detection rates curves with respect to the value of λ.

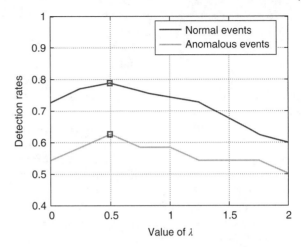

Table 9.8 Confusion matrices and execution times of Xerox Stop Sign data set.

Run time	LRSP		SSP	
	37 s		159 s	
	Normal (%)	Anomaly (%)	Normal (%)	Anomaly (%)
Normal	88.2	25.0	91.2	25.0
Anomaly	11.8	75.0	8.8	75.0

For the Xerox Stop Sign data set, the training dictionary contains nine normal event classes (containing eight trajectories each) and one anomalous trajectory class (containing four trajectories). An independent set of 34 normal trajectories and eight anomalous trajectories are used to test our approach. Table 9.8 shows the execution times of LRSP and SSP. We can see that the proposed LRSP method runs much faster than the SSP with a small loss in detection rates. This is expected because ESP has the benefit of prelabeled event classes. In the structured scenarios, the performance of SSP in fact serves as the practical upper bound for LRSP.

9.5 Conclusion and Future Research

The problem of anomaly detection is encountered in a broad array of applications wherein it is important to detect unusual patterns in data generated or acquired by machines. In this chapter, we have attempted to provide a comprehensive overview of techniques focusing on video captured in transportation settings. Various models are presented that place different requirements on the availability of labeled and structured data, so that the reader may select the appropriate methodology for a given application setting. In particular, a family of sparsity models are elaborated upon as a recent promising approach to the problem.

In general, supervised and structured techniques are likely to yield high anomaly detection performance; however, the requirement for data annotation may not always be practical. On the other hand, unsupervised and unstructured techniques attempt to tackle a more challenging problem, but would be practically more viable. We believe there is a lot of scope for further investigation in this second setting. While the detection results presented in Section 9.3 are promising, there is still significant room for improvement, especially for data captured in real-world scenarios with complex human and traffic behavior in the presence of many noise sources. A related direction warranting investigation is the ability to incrementally and automatically update an anomaly detection model trained for one time instance and geographic location to make accurate inferences in a different time/space instance where there may be a paucity of labeled data, at least initially. Recent progress in domain adaptation and online learning within the field of computer vision could be promisingly leveraged to address these challenges.

Most current methods define anomalies in a purely statistical sense. To the human, an anomaly can also bear semantic interpretation. In the Boston marathon bombings of 2013, a very specific type of human behavior and movement warranted special attention, driven purely by the semantic context of the attack. A systematic means to unify semantic rules with statistical models for anomaly detection would be an interesting and fruitful step forward.

Finally, many sophisticated anomaly detection models available today are computationally too prohibitive for real-time inference in practical scenarios. On the one hand, one can argue that Moore's law comes to the aid to accelerate hardware performance. However on the flip side, there is a trend to move more of the computing to the edge of a network, so that an increasing fraction of video processing and analysis takes place within the capture device in order to save network bandwidth. Furthermore, the capture devices themselves are becoming more portable and miniaturized. (For instance, at public events, a large fraction of video footage is captured by smartphone cameras.) There is thus an ongoing need to develop computationally tractable inference algorithms.

References

1 Saligrama V, Konrad J and Jodoin P 2010 Video anomaly identification. *IEEE Signal Processing Magazine* 27(5), 18–33.

2 Piciarelli C, Micheloni C and Foresti G 2008 Trajectory-based anomalous event detection. *IEEE Transactions on Circuits and Systems for Video Technology* 18(11), 1544–1554.

3 Xiang T and Gong S 2008 Video behavior profiling for anomaly detection. *IEEE Transactions on Pattern Analysis and Machine Intelligence* 30(5), 893–908.

4 Mo X, Monga V, Bala R and Fan Z 2014 Low rank sparsity prior for robust video anomaly detection. In *IEEE International Conference on Acoustics Speech and Signal Processing*, May 2014, Florence.

5 Li C, Han Z, Ye Q and Jiao J 2011 Abnormal behavior detection via sparse reconstruction analysis of trajectory. In *Proceedings of the IEEE Conference on Image and Graphics*, Hefei, Anhui, China, August 12–15, 2011, pp. 807–810.

6 Zhao B, Fei-Fei L and Xing E 2011 Online detection of unusual events in videos via dynamic sparse coding. In *Proceedings of the IEEE Conference on Computer Vision Pattern Recognition*, Colorado Springs, June 20–25, 2011, pp. 3313–3320.

7 Simon C, Meessen J and De Vleeschouwer C 2010 Visual event recognition using decision trees. *Multimedia Tools and Applications* 50(1), 95–121.

8 Vaswani N, Roy-Chowdhury A and Chellappa R 2005 Shape activity: a continuous-state hmm for moving/deforming shapes with application to abnormal activity detection. *IEEE Transactions on Image Processing* 14(10), 1603–1616.

9 Pruteanu-Malinici I and Carin L 2008 Infinite hidden Markov models for unusual-event detection in video. *IEEE Transactions on Image Processing* 17(5), 811–822.

10 Han M, Xu W, Tao H and Gong Y 2004 An algorithm for multiple object trajectory tracking. In *Proceedings of the IEEE Conference on Computer Vision Pattern Recognition*, Washington, DC, June 29–July 1, 2004, 1, 864–871.

11 Mo X, Monga V, Bala R and Fan Z 2014 Adaptive sparse representations for video anomaly detection. *IEEE Transactions on Circuits and Systems for Video Technology* 24(4), 631–645.

12 Yilmaz A, Javed O and Shah M 2006 Object tracking: a survey. *ACM Computing Surveys* 38(4), 1–45.

13 Stauffer C and Grimson W 2000 Learning patterns of activity using real-time tracking. *IEEE Transactions on Pattern Analysis and Machine Intelligence* 22(8), 747–757.

14 Laptev I 2005 On space–time interest points. *International Journal of Computer Vision* 64(2–3), 107–123.

15 Williams G, Tuytelaars T and Gool L 2008 An efficient dense and scale invariant spatio-temporal interest point detector. In *Proceedings of the IEEE European Conference on Computer Vision*, Marseille France, October 12–18, 2008, pp. 650–663.

16 Scovanner P, Ali S and Shah M 2007 A 3-dimensional sift descriptor and its application to action recognition. In *MM '07 Proceedings of the 15th ACM International Conference on Multimedia*, Augsburg, Germany, September 24–29, pp. 357–360.

17 Klasser A, Marszalek M and Schmid C 2008 A spatio-temporal descriptor based on 3D-gradients. In *BMVC 2008—19th British Machine Vision Conference*, September 2008, pp. 275:1–10, British Machine Vision Association, Leeds.

18 Dollar P, Rabaud V, Cottrell G and Belongie S 2005 Behavior recognition via sparse spatiotemporal features. In *Proceedings of the 2nd Joint IEEE International Workshop on Visual Surveillance and Performance Evaluation of Tracking and Surveillance (VS-PETS)*, October 15–16, 2005, pp. 65–72, IEEE, Department of Computer Science and Engineering, University of California, San Diego, La Jolla, CA.

19 Yeffet L and Wolf L 2009 Local trinary patterns for human action recognition. In *Proceedings of the IEEE Conference on Computer Vision*, Kyoto, Japan, September 29–October 2, 2009, pp. 492–497.

20 Dalal N and Triggs B 2005 Histogram of oriented gradients for human detection. In *Proceedings of the IEEE Conference on Computer Vision Pattern Recognition*, San Diego, June 20–26, 2005, 1, 886–893.

21 Lucas B and Kanade T 1981 An iterative image registration technique with an application to stereo vision. In *Proceedings of the 7th International Joint Conference on Artificial Intelligence, 1981*, Vancouver, BC, August 24–28, 1981, pp. 674–679.

22 Wang H, Klaser A, Schmid C and Liu CL 2013 Dense trajectories and motion boundary descriptors for action recognition. In *Proceedings of the IEEE Conference on Computer Vision Pattern Recognition*, Portland, June 23–28, 2013, pp. 3551–3558.

23 Utasi A and Czuni L 2008 HMM-based unusual motion detection without tracking. In *Proceedings of the IEEE Conference on Pattern Recognition*, Tampa, FL, December 8–11, 2008, pp. 1–4.

24 Wright J, Yang A, Ganesh A, Sastry S and Ma Y 2009 Robust face recognition via sparse representation. *IEEE Transactions on Pattern Analysis and Machine Intelligence* 31(2), 210–227.

25 Baraniuk R 2007 Compressive sensing [lecture notes]. *IEEE Signal Processing Magazine* 24(4), 118–121.

26 Mo X, Monga V, Bala R and Rodrguez-Serrano JA 2013 Practical methods for sparsity based video anomaly detection. In *IEEE International Conference on Intelligent Transportation Systems*, The Hague, the Netherlands, October 6–9, 2013, pp. 955–960.

27 Wang X, Ma X and Grimson W 2009 Unsupervised activity perception in crowded and complicated scenes using hierarchical Bayesian models. *IEEE Transactions on Pattern Analysis and Machine Intelligence* 31(3), 539–555.

28 Tropp J and Gilbert A 2007 Signal recovery from random measurements via orthogonal matching pursuit. *IEEE Transactions on Information Theory* 53(12), 4655–4666.

29 Srinivas U., Mousavi H., Monga V., Hattel A. and Jayarao B. 2014 SHIRC: simultaneous sparsity model for histopathological image representation and classification. *IEEE Transactions on Medical Imaging* 33(5), 1–17.

30 Zhang Z, Zha H and Simon H 2002 Low-rank approximations with sparse factors I: basic algorithms and error analysis. *SIAM Journal on Matrix Analysis and Applications* 23(3), 706–727.

31 Candès EJ, Li X, Ma Y and Wright J 2011 Robust principal component analysis? *Journal of the ACM* 58(3), 11:1–11:37.

32 Mo X, Monga V, Bala R and Fan Z 2014 Low Rank Sparsity Prior for Robust Video Anomaly Detection. In *Proceedings of the IEEE International Conference on Acoustics Speech and Signal Processing*, May 2014, Florence, Italy.

33 Chandrasekaran V, Sanghavi S, Parrilo P and Willsky A 2011 Rank-sparsity incoherence for matrix decomposition. *SIAM Journal on Optimization* 21(2), 572–596.

34 Liu Z and Vandenberghe L 2010 Interior-point method for nuclear norm approximation with application to system identification. *SIAM Journal on Matrix Analysis and Applications* 31(3), 1235–1256.

35 Yang A, Sastry S, Ganesh A and Ma Y 2010 Fast L1-minimization algorithms and an application in robust face recognition: a review. In *Proceedings of the IEEE Conference on Image Processing*, Hong Kong, September 2010, pp. 1849–1852.

36 Malioutov D, Cetin M and Willsky A 2005 Homotopy continuation for sparse signal representation. In *Proceedings of the IEEE Conference on Acoustics, Speech, and Signal Processing*, Philadelphia, March 2005, 5, 733–736.

37 Knott G 2000 *Interpolating Cubic Splines*, Birkhäuser, Basel.

38 CAVIAR Datasets. Caviar: EC Funded CAVIAR project/IST 2001 37540. http://homepages.inf.ed.ac.uk/rbf/CAVIAR/ (accessed October 17, 2016).

39 AVSS2007. AVSS: i-Lids dataset for AVSS 2007. 2007. http://www.eecs.qmul.ac.uk/~andrea/avss2007_d.html (accessed October 17, 2016).

40 Public Dataset of Traffic Video (PDTV). *Public Dataset of Traffic Video (PDTV)* n.d. http://www.tft.lth.se/video/co_operation/data_exchange/

Part II

Imaging from and within the Vehicle

10

Pedestrian Detection

Shashank Deshpande and Yang Cai

Carnegie Mellon University, Pittsburgh, PA, USA

10.1   Introduction

Pedestrian detection is critical to intelligent transportation systems, ranging from autonomous driving to infrastructure surveillance, traffic management, and transit safety and efficiency—even law enforcement. Pedestrian detection involves many types of sensors, such as closed-circuit television (CCTV) cameras, thermal imaging devices, near-infrared (NIR) imaging devices, and onboard RGB cameras. There is also a broad spectrum of pedestrian detection algorithms based on IR signatures, shape features, gradient features, machine learning, or motion features. There have been comprehensive reviews of pedestrian detection algorithms, for example, [1]. In this chapter, we intend not to duplicate prior reviews, but rather explore new avenues. We focus on the relationship between the various algorithms, recent progress made in affordable NIR sensors, and light detection and ranging (LiDAR), and camera fusion–based pedestrian detection algorithms.

10.2   Overview of the Algorithms

For decades, security, robotics, and defense communities have developed a forest of pedestrian detection algorithms. Each of them has its basic assumptions, effectiveness, and limitations. How does one see the forest above the individual trees? We begin with imaging devices and their related algorithms. Figure 10.1 shows a flowchart of the overall algorithm selection process. The first question is "Is the camera a thermal camera?" If it is a thermal camera, then we may simply use the intensity threshold of body temperature to detect humans. The second question is "Is it a depth imaging sensor such as a LiDAR?" If it is, then we can extract the depth and three-dimensional (3D) features for pedestrian detection. If the sensor is a visible RGB or monochrome NIR camera, then we ask the question "Does the camera move?" If the camera is stationary, then we can use a background subtraction method to extract the pedestrian shape. Otherwise, we must use either motion-based models such as optical flow, or shape-based methods such as histogram of oriented gradient (HOG) or generalized Hough transform (GHT). One could also use learning-based

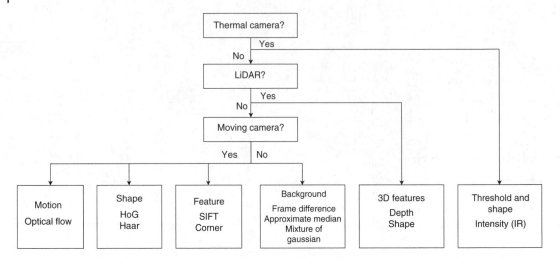

Figure 10.1 Overview of the pedestrian detection algorithms as described in Ref. [2].

methods such as Haar-like wavelet features, or feature point-based methods such as scale-independent feature transform (SIFT) or corners.

Autonomous driving pushes pedestrian detection algorithms to the limit because it requires responses to be in real time. For a pedestrian detection algorithm to be useful for an autonomous car, the system must be able to detect pedestrians as soon as one enters the field of view of the camera. This requires the capacity to detect deformable and noisy pedestrian shapes at an individual frame basis. Recently, LiDAR- and camera fusion-based algorithms have been promising in completing complex object classification tasks, which can also be applied to pedestrian detection. In light of this, we present a sensor fusion algorithm in detail along with an example.

10.3 Thermal Imaging

A pedestrian on the road can be observed as a source of image signals for two different imaging modalities: visible RGB camera (similarly, monochrome NIR) or thermal imaging camera. Pedestrian detection using thermal imagery usually involves a preprocessing stage, including the following elements: morphological operations to clean up the human shape (e.g., filling in holes), figure background segmentation to extract the human shape, and aspect ratio filtering to distinguish a human from another moving entity, such as a dog. After preprocessing, a feature descriptor is used. The most popular feature descriptors are HOGs [3] and the discriminatively trained deformable part model [4]. The final output stage involves a binary classifier: pedestrian or not. The support vector machine (SVM) [5] or AdaBoost-based classifiers [6] can be used here. Recently, various nighttime pedestrian detection systems have combined both thermal imagery and visual imagery, showing better detection rates [7, 8].

Thermal imaging–based systems are robust in detecting pedestrians at nighttime. However, they are noisy during the daytime, especially in hot environments. In the automotive industry, companies are beginning to equip vehicles with night vision systems for pedestrian detection. For example, the Night View Assist PLUS system from Mercedes-Benz can detect and recognize people and animal at a range of up to 160 m ahead of the vehicle.

10.4   Background Subtraction Methods

Given a set of video frames (assuming the camera is stationary), how do we extract the moving foreground and remove the static background? A binary blob is the simplest visual form of a shape description. In fact, the blob is so important that intelligence analysts often refer to video analytics as "blobology." Here, we introduce the following three basic background subtraction methods: frame subtraction, approximate median, and Gaussian mixture model (GMM).

To obtain binary blobs, we need to do some preprocessing. First, we reduce the resolution of the frames to speed up computation time, since it can be a lengthy process and it is not necessary to have high-resolution images for this task. As a rule of thumb, reducing the resolution by a factor of 5–10 is reasonable. Next, we convert the original image into a grayscale image if acquired by an RGB camera (as opposed to NIR, which is inherently grayscale). We then use an appropriate grayscale threshold to produce a binary image and fill in any interior holes left in the shape.

10.4.1   Frame Subtraction

Frame subtraction is the simplest way to extract a moving object from the video. Assuming the camera is fixed, p_i is the pixel value of the current frame and p_{i-1} is the pixel value of the previous frame. If their pixel difference is larger than the threshold T, then the pixel is the moving foreground. Otherwise, the pixel is the background.

$$| p_i - p_{i-1} | > T$$

The results of a background subtraction are binary, creating a binary image, wherein foreground figures are blobs as illustrated in Figure 10.2.

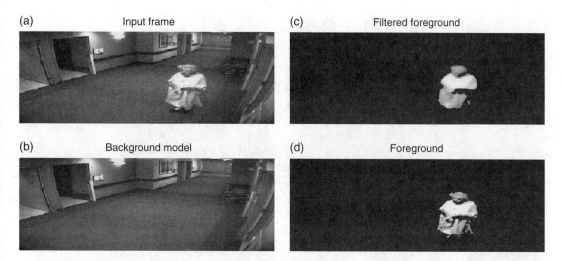

Figure 10.2 Background subtraction: (a) original video, (b) background, (c) filtered foreground, and (d) foreground. We can see that the extracted foreground figure has shadow attached because the shadow is also a moving object.

The pseudo code for frame subtraction is as follows:

```
read first_frame
prev_frame = first_frame
while (frames remaining)
current_frame = read next_frame
        for (each pixel P_i)
                delta = | current_frame - prev_frame |
                if(delta[P_i] > threshold)
                                foreground_frame[P_i] = current_frame[P_i]
                elseif (delta[P_i] < threshold)
                    foreground_frame[P_i] = 0
                end
        end
        prev_frame = current_frame
        filter foreground_frame
end
```

However, frame subtraction is sensitive to noise in the video, and is very sensitive to the threshold T. In addition, it is sensitive to the speed of the foreground objects and the camera frame rate. If the object was moving too fast or if the frame rate was too low, then results would be poor.

10.4.2 Approximate Median

The approximate median algorithm makes a more advanced assumption on the image background. Using the approximate median algorithm, we assume the background is somewhat consistent over the short term, but could still change over longer periods of time. This assumption helps prevent short-term noise, while still being able to accommodate a dynamic scene. This algorithm is called "approximate median" because, rather than storing a large number of frames in memory, we simply add or subtract 1 when the scene changes, which is much less computationally expensive, and produces similar results. Figure 10.3 shows an example of changes in pixel values occurring due to the passing man in a video sequence.

Figure 10.3 Pixels of a video sequence processed for background/foreground detection, capturing changes occurring due to the passing man.

Note that long-term changes, such as the changes in shadows over time, result in pixel value changes that are below the threshold. These changes are integrated into the median background value without ever entering the foreground scene.

The pseudo code for the approximate median algorithm is as follows:

```
read first frame
median_frame = first frame     % Assume the first frame is the median
while (frames remaining)
        current_frame = read next_frame
        for (every pixel P_i)
                delta = | current_frame - median_frame|
                if (current_frame[P_i] > median_frame[P_i])
                        median_frame[P_i] + 1
                else
                        median_frame[P_i] - 1
                end
                if (delta > threshold)
                        foreground_frame[P_i] = current_frame[P_i]
                else
                        foreground_frame[P_i] = 0
                end
        end
end
```

10.4.3 Gaussian Mixture Model

The GMM can be expressed as follows:

$$F(x) = \sum_{i=1}^{k} w_i \cdot e^{\frac{-(x-\mu)^2}{\sigma^2}}$$

Here, u is the mean of the background pixels values in each Gaussian function, w_i is the weight, and σ is the standard deviation of each component (where higher weight and lower σ mean higher confidence). There are typically three to five Gaussian components per pixel. Unfortunately, GMM needs more frames for training, and it is more computationally expensive than the two aforementioned methods. According to our test results, GMM does not generate better segmentation results than the approximate median method. Therefore, GMM is not recommended as the first choice. Approximate median often generates decent foreground segmentation for outdoor and indoor videos. It performs better than the frame subtraction and GMM methods in many cases [3].

10.5 Polar Coordinate Profile

Contours are the edges along the exterior of an object that can be obtained from binary blobs. Usually, blobs contain holes and noisy edges, and small spurious blobs may also be present in the image. To remove holes in the blobs, we can use morphological closing, which is a morphological dilation followed by an erosion. To remove small spurious blobs, we can use morphological opening, which is a morphological erosion followed by a dilation. Several rounds of opening and closing with different structuring elements may be needed to obtain a clean blob. Note that these

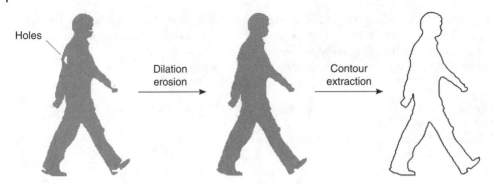

Figure 10.4 An example of the morphological operations to fill the holes in the object for better contour extraction.

morphological operations do change the blob shape slightly, especially the fine details of the perimeter as illustrated in Figure 10.4.

The main advantage of the polar coordinate profile representation method is that it can transform a 2D feature space to a 1D feature space, which simplifies many problems. However, it may create other problems such as modifying the shape of the object slightly. Here, we apply it to represent human forms as the contour of a blob. The polar coordinate profile is useful in video analytics for multiple reasons: (i) scale invariance—size does not matter; (ii) rotation invariance—any rotated version of the contour should have no impact on the recognition; and (iii) shift invariance—the location of the shape within an image should not affect performance. The polar profile is also called the centroid contour distance curve (CCDC).

How do we find the centroid? In many cases, we can draw a rectangular bounding box around the shape and place the centroid at the center of the bounding box. If we have a prior knowledge of the shape, then we may use that knowledge to locate the centroid. An example of this would be placing the centroid near the navel of a human blob. Next, we can plot the polar coordinate profile. Each point along the contour is defined by its distance from the centroid, r, and its angle around the boundary from some arbitrary reference, θ. By default, we normally begin at 0° in the clockwise direction. We may also normalize the boundary to [1,0] range for scale invariance. Human shapes are very complex and diverse. Therefore, for some values of θ, there may exist a number of different values of r. A single-valued polar profile can be generated by choosing the minimum, maximum, or average of the different values of r. Figure 10.5 shows a 2D shape and its polar coordinate profile (r, θ) plot.

How can we match polar coordinate profiles extracted from a scene with those from a training set? The matching process can be done by sliding the test profile along the candidate profiles and finding a significant fit. Given a polar coordinate profile, we have at least three methods for matching it with known profile templates: nearest neighbor, dynamic time warping (DTW), and Fourier descriptors (FDs). Of the three methods mentioned here, nearest neighbor is the simplest. For θ from 0 to 360°, we summarize the Euclidian distance between the template profile and the polar coordinate profile in terms of r. The profile with the minimal distance is a match. DTW incorporates feature point alignment during the matching process. DTW aligns peaks to peaks and valleys to valleys. As we discussed in Chapter 6, DTW is computationally expensive. However, with some constraints, such as search window, slope, ending point, and monotonicity, approximate solutions can be obtained within a reasonable amount of time. FDs convert the 1D polar coordinate profile into a Fourier series. We can use fast Fourier transform (FFT) to make the conversion. After the FFT, the contour

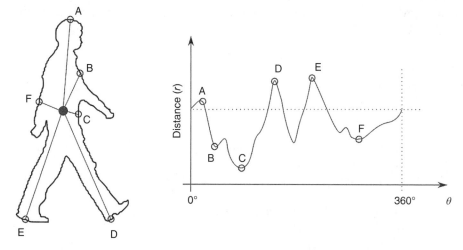

Figure 10.5 A 2D shape and its polar coordinate profile (r,θ) plot. Here, min r values are chosen.

is represented by an infinite series of Fourier coefficients. What is the physical meaning of these coefficients? The high-frequency terms describe microscopic details in the profile, while the low-frequency terms describe the metaphor, or macroscopic shape of the profile without the details. In light of this, we can use a low-pass filter to remove the curvy details while keeping the high-level description of the profile. According to Zahn et al., the first 10 Fourier coefficients should be enough to represent many 2D shapes for problems such as recognizing handwritten numbers.

10.6 Image-Based Features

10.6.1 Histogram of Oriented Gradients

HOGs is a shape feature–based pedestrian detection algorithm that works on individual images. Assuming that we divide an image into small grids and calculate the maximal gradient and its orientation for each cell, we may find that the oriented gradients roughly follow the contour of a pedestrian. For example, if the area of a pedestrian is darker than the background, then the HOGs would portray a rough contour of a human shape with their normal vectors pointed toward the outward direction. Conversely, if the area of a pedestrian is lighter than the background, then the HOGs would also constitute a human shape with their normal vectors pointed toward the inward direction. This assumes that the intensity of a pedestrian shape and the background are relatively uniform, which is often true in a transportation system environment, such as a highway, tunnel, street intersection, and so on. The HOG descriptor [3] was first introduced by Dalal and Triggs. A typical HOG descriptor is as follows:

1) Divide m cell grids (8×8 pixels) in a detection window.
2) Compute gradients for each pixel.

$$\nabla I = \left[\frac{\partial I}{\partial x} \cdot \frac{\partial I}{\partial y}\right]^{T} \quad \| \nabla I \| = \sqrt{\left(\frac{\partial I}{\partial x}\right)^{2} + \left(\frac{\partial I}{\partial y}\right)^{2}} \quad \theta = \tan^{-1}\left(\frac{\partial I}{\partial x} \bigg/ \frac{\partial I}{\partial y}\right)$$

3) Create a weighted HOGs for each cell (e.g., use nine bins from 0 to 180°).

4) Normalize contrast, given a small constant ε, let v be the nonnormalized vector containing all histograms in a given block.

$$v_i \rightarrow v_i / \sqrt{\| v_i \|^2 + \varepsilon^2}$$

5) Make a feature vector.

$$v = [v_1, v_2, \ldots]$$

HOG is in fact a learning algorithm. In order to recognize a pedestrian, we must first train the model with a classifier, for example, k-nearest neighbor or SVM. We also need to crop pedestrian images and nonpedestrian images to create a training sample set. We want the algorithm to learn what kind of HOG feature vectors constitute a pedestrian shape and which ones do not. Once the algorithm has been trained, it can be used to detect pedestrians by sliding the detection window from the top left of the image to the bottom right corner row after row. If we do not know the dimension of the pedestrians in the image, then HOG is typical run many times with different sizes of the detection window, which results in slowing of the detection speed.

Besides scalability, HOG has other limitations and shortcomings. For example, HOG typically fails when there are occlusions between pedestrians. When the shapes of pedestrians overlap each other, the occluded shape may be dramatically different from the known templates of a pedestrian shape. Also, the performance of HOG suffers when there are both illumination variation on the pedestrian and shadows cast by the pedestrian. Other challenges faced by pedestrian detection algorithms include deformation of human shapes (e.g., walking), unconstrained backgrounds, complex illumination patterns, and a variety of clothing. The deformation problem can be solved by applying a deformable parts model (DPM). Occlusion handling was a serious roadblock for pedestrian detection algorithms until the advent of deep feature learning for occlusion handling [9]. We discuss these two major problems in sections 10.6.2 and 10.6.3.

10.6.2 Deformable Parts Model

Many variations and improvements over the original HOG descriptor for pedestrian detection have been proposed. One key advancement is the discriminatively trained deformable part based model (latent SVM). While there are various flavors, there are three core components comprising a parts-based model for pedestrian detection and object detection in general. The DPM technique is known to be computationally expensive, and takes much more time compared to other global feature representation-based detection methods. Recent methods [10, 11] using DPM achieving real-time performance have been reported.

10.6.3 LiDAR and Camera Fusion–Based Detection

LiDAR or laser imaging is a critical device for autonomous or semiautomatic vehicles to detect pedestrians on the road. Pedestrian detection by fusing LiDAR and camera data is perhaps one of the most heavily explored multisensor fusion techniques in the past decade. The purpose of such a fusion is to exploit the redundancy and complementary characteristics of the two sensors for improving the reliability and accuracy of the detection system. The aim of research in this field has been to explore how much valuable information can be extracted from both LiDAR sensors and cameras for pedestrian detection and the ways to combine the information that can lead to better classifier accuracy. Here, we outline the approach for LiDAR and camera fusion for pedestrian detection, discuss the most recent state-of-the-art algorithms, and look at a simple implementation that demonstrates the strengths of such fusion.

The description of the outline (shown in the Figure 10.6) for the general flow followed by the algorithms employed in pedestrian detection by fusing LiDAR and vision-based features has been studied

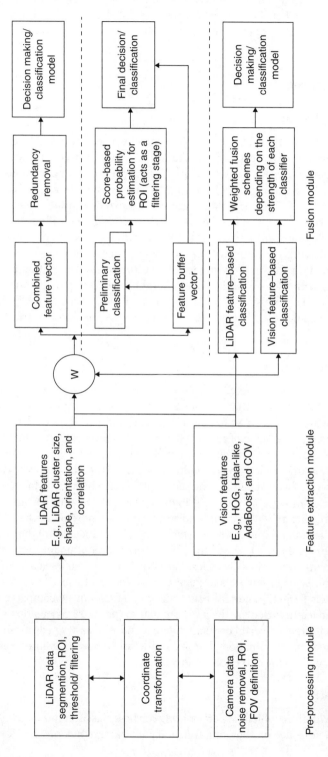

Figure 10.6 Overview of the fusion-based pedestrian-based methods.

very closely in Ref. [12]. Fusing LiDAR and camera data not only helps in improving the classification accuracy, but it can also be beneficial in extracting more information like distance of the pedestrian from the vehicle, time to impact, approximating the intent of the pedestrian, and so on.

10.7 LiDAR Features

10.7.1 Preprocessing Module

For the scan of the scene delivered by the LiDAR sensor, we need to discuss several preprocessing steps prior to discussing extraction of feature vectors and subsequent pedestrian detection/ classification. The general tasks performed in LiDAR preprocessing include data prefiltering, coordinate transformation, data point clustering, and region-of-interest (ROI) extraction. Before the data prefiltering stage, LiDAR data points are synchronized with the camera input. This step is a key to ensure consistency in LiDAR and camera inputs. The data prefiltering smoothes the noisy LiDAR data points which can occur due to abnormal surfaces, error in data collection, and so on. Once, the data prefiltering is performed, LiDAR data generally undergoes a polar-to-Cartesian coordinate transformation for projection and efficient understanding of data point distribution. Data point clustering with ROI extraction is a method for clustering data points corresponding to each object and to likely pedestrians. Similarly, camera data preprocessing also involves noise removal, image enhancement, definition of field of view (FoV) and ROI extraction corresponding to likely pedestrians.

10.7.2 Feature Extraction Module

LiDAR-based features can be extracted both in polar and Cartesian coordinate systems. Extensive study of different LiDAR features that can be extracted and the implications of those features for pedestrian or people detection can be found in Refs. [13] and [14]. LiDAR data strongly brings out the height, width, and relative shape of an object. Thus, the typical extracted LiDAR features revolve around the statistical characters of the distribution of LiDAR data points such as standard deviation, circularity, and mean boundary length and width. Since LiDAR data accurately estimates the longitudinal distance between the obstacle and the sensor, it has also been used for speed-based features. For example, obstacles moving at relatively high speeds can be filtered out as nonpedestrians. Although, the features are relatively simple to extract and comprehend, their distribution corresponding to pedestrian and nonpedestrian obstacles have a significant difference.

Vision-based features are more inclined toward capturing predominant shape structures that can differentiate pedestrians from other obstacles. Variations of HOG, deformable HOG, and Haar-like features have held the place of favorites in pedestrian detection for over a decade of study. Recent surveys such as in Refs. [15–17] and [18] have enlisted and effectively compared the advantages of different vision-based features that are the most popular for pedestrian detection algorithms. Vision-based features are definitely powerful enough to detect and/or classify pedestrians, but the environmental limitations such as illumination changes and weather have prevented vision-based models from achieving ideal pedestrian detection performance.

10.7.3 Fusion Module

This module deals with the problem of determining the best classifier model combination to ensure a good exploitation scheme for the complimentary characteristics of LiDAR- and vision-based features. Extraction of meaningful and useful information from the fusion of heterogeneous data is not straightforward. We describe three ways that fusion may be performed:

1) The features extracted both from vision and LiDAR data are combined together to form a single feature vector per data instance. The process of combining the two vectors can itself have a smart method such as a weighted combination. For example, if the environmental conditions favor vision-based features (clear sunny day, etc.), then we can assume a higher weight for vision features. This combined feature vector undergoes a processing stage where redundant features are removed. Dimensionality reduction techniques such as linear discriminant analysis (LDA) or principal component analysis (PCA) are generally used to reduce complexity by feature analysis. Classifiers like SVM, support vector regression (SVR), or GMM then map the N-dimensional feature vector in the feature space to predicted decisions.

2) First, the detection model is built using either vision- or LiDAR-based features. The purpose of this model is to act as a filtering stage which detects high probability windows that are more likely ROI (containing pedestrians) segments. Thus, this high-level filtering ensures that the more dominant feature vector, either LiDAR- or vision-based, has reduced false positives. For example, if LiDAR features are extracted to identify median boundary length, then the classification model can confuse a pole as a pedestrian. In this case, vision-based features can effectively filter out poles as non-ROIs. A scheme such as this explicitly exploits the complimentary properties of LiDAR and vision features.

3) LiDAR and vision features are used to separately construct two classification models. These classification models yield a probability of confidence of detection. A fusion strategy combines the information provided by each classifier to provide a final decision. The fusion strategy also takes into account the environmental conditions to weigh the likelihood of the probability that is output by each classifier. Such a scheme treats both LiDAR- and vision-based classification as two separate segments until the point where their decision outputs are fused. This late-fusion approach enables tweaking feature extraction or classification stages in each segment without concern about the impact on the other segment.

Few recent studies have explored pedestrian detection models solely based on LiDAR sensor data. Zhao et al. [19] describe an efficient way for object modeling and feature extraction based on the appearance of the LiDAR data stream. The strength of the method lies in the ability to differentiate a single pedestrian from a group of pedestrians. The paper reports an accuracy of 93% for pedestrian detection. Oliveira and Nunes [20] propose a featureless approach for detecting objects using LiDAR. By treating partial segmentation, the method is able to efficiently deal with the problem of object occlusion, which introduces a large error when dealing with LiDAR features. This approach introduces a flexible pedestrian detector model which can prove to be a very useful technique in highly dynamic environments. Whereas, Premebida et al. [13] describe the strengths of different LiDAR features extracted by comparing the receiver operating characteristic (ROC) plots and calculating their area under the curve (AUC). The paper also compares different classification methods and their impacts. This work serves as an effective one stop site for understanding the classical LiDAR feature extraction and classification based pedestrian detection model.

As described in Ref. [12], the integration of a LiDAR sensor and a camera, to exploit the redundant and complementary properties of the two data types for improving the reliability and accuracy of pedestrian detection algorithms, has gained the attention of researchers in the field of intelligent vehicles (IVs), driver assistance systems, and mobile robotics in the past few years. Although the multisensor fusion approach promises a better detection model, it is a challenging task. Identifying the underlying symbiotic properties of LiDAR and camera data, and using an appropriate fusion scheme to exploit the best of both techniques, gives rise to a large set of possible combinations. Exploring these possibilities to develop the most useful application-specific algorithm has been the crux of the research for the past decade. Although a significant amount of effort and time has gone into exploring the best possible fusion schemes, there are many topics in this field that are not fully exploited. Recent approaches have started to examine fusing more sensors such as radar, inertial measurement unit (IMU), and GPS in the detection model to study the impact on the accuracy and robustness of the fusion scheme.

Table 10.1 surveys some interesting recent contributions to pedestrian detection using LiDAR and visible-spectrum camera fusion approach.

In Section 10.7.4, we describe a simple implementation of LiDAR and camera fusion for pedestrian detection. The intent of this demonstration is to show the possibility of such a fusion scheme, discuss the complementary properties of these sensors, and highlight their impact on the accuracy of pedestrian detection. We first discuss the dataset that we have used and the overview algorithm. For our experiments, we have made use of laser and image pedestrian detection (LIPD) dataset in urban environment, available online at http://home.isr.uc.pt/~cpremebida/dataset.

10.7.4 LIPD Dataset

The LIPD dataset was collected in the Coimbra University/ISR Campus zone. A Yamaha-ISR electric vehicle (ISRobotCar), equipped with a multilayer automotive laser scanner Alasca-XT from Ibeo; a TopCon HyperPro GPS device in RTK mode; one IMU Mti from Xsens; and a monocular Guppy camera were used for the data collection. Table 10.2 lists the specification of LiDAR and camera sensors used for data collection.

The details for system calibration and data collection setup are explained in great detail in Ref. [26]. The camera intrinsic parameters calculated using Ref. [27] have been listed in Table 10.3.

The collected LiDAR data needs to be transformed into Cartesian coordinates for projection onto the image plane. The transformation matrix for perspective projection (calculation discussed in Ref. [26]) is as follows:

$$\text{Transformation matrix} = \begin{bmatrix} 0.999 & -0.014 & -0.009 & 11.92 \\ 0.014 & 0.999 & 0.026 & -161.26 \\ 0.009 & -0.027 & 0.999 & 0.78 \\ 0 & 0 & 0 & 1 \end{bmatrix}$$

The LIPD dataset contains classification and detection datasets. For the purpose of this implementation, we have used only the detection dataset. In total, 14,367 raw scans in the laser training subset of the detection dataset were used to study the threshold for FOV and object labeling, and 19,604 training images from the image training subset were used for training HOG features.

10.7.5 Overview of the Algorithm

Figure 10.7 illustrates the flow of the algorithm. The LIPD dataset provides a timestamp for synchronization of LiDAR and camera data. In the LiDAR module, the data point distribution is first used to further define the FOV. This is done simply by defining a threshold for the permissible values. The LiDAR data points in the FOV are segmented by a clustering method to extract the boundaries of different obstacles. The segments are labeled as pedestrian-like and nonpedestrian obstacles by inspecting the shape of the LiDAR data point distribution in each obstacle segment. The segments

Table 10.1 Fusion-based method.

Reference	LiDAR features	Vision features	Classification model	Comments
[21]	• Number of points • Standard deviation • Mean average deviation • Width/height • Linearity • Circularity and radius • Boundary length • Mean curvature • Mean angular difference • Kurtosis • PCA-based shape factors • N-binned histogram • Boundary regularity	HOG features	C-SVM for both light detection and ranging (LiDAR) and vision features. Conditional probability is defined for fusing the detection probability of each module for final decision.	In the proposed method, the object's position is detected by the LiDAR (structure information), and the vision-based system (appearance information) classifies pedestrians. A Bayesian modeling approach is used for developing the fusion rule. The resulting human detection consists in rich information that takes into account the distance of the cluster and the confidence level of both detection methods.
[22]	• LiDAR data point projection–based clusters are subjected to thresholding to obtain the ROI. • Shape of the LiDAR projection cluster. • Speed estimated by the mean of the Kalman filter with a constant velocity model.	Haar-like features	AdaBoost classifier	The Bayesian formalism offers the possibility to integrate correlated data in the formalism as soon as the conditional likelihoods between a current observation and the previous are modeled. Thus, the model is more robust. The estimated speed and the vision score–based likelihood are fused in a Bayesian model using autoregressive (AR) formalism.
[23]	• Shape of the LiDAR projection cluster. This is estimated using poly line approximation and line merging.	Haar-like features	Obstacle classification is performed based on the LiDAR projection shape. AdaBoost classification is used for trained Haar-like features.	LiDAR data distribution is used for identifying different obstacles. Obstacles are classified into pedestrians and other obstacles based on the shape of the LiDAR point distribution. The paper also differentiates between static and moving objects to improve the accuracy of object classification. The obstacles classified as pedestrians or pedestrian-like are the ROIs. An AdaBoost classifier trained on the Haar-like features makes the final decision.
[24]	• A course to fine segmentation employing β skeleton method, provides segments (partial and/or full segments). • Segments are then labeled, and scored by a Procrustes analysis.	• HOG • Local receptive features	SVM and MLP classifiers, fuzzy integral–based fusion model.	The region of interest (ROI) is found by looking at the LiDAR data distribution and an image classifier is used to label the projected ROI. The proposed system deals with partial segments; it is able to recover depth information even if the LiDAR-based segmentation of the entire obstacle fails, and the integration is modeled through contextual information. The fusion approach is based on the "semantic" information.
[25]	Using an extended Kalman filter, local tracking techniques are used in LiDAR and the radar system is fused to form a global tracking approach.	Gabor features	Maximum-likelihood estimation to fuse multisensor tracks	The paper describes a fusion strategy using LiDAR and radar information for on-road object detection with emphasis on a local and global tracking approach.

Table 10.2 Sensor specifications.

Sensor type	Specifications
LiDAR	Lbeo (Alasca XT model)
	Vertical res: four layers [−1.6; −0.8; 0.8; 1.6°]
	Frequency: 12.5 Hz
	Range: 0.3–200 m
	Used field of view: 120°
Camera	FireWire Allied (Guppy)
	Sensor: CCD (Bayer color format)
	Lens: 6.5 mm (CS-mount)
	Resolution: 640 × 480 (30f/s)

Table 10.3 Intrinsic camera parameters.

Focal length	f_c = |623.732; 624.595|
Principal point	cc = |360.876; 256.895|
Distortion coefficients	K_c = |−0.304; 0.106; −0.00032; −0.000027; 0.0|

labeled as pedestrian-like are projected onto the image plane. The region in the image plane where projected LiDAR data points are present is considered the filtered ROI.

On the ROI in the image plane labeled by the LiDAR module, HOG features are extracted for testing. HOG features are extracted from the training set for training the SVM classifier. The classifier model maps the testing HOG features to a decision label for final classification. LiDAR range data for the true label, that is, pedestrian segment, is then projected to obtain the distance for impact information. This fusion scheme successfully reduces the false positives generally detected in LiDAR-based techniques and false negatives in vision-based approaches, thereby increasing the overall detection accuracy of a pedestrian detection system. In Sections 10.7.6 and 10.7.7, we explore in detail both LiDAR and vision modules. We also discuss the results and advantages of a fusion scheme.

10.7.6 LiDAR Module

We follow four stages in the LiDAR module for ROI extraction: (i) preprocessing, (ii) thresholding, (iii) Segmentation, and (iv) ROI labeling, as shown in Figure 10.8.

1) Preprocessing: The preprocessing stage is responsible for prefiltering and coordinate transformation. As described in Ref. [26], the acquired raw LiDAR is defined in $\mathbb{R}^3$ space. Let's denote a single range point i from the laser data points as p_i defined by $(r_i, \alpha_i, \Theta_i)$. This can be visualized in the Cartesian plane as shown in Figure 10.9. Here, $\Theta_i \in \{-1.6°, -0.8°, 0.8°, 1.6°\}$ represents four vertical layers of the LiDAR sensor. The (x_i, y_i, z_i) *coordinates* for projection can be computed using $(r_i, \alpha_i, \Theta_i)$ values per laser data point. Figure 10.10a shows the LiDAR scan for the sample image shown in Figure 10.10b. LiDAR scans include information of the surroundings that might not be visible to the camera. Hence, we first filter out data points that are out of the scope of image plane as these data are irrelevant for processing.

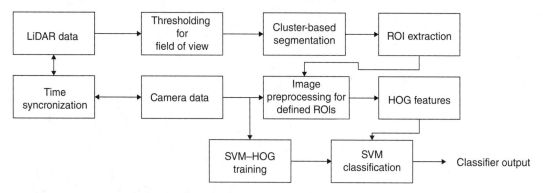

Figure 10.7 Overview of the algorithm flow followed in our implementation.

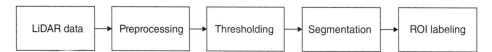

Figure 10.8 Stages for LiDAR module.

Figure 10.9 Cartesian and polar coordinate representation of a point p_i.

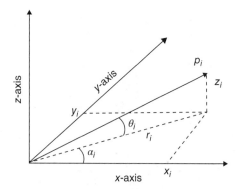

2) Thresholding: The thresholding stage is responsible for defining the FOV. This step needs to be used as required by the given application. We define the FOV considering hints such as pedestrians 20 m or further are not an immediate threat. Hence, the FOV is defined between 1 and 20 m from the LiDAR sensor. Figure 10.10c shows a sample FOV described by thresholding.

3) Segmentation: The segmentation stage is responsible for clustering of data points that share similar properties. Reference [28] presents different clustering and segmentation methods and their comparison in great detail. For our implementation, we have used *jump distance clustering* for segmentation. Jump distance clustering is a widely used method for 2D laser range data in mobile robotics (see Ref. [28] for an overview). It is fast and relatively simple to implement. The basic idea here is that if the Euclidean distance between two adjacent data points exceeds a given threshold, a new segment is generated. All the data points with Euclidean distance below the threshold are grouped together into a segment. Figure 10.10d shows the result of such a segmentation method.

4) ROI labeling: Once different segments are clustered together, we examine the distribution within each segment. Segments corresponding to pedestrian-like objects have a linear, tightly held distribution, whereas, segments corresponding to vehicles or buildings have a broader

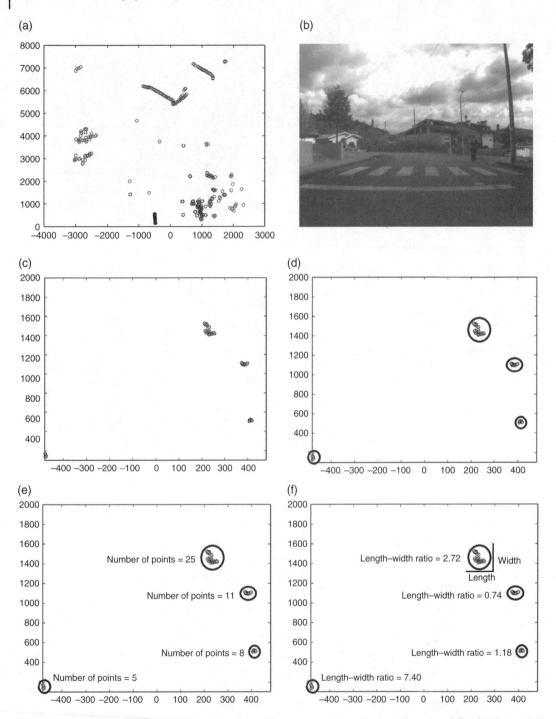

Figure 10.10 (a) LiDAR range points in Cartesian plane, all layers grouped together (*x*- and *y*-axis as defined in Figure 10.9); (b) the image corresponding to the LiDAR scan (*x*- and *y*-axis corresponding to image plane); (c) range points outside a threshold (FOV) in *x*- and *y*-axis are discarded; (d) segmentation using the jump distance clustering method; (e) number of data points per segment; (f) the length-to-width ratio calculation of each segment; (g) segment labels; (h) ROI in LiDAR plane; (i) corresponding ROI in image plane; and (j) bounding box displaying the ROI in the image.

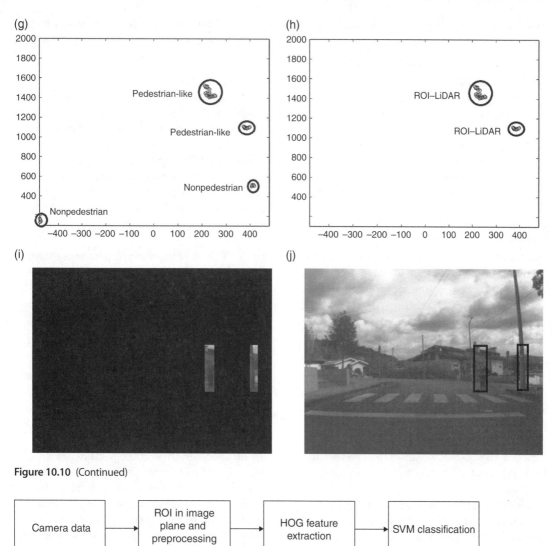

Figure 10.10 (Continued)

Figure 10.11 Stages for vision module.

distribution. We define a threshold for the number of data points a pedestrian-like segment should contain (Figure 10.10e). Among these segments, we calculate the length-to-width ratio of the LiDAR data distribution as can be seen in Figure 10.10f, and we label segments as pedestrian-like if the width-to-length ratio falls within a predefined range that is learned from the training data. The window defined by these labeled segments are the ROIs in the image plane (Figure 10.10g–j).

10.7.7 Vision Module

The segments labeled by the LiDAR module as pedestrian-like are mapped to the image plane to define the ROI for the vision module. Figure 10.11 provides a block diagram representation of the steps performed in the vision module. As a preprocessing step, we convert the ROI into a grayscale image and perform histogram equalization for contrast enhancement.

a) HOG feature extraction

HOGs for human detection was proposed by Dalal and Triggs [3] in 2005. HOG is a set of features extracted from a dense and overlapping grid of normalized local histograms of image gradient orientations over image windows. To compose HOG, the cell histograms of each pixel within the cell cast a weighted vote, according to the gradient L2 norm, for an orientation-based histogram channel. Rectangular cell HOG features are computed from a $g \times g$ grid, which defines the number of cells in the detection window. We have used a 3×3 grid in our implementation. Each cell contains λ bins of the oriented histograms ($\lambda = 9$ in our implementation). Reference [29] describes in detail the selection of the grid size and the number of bins, and their impact on HOG features for human detection. The nine histograms with nine bins were then concatenated to make an 81-dimensional feature vector.

b) SVM classification

SVMs are based on the statistical theory of learning. SVM is very sensitive to parameter tuning. In our implementation, we have used a radial basis function (RBF) kernel with margin parameter $C = 3000$ and a scaling factor (sigma) = 0.1 for tuning the SVM training model. We trained the model for the HOG features extracted on the training set in the LIPD dataset. Figure 10.12 shows samples of positive and negative training data.

10.7.8 Results and Discussion

The proposed sample implementation was evaluated in terms of accuracy and ROC curves, discuss the performance of each module (LiDAR and vision) separately, and compare their performance against the fusion scheme.

10.7.8.1 LiDAR Module

Figure 10.13 shows some of the pedestrian-like and nonpedestrian segments classified by the LiDAR module. A total of 27,309 segments were observed in the FOV for 4823 test image frames (5–9 segments per frame). Out of the 3417 actual pedestrian segments, 3354 were classified as pedestrian-like; whereas out of 23,892 nonpedestrian segments, 4783 segments were classified as pedestrian-like. Table 10.4 summarizes these numbers in the form of a confusion matrix. Since the interest in this stage was to filter out the segments that definitely don't have pedestrians, we observe from the confusion matrix that the true negative rate is negligible, whereas the false positive rate is higher due to the detection of pedestrian-like segments.

10.7.8.2 Vision Module

While dealing independently with the vision module, within a predefined FOV, we consider a sliding window of 80×160 pixels as shown in Figure 10.14. We consider a $20\,pixels$ horizontal motion per iteration for the sliding window.

Figure 10.15 shows several of the windows classified as pedestrians using the HOG–SVM classifier. A total of 139,867 windows were observed (29 windows per frame). Out of which 3417 windows were pedestrian. Table 10.4 summarizes the confusion matrix for the vision-based module. Here, we can observe that the number of false positives is much lesser than what we observed with LiDAR module classification. However, in the case of drastic environmental changes, such as weather or illumination, vision-based classification is typically not as accurate as observed in the present study.

Figure 10.12 Samples of the training data: (a) positive training images and (b) negative training images.

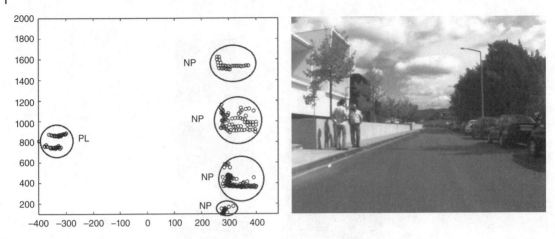

Figure 10.13 An illustration of segment labeling and its corresponding image plane objects. Here, NP, nonpedestrian segment; PL, pedestrian-like segment.

Table 10.4 Confusion matrix for results from LiDAR, visual, and fusion methods.

	LiDAR Pedestrian	Module Nonpedestrian	Vision Pedestrian	Module Nonpedestrian	Fusion Pedestrian	Scheme Nonpedestrian
Pedestrian	3354	63	3249	168	3309	108
Nonpedestrian	4783	19,109	1723	134,727	84	4117

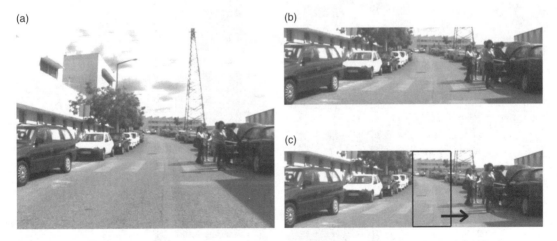

Figure 10.14 (a) The original frame, (b) cropped FOV, and (c) sliding window in the FOV of a predefined size.

The segments classified as pedestrian-like in the LiDAR module are the ROI windows for the vision module in the fusion scheme. In the LiDAR module, a total of 8940 segments are marked as pedestrian-like segments. A window of 80×160 around these segments are the ROIs for HOG–SVM classifier. Table 10.4 summarizes the confusion matrix for the fusion scheme. We can observe that by

Figure 10.15 Examples of HOG–SVM-based vision module results for pedestrian detection. Images at the top are the input images and the images at the bottom are the windows identified as containing pedestrians.

Figure 10.16 Examples of fusion scheme–based pedestrian detection. Along with detection, we have also measured the distance for impact.

using the fusion scheme, both the number of false positives and true negatives have been drastically reduced. Also, by using the LiDAR module filtering, we have reduced the number of windows being processed in the vision module from 139,867 to 7618. This is a huge reduction in computation time. Other advantage of using the fusion scheme includes the ability to reduce false positives in difficult weather and illumination conditions, since the LiDAR module is not significantly sensitive to environmental variations. Once the pedestrian windows are classified, we can also use the LiDAR range data to obtain the distance for crossing and estimate the time for impact assuming a constant velocity model. This aids in alerting the driver with sufficient time for response. Figure 10.16 shows several still frame results for the fusion scheme.

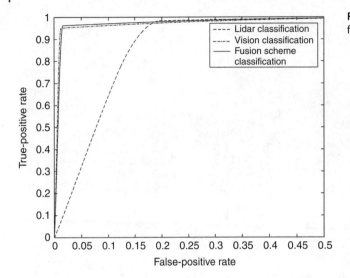

Figure 10.17 The ROC of LiDAR, vision, and fusion methods.

10.8 Summary

Pedestrian detection has been a challenging task for vision systems due to occlusion, deformation, and complexity of foreground and background data. Thermal imaging and NIR structured lighting are relatively simple to implement, and are reliable methods to use at night or in indoor environments. For CCTV applications, approximated median is more accurate than frame subtraction and faster than GMM. Our empirical results show that approximated median performs well in detecting pedestrians in outdoor and indoor environments. The polar coordinate profile can map the 2D shape description to 1D models, which simplifies pattern-matching algorithms. HOG enables individual image-based detection, based on the "cell-like" HOGs. However, it has several limitations, including issues with scalability, occlusion, deformation, unconstrained backgrounds, complex illumination patterns, and a wide variety of articulated poses and clothing. The deformation problem can be solved by applying a DPM. However, occlusion handling is a serious problem for pedestrian detection algorithms. LiDAR-based methods can reduce the false positives and computational time for pedestrian detection. By fusing LiDAR and visual camera data, we can improve the reliability and accuracy of the overall detection rate. Our future work includes incorporating deep knowledge about human shapes, dynamics, and multiple sensors in pedestrian detection.

References

1 P. Dollar, C. Wojek, B. Schiele, P. Perona, "Pedestrian Detection: A Benchmark," in IEEE Conference on Computer Vision and Pattern Recognition 2009, CVPR, Miami, FL, June 2009.
2 Y. Cai, "Cognitive Video Lecture Handout," ECE Graduate Course 18-799K, Spring, 2016.
3 N. Dalal, B. Triggs, "Histograms of Oriented Gradients for Human Detection," in Proceedings of the Conference on Computer Vision and Pattern Recognition, San Diego, California, USA, pp. 886–893, 2005.
4 P. Felzenszwalb, D. McAllester, D. Ramanan, "A Discriminatively Trained, Multiscal, Deformable Parts Model," in IEEE Conference on Computer Vision and Pattern Recognition, 2008, Anchorage, AK, USA. IEEE, 2008.

5 B. Boser, I. Guyon, V. Vapunik, "A Training Algorithm for Optimal Margin Classifiers," in Proceedings of the 5th Annual Workshop on Computational Learning Theory (COLT'92), Pittsburgh, 1992.

6 R. Schapire, Y. Freund, "Experiments with a New Boosting Algorithm," in Proceedings of the 13th International Conference on Machine Learning, Bari, Italy, 1996.

7 V. John, S. Mita, Z. Liu, B. Qi, "Pedestrian Detection in Thermal Images Using Adaptive Fuzzy C-Means Clustering and Convolutional Neural Networks," in 14th IAPR International Conference on Machine Vision Applications, Tokyo, Japan, 2015.

8 B. Qi, V. John, Z. Liu, S. Mita, "Use of Sparse Representation for Pedestrian Detection in Thermal Images," in IEEE Computer Vision and Pattern Recognition Workshops, Columbus, OH, 2014.

9 W. Ouyang, X. Wang, "A Discriminative Deep Model for Pedestrian Detection with Occlusion Handling," in IEEE International Conference on Computer Vision and Pattern Recognition, Providence, RI, 2012.

10 H. Cho, P. Rybski, A. Bar-Hillel, W. Zhang, "Real-Time Pedestrian Detection with Deformable Part Models," in IEEE Intelligent Vehicles Symposium, Madrid, Spain, 2012.

11 M. Pedersoli, J. Gonzalez, X. Hu, X. Roca, "Toward Real-Time Pedestrian Detection Based on a Deformable Template Model," in IEEE Transactions on Intelligent Transportation Systems, Vol. 15, No. 6, December 2014.

12 C. Premebida, O. Ludwig, U. Nunes, "LiDAR and Vision-Based Pedestrian Detection System," *Journal of Field Robotics*, vol. 26, pp. 696–711, 2009.

13 C. Premebida, O. Ludwig, U. Nunes, "Exploiting LiDAR-Based Features on Pedestrian Detection in Urban Scenarios," in Proceedings of the 12th International IEEE Conference on Intelligent Transportation Systems, St. Louis, MO, 2009.

14 K. Arras, O. Mozos, W. Burgard, "Using Boosted Features for the Detection of People in 2D Range Data," in IEEE International Conference on Robotics and Automation, Rome, Italy, 2007.

15 D. Geronimo, A. Loper, A. Sappa, "Computer Vision Approaches to Pedestrian Detection: Visible Spectrum Survey," in Proceedings of the 3rd Iberian Conference on Pattern Recognition and Image Analysis, Part I, Girona, Spain, 2007.

16 M. Enzweiler, D. Gavrila, "A Multilevel Mixture-of-Experts Framework for Pedestrian Classification," *IEEE Transactions on Image Processing*, vol. 20, no. 10, pp. 2967–2979, 2011.

17 M. Perez, "Vision-Based Pedestrian Detection for Driving Assistance," Multidimensional DSP Project, Spring 2005. http://users.ece.utexas.edu/~bevans/courses/ee381k/projects/spring05/perez/LitSurveyReport.pdf (accessed October 17, 2016).

18 T. Gandhi, M. Trivedi, "Pedestrian Collision Avoidance Systems: A Survey of Computer Vision Based Recent Studies," in Proceedings of the IEEE ITSC, Toronto, Canada, 2006.

19 H. Zhao, Q. Zhang, M. Chiba, R. Shibasaki, J. Cui, H. Zha, "Moving Object Classification Using Horizontal Laser Scan Data," in IEEE International Conference on Robotics and Automation, Kobe, Japan, 2009.

20 L. Oliveira, U. Nunes, "Context-Aware Pedestrian Detection Using LiDAR," in IEEE Intelligent Vehicles Symposium, 2010, San Diego, CA.

21 L. Spinello, R. Siegwart, "Human Detection Using Multimodal and Multidimensional Features," in IEEE International Conference on Robotics and Automation, Pasadena, CA, 2008.

22 L. Pangop, R. Chapuis, S. Bonnet, S. Cornou, F. Chausse, "A Bayesian Multisensor Fusion Approach Integrating Correlated Data Applied to a Real-Time Pedestrian Detection System," in IEEE IROS2008 2nd Workshop on Perception, Planning and Navigation for Intelligent Vehicles, Nice, France, 2008.

23 A. Broggi, P. Cerri, S. Ghidoni, P. Grisleri, H. Jung, "A New Approach to Urban Pedestrian Detection for Automatic Braking," *IEEE Transactions on Intelligent Transportation Systems*, vol. 10, no. 4, pp. 594–605, 2009.

24 L. Oliveira, U. Nunes, P. Peixoto, M. Silva, F. Moita, "Semantic Fusion of Laser and Vision in Pedestrian Detection," *Pattern Recognition*, vol. 43, no. 10, pp. 3648–3659, 2010.

25 H. Cheng, N. Zheng, X. Zhang, J. Din, H. Wetering, "Interactive Road Situation Analysis for Driver Assistance and Safety Warning Systems: Framework and Algorithms," *IEEE Transactions on Intelligent Transportation Systems*, vol. 8, no. 1, pp. 157–167, 2007.

26 C. Permebida, "Pedestrian Detection Using Laser and Vision," A Dissertation to Department Electrical and Computer Engineering of COIMBRA University under the supervision of U.Nunes

27 Y. Bouguet, "Camera calibration toolbox for matlab |on-line|" at https://www.vision.caltech.edu/bouguetj/calib_doc/, 2007 (accessed September 28, 2016).

28 C. Premebida, U. Nunes, "Segmentation and Geometric Primitives Extraction from 2D Laser Range Data for Mobile Robot Applications," in Proccedings 5th National Festival of Robotics, Scientific Meeting ROBOTICA, Coimbra, Portugal, 2005

29 O. Ludwig Jr, D. Delgado, V. Goncalves, U. Nunes, "Trainable Classifier-Fusion Schemes: An Application to Pedestrian Detection," in Proceedings of the 12th International IEEE Conference on Intelligent Transportation Systems, St. Louis, MO, 2009

11

Lane Detection and Tracking Problems in Lane Departure Warning Systems

Gianni Cario, Alessandro Casavola and Marco Lupia

Department of Informatics, Modeling, Electronics and Systems Engineering (DIMES), University of Calabria, Rende, Italy

11.1   Introduction

In this chapter, we focus on lane detection (LD) and lane tracking (LT) problems arising in lane departure warning systems (LDWSs). LWDSs refer to specific forms of advanced driver assistance systems (ADASs) designed to help the driver to stay in the lane, by warning her/him in sufficient advance that an imminent and possibly unintentional lane departure is going to take place so that she/he can take the necessary corrective measures. These systems work by estimating the current vehicle position and direction within the lane by identifying the lane demarcating's stripes from the processing video or images of the oncoming road sections taken by a camera mounted on the front windshield of the vehicle (Figure 11.1). According to TRACE [1], a study on traffic accident causation in Europe, a high penetration of LDWSs will produce the following:

- A 25% potential reduction of frontal and run-off-road collisions
- A 25% potential reduction of severity of injuries associated with frontal collisions and reduction of 15% in run-off-road collisions
- A 60% potential reduction of lateral collisions

Some interesting contributions to LDWS development can be found in McCall and Trivedi [2] and Lee [3]. An interesting approach is the so-called time-to-lane-crossing (TLC)-based method, first proposed by Godthelp et al. [4], where an alarm is triggered when the TLC—the time interval by which a lane crossing is supposed to occur under current vehicle speed and direction—is below a specified threshold. Such systems typically use acoustic or vibration warnings, the latter applied to the driving seat or the steering wheel. In general, TLC-based methods provide earlier warnings than roadside rumble stripes (RRSs), because alarms are triggered in sufficient advance before the driver really gets into danger.

A standard configuration of an LDWS is depicted in Figure 11.2, where it is assumed that the vehicle is equipped with a camera mounted on the windshield and angular speed sensors mounted on the rear wheels.

The main devices and functions in a standard LDWS are as follows:

- Camera(s)
- Longitudinal and angular rear wheel speed sensors

Computer Vision and Imaging in Intelligent Transportation Systems, First Edition.
Edited by Robert P. Loce, Raja Bala and Mohan Trivedi.
© 2017 John Wiley & Sons Ltd. Published 2017 by John Wiley & Sons Ltd.
Companion website: www.wiley.com/go/loce/ComputerVisionandImaginginITS

- LD and LT software modules
- Lane departure warning module

The camera mounted on the windshield provides images of the oncoming road sections from which the lane stripes are detected and the car position and heading estimated. The speed sensors, mounted on the rear wheels, measure angular speed. Such information is sufficient to allow one to compute the longitudinal and lateral speed of the vehicle and to estimate the current TLC. Each single frame of the camera stream is analyzed to identify the lane stripes, and an LT module is used to reduce the discrepancy in the reconstruction of the lane stripes between two successive frames. Finally, a warning logic is used to eventually issue an alarm if the computed TLC is shorter than a prescribed threshold.

11.1.1 Basic LDWS Algorithm Structure

A typical LDWS algorithm can be broadly divided into three main functional blocks (Figure 11.3) as follows:

1) LD: recognition of the two stripes that bound the roadway lane in a single image frame
2) LT: recognition and association of the stripes recognized in consecutive frames of the image streaming
3) Lane departure: signaling of exceeding the limits of the lane

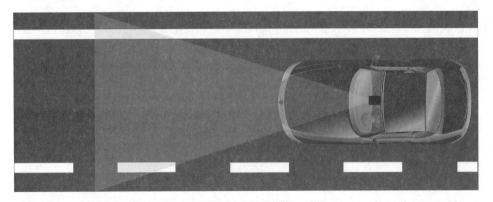

Figure 11.1 Lane departure warning systems.

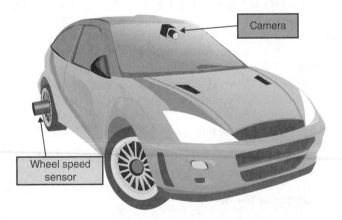

Camera

Wheel speed sensor

Figure 11.2 LDWS overview.

Figure 11.3 LDWS algorithm structure.

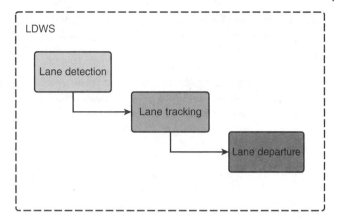

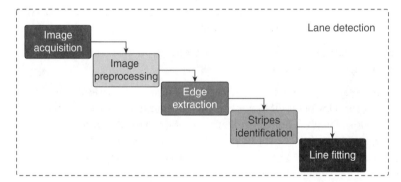

Figure 11.4 Lane detection.

11.2   LD: Algorithms for a Single Frame

The typical steps necessary for the identification of the lane stripes within a single image frame (LD) are listed in detail in Figure 11.4.

Usually, once having acquired an image frame from the camera, some preprocessing steps are necessary to improve the quality and possibly perform some corrections. Then, the preprocessed image undergoes filtering steps necessary to extract the contours and identify the pixels belonging to the stripes of demarcation of the roadway's lane. The last stage is the fitting of the pixels of the stripes with a model to obtain a mathematical representation of the stripes within the image. Details on these tasks are provided in Sections 11.2.1, 11.2.2, 11.2.3, and 11.2.4.

11.2.1   Image Preprocessing

Image preprocessing is performed to simplify and make more effective the subsequent extraction of the lane stripes and the identification of the lane characteristics. The use of filtering techniques allows one to do the following:

- Improve the image contrast
- Reduce the image noise
- Improve the separation of the foreground objects from the background

Frequently used techniques for the aforementioned purposes include gray-level optimization and image smoothing [5], described in Sections 11.2.1.1 and 11.2.1.2.

11.2.1.1 Gray-Level Optimization

The simplest type of image enhancement is the pixel-wide adjustment of intensity values. One such operation is **contrast stretching** that remaps pixel values to utilize the entire set of available values. The enhancement process can be denoted by

$$G(x,y) = T\big[I(x,y)\big],$$ (11.1)

where $I(x,y)$ is the gray value of the (x,y)th pixel of the input image, $G(x,y)$ is the gray value of the (x,y)th pixel of the enhanced image, and T is the transformation function. How the gray levels are changed depends on the remapping function used. For example,

- $G(x,y) = [I(x,y)]^{0.5}$ increases the brightness of all pixels. However, because of its nonlinearity, it increases the intensity of low brightness pixels much more than the brightness pixels.

- $G(x,y) = [I(x,y)]^{1/\gamma}$, also known as gamma correction. It allows one to increase or decrease the brightness and contrast depending on the value of γ.

In addition, it is possible to use adaptive techniques such as histogram equalization to optimize gray level. Histogram equalization rescales pixel intensities to produce an image in which the pixel brightness values are distributed more evenly. The resulting image tends to exhibit greater contrast. The mathematical formulation is as follows: Let I be the original image represented as a $M \times N$ matrix of integer pixel intensities ranging from 0 to $L-1$. L is the number of possible intensity values. Let p denote the normalized histogram of I with a bin for each possible intensity. So,

$$p_n = \frac{\text{number of pixels with intensity } n}{\text{total number of pixels}} \quad n = 0,1,\ldots,L-1.$$

The histogram equalized image I_{HE} will be defined by

$$I_{\text{HE}}(x,y) = (L-1) \sum_{n=0}^{I(x,y)} p_n$$ (11.2)

11.2.1.2 Image Smoothing

Smoothing filters are used to reduce noise and to blur the image. Usually, the images are blurred during the preprocessing phases in order to eliminate small details or small jumps in lines and curves before extracting large objects. The noise reduction can be carried out by blurring the image with a linear or a nonlinear filter.

The action of a linear smoothing or lowpass filter is simply the averaging of the pixels' brightness contained within the mask filter. The output brightness value of a pixel is obtained by averaging the intensity of all pixels in a defined neighborhood. Averaging is performed as follows: Given original image I and a local neighborhood of size $(2m+1) \times (2m+1)$ about location (x,y). Then, the resulting image I_S is given by

$$I_S(x,y) = \sum_{i=-m}^{m} \sum_{j=-m}^{m} h(i,j) I(x+i, y+j)$$ (11.3)

Because random noise typically consists of sharp transitions in the gray level, the most common application of such filters is in noise reduction.

However, considering that edge contours are characterized by abrupt transitions in gray level, the smoothing action could have undesirable effects in the identification of the edges. It is therefore necessary in practice to find a suitable trade-off between noise reduction and excessive blurring of edge features.

11.2.2 Edge Extraction

The edge, or contour, of an object represents the line of separation between the object and background, or between one object and another. Edge-based segmentation algorithms seek a connected discontinuity in some pixel characteristic, for example, the level of gray, the color, and the spatial regularity [6].

Most detectors of discontinuities make use of derivative operators. These operators are applied directly to the image (gray levels or Red/Green/Blue levels) or to some derived variable, obtained by applying a suitable transformation. The points of discontinuity are subsequently detected and concatenated so as to obtain closed contours that delimit distinct areas. The first and second derivatives differ significantly from zero only in correspondence of dark-to-light and light-to-dark transitions, as shown in Figure 11.5.

- The first derivative of the profile is positive at a dark-to-light transition and negative at a light-to-dark transition, and zero in areas where the gray level remains almost constant.
- The second derivative is positive in the vicinity of the dark side of a boundary, negative on the light side, and zero where the gray level is almost constant, or varying at a constant rate. It passes through zero (zero crossing) in correspondence with the transitions.

In general, for an image described by gray levels $I(x,y)$, an edge is defined from the gradient of brightness function as follows:

$$\nabla I(x,y) = \left(\frac{\partial I}{\partial x}, \frac{\partial I}{\partial y} \right)^T \approx \left(D_x, D_y \right)^T \tag{11.4}$$

Figure 11.5 First and second derivates around an edge.

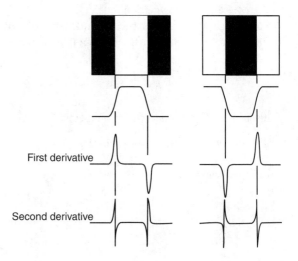

First derivative

Second derivative

Here, D_x and D_y represent partial first derivatives in the x and y directions, respectively, and edge metrics are given by the following:

- Magnitude: $|\nabla I(x,y)| \approx D_x + D_y$
- Orientation: $\theta(x,y) = \tan^{-1}(D_y / D_x)$

The partial derivatives D_x and D_y are approximated using digital derivative operators. Common first-order derivative operators use two kernels. One for changes in the vertical axis and the other for horizontal axis. The convolution of the kernels and the original image provide an approximation of the partial derivatives.

- Sobel

$$D_x = \begin{bmatrix} 1 & 0 & -1 \\ 2 & 0 & -2 \\ 1 & 0 & -1 \end{bmatrix}, \ D_y = \begin{bmatrix} 1 & 2 & 1 \\ 0 & 0 & 0 \\ -1 & -2 & -1 \end{bmatrix}$$

- Prewitt

$$D_x = \begin{bmatrix} -1 & 0 & 1 \\ -1 & 0 & 1 \\ -1 & 0 & 1 \end{bmatrix}, \ D_y = \begin{bmatrix} -1 & -1 & -1 \\ 0 & 0 & 0 \\ 1 & 1 & 1 \end{bmatrix}$$

- Roberts

$$D_x = \begin{bmatrix} 1 & 0 \\ 0 & -1 \end{bmatrix}, \ D_y = \begin{bmatrix} 0 & 1 \\ -1 & 0 \end{bmatrix}$$

Figure 11.6 shows an example roadway image and Figure 11.7 shows a first-derivative operator applied to the image.

11.2.2.1 Second-Order Derivative Operators

11.2.2.1.1 Gaussian's Laplacian

A common way to obtain the second derivative of a function $f(x,y)$ at a point is to calculate the Laplacian at that point. Given $f(x,y)$, its Laplacian is defined as follows:

$$L(x,y) = \nabla^2 f = \frac{\partial^2 f}{\partial x^2} + \frac{\partial^2 f}{\partial y^2} \tag{11.5}$$

A discrete approximation for the Laplacian can be written as

$$L(x,y) = 4f(x,y) - f(x-1,y) - f(x+1,y) - f(x,y-1) - f(x,y+1), \tag{11.6}$$

with kernel representation

$$L = \begin{bmatrix} 0 & -1 & 0 \\ -1 & 4 & -1 \\ 0 & -1 & 0 \end{bmatrix}.$$

Figure 11.6 Original image.

Figure 11.7 Output from Sobel operator.

Note that the Laplacian is overly sensitive to noise (as a second-order derivative operator) and unable to detect the direction of the boundary (as a separate scalar). For these reasons, the Laplacian is rarely used alone for edge-detection purposes. One solution, proposed by Marr and Hildreth [7], uses the Laplacian in connection with a smoothing filter based on Gaussians. The resulting operator is referred to as the Laplacian of Gaussian (LoG) [8].

$$\nabla^2 h = -\frac{r^2 - \sigma^2}{2\sigma^2} e^{\left(-\frac{r^2}{2\sigma^2}\right)} \qquad (11.7)$$

The convolution of the image with the given operator is equivalent to filtering the image with a Gaussian smoothing filter, to reduce the effects of the noise, and then applying the Laplacian operator on the result. However, for implementation on a digital computer, the continuous operator must be approximated by a discrete mask whose elements are obtained by sampling the continuous LoG on a discrete grid [9].

Then, the localization of the contours requires a search for the zero crossings of the second derivative of the image.

Figure 11.8 shows an example of LoG operator applied to the image of Figure 11.6.

11.2.2.2 Canny's Algorithm

The derivative operators considered thus far have a heuristic origin. On the contrary, an analytical approach was adopted by Canny [10], who has studied in detail the behavior of the gradient operator applied to a noisy image.

Canny set three quality criteria for an edge-detection algorithm and determined analytically the optimal filter against these three criteria. The edge detection is performed by convoluting the noisy image I with a filter impulsive response h, chosen so as to meet the following three criteria:

1) Ability to detect the edge
2) Ability to locate the edge
3) Uniqueness of the solution

An edge is identified in correspondence to a local maximum of the convolution $I * h$.

Figure 11.9 shows an example of Canny operator applied to the image of Figure 11.6.

Figure 11.8 Output from Laplacian of Gaussian operator.

Figure 11.9 Output from Canny operator.

11.2.2.3 Comparison of Edge-Detection Algorithms

The various techniques for edge extraction described before have been compared both from a computational and qualitative point of view. In particular, for this second aspect, the Pratt's figure of merit (FOM) [11] was used, which computes an error between the locations of estimated versus true gradient.

$$\text{FOM} = \frac{1}{I_N} \sum_{i=1}^{I_A} \frac{1}{1 + ad^2(i)} \tag{11.8}$$

Here,

- $I_N = \max(I_I, I_A)$, with I_I and I_A, respectively, the number of exactly identified points on the edge.
- a is a scale factor, usually set to 1/9.
- $d(i)$ is the distance (in pixel units) between the identified ith pixel of the edge and its exact position.

The ideal image (Figure 11.10) is used as a benchmark to assess the results achieved by the various edge-detection algorithms and is obtained by manually marking the position of the ideal edge pixels in an area of interest of the original image. From the results of Table 11.1, it seems that the Sobel or the Prewitt operators may be the most suitable algorithms for real-time application with stringent execution times. In fact, these operators show the highest performance, both in terms of computational times and FOM index.

11.2.3 Stripe Identification

The next step in the LD pipeline is stripe identification, whose main functions are depicted in Figure 11.11. Two main approaches are the method of the edge distribution function (EDF) and the Hough transform, described Sections 11.2.3.1 and 11.2.3.2.

(a)

(b)

Figure 11.10 FOM: ideal image: (a) original image and (b) ideal image.

Table 11.1 Comparisons among edge detectors.

Edge detector	Execution time (s)	FOM
Sobel	0.015	0.315
Prewitt	0.015	0.311
Roberts	0.018	0.275
Canny	0.193	0.217
LoG	0.045	0.252

11.2.3.1 Edge Distribution Function

The EDF $F(d)$ is defined as the histogram of the gradient magnitude with respect to its orientation [3].

Note in Figure 11.12 how, on the road surface, the local maxima of $F(d)$ correspond to points on the lane stripes.

11.2.3.2 Hough Transform

The Hough transform [12] is a technique that recognizes specific configurations of points in an image, such as line segments, curves, or other patterns. The basic principle is that the form sought may be expressed through a known function depending on a set of parameters. A particular instance of the form sought is therefore completely specified by the values taken by such a set of parameters.

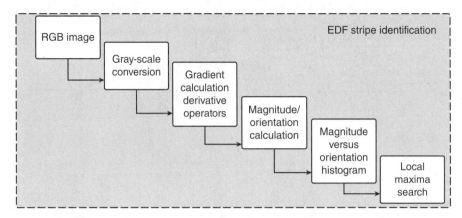

Figure 11.11 Stripes identification via edge distribution function.

Figure 11.12 Edge distribution function.

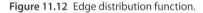

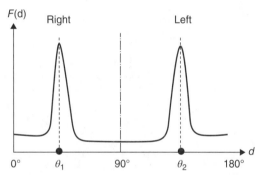

For example, by taking as a representation of straight lines, the equation $y = ax + b$, any straight line is completely specified by the value of the parameters (a, b). Equivalently, if one takes a different type of representation, say $\rho = x\cos\theta + y\sin\theta$, the straight line is completely specified by the pair (ρ, θ).

Thus, given a class of patterns of interest and its mathematical representation, one can consider a transformation from the image plane to the parameter plane. Let us take as an example the cases of interest depicted in Figure 11.13. A single point in the image plane can be represented by the intersection of a collection of two or more straight lines passing through that point, that map to a collection of points falling on a unique curve in parameter space. Conversely, a straight line in the image plane corresponds to an intersection point of multiple curves in the parameter plane.

For the determination of the intersection points in the parameter plane, and therefore of the corresponding straight lines in the image plane, consider a discretization of the parameters (ρ, θ). This allows one to represent the plane on an accumulation matrix $H(m, n)$ whose row and column indexes correspond to quantized values of ρ and θ. Then, for each point $P(x, y)$ in the image, the following steps are executed:

- For $\theta_n \in [-\pi/2, \pi/2]$ with step $d\theta$,
  - Evaluate $\rho(n) = x\cos(\theta_n) + y\sin(\theta_n)$
  - Get the index m corresponding to $\rho(n)$
  - Increase $H(m, n)$ by 1.

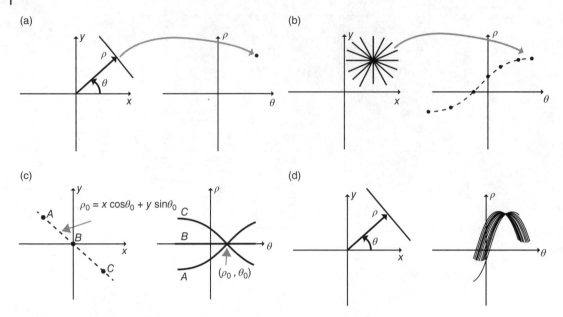

Figure 11.13 Hough transform: (a) image plane—parameters plane transformation; (b) single-point transformation; (c) three-point transformation; and (d) straight-line transformation.

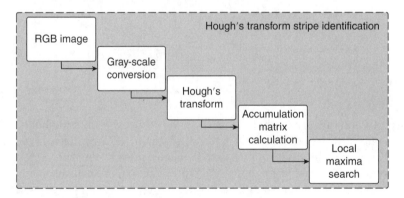

Figure 11.14 Straight-line identification via Hough transform.

This process is repeated for all points in the image, and local maxima of $H(m, n)$ are extracted that denote the straight lines in the image. The overall process is depicted in Figure 11.14. Several parameters such as the resolution of the accumulation matrix are tuned heuristically based on the application and characteristics of the captured images.

11.2.4 Line Fitting

An image typically consists of a set of tokens, which may be pixels, isolated points, sets of contour points, etc., and one may want to determine whether or not they belong to a simple prescribed form. For example, one could be interested to aggregate different tokens because together they form a circle. The search for such groups is known as *fitting*.

The fitting process, in general, uses a model to obtain a representation that highlights the image structures of interest and there are a great variety of possible strategies for fitting. The problem of fitting can be made more efficient if one knows that certain forms are present in the image and a good mathematical model to describe them is available. The most used models in lane stripe fitting problems include the following:

- Linear
- Polynomial, second or third order
- Linear parabolic (LP)
- Clothoid
- Spline

We elaborate next on linear and LP fitting in Sections 11.2.4.1 and 11.2.4.2.

11.2.4.1 Linear Fitting

Linear fitting can be performed either by using the Hough transform directly, which returns the parameters of the straight lines in the image as previously described, or by interpolation or least-squares methods. The use of the least-squares method assumes that all points that belong to a particular line are known and that only the parameters of the mathematical representation of the line need to be identified. We can represent a line in the image as a collection of points given in the Cartesian coordinate (x, y), satisfying the relationship $ax + by + c = 0$. Least-squares fitting consists of determining the triplet (a, b, c) that minimizes the usual quadratic fitting criterion.

11.2.4.2 LP Fitting

Another method used for lane stripes determination is based on LP fitting [13]. The basic idea is to perform a linear fitting in the near field of the image and a quadratic fitting in the far field (Figure 11.15a) to better accommodate perspective distortion of road curves when it is required to fit a large section of road ahead. A fixed threshold x_m that separates the two image fields is selected and the curve that represents the stripe is given by

$$f(x) = \begin{cases} a + b(x - x_m) & x > x_m \\ a + b(x - x_m)x + c(x - x_m)^2 & x \leq x_m \end{cases} \tag{11.9}$$

In order to accomplish such an LP fitting, the first step is to identify the two areas of interest (lane boundary region of interest or LBROI) within the current frame (Figure 11.15b). This process can be made easier by exploiting the same division used at previous frame or by an initial calibration phase performed at system startup.

The parameters a, b, c of Equation 11.9 are determined by an LP fitting minimizing the weighted quadratic error E_{lp} as follows:

$$E_{\text{lp}} = \sum_{i=1}^{m} M_{n_i} \left[y_{n_i} - f(x_{n_i}) \right]^2 + \sum_{j=1}^{n} M_{f_j} \left[y_{f_j} - f(x_{f_j}) \right]^2 \tag{11.10}$$

Here, (x_{n_i}, y_{n_i}), $i = 1, \ldots, m$, represent the mth pixel coordinates, and M_{n_i} the corresponding weights in the near field (only for the pixels $\neq 0$). Similarly, (x_{f_j}, y_{f_j}) and M_{f_j}, $j = 1, \ldots, n$, represent the same quantities for the n pixels in the far field.

(a)

(b)

Figure 11.15 Linear-parabolic fitting: (a) Near- and far-field image separation and (b) lane boundary region of interest (LBROI).

The E_{lp} error is minimized analytically by finding a solution to the following linear system of equations:

$$A^T W A c = A^T W b. \tag{11.11}$$

Here,

$$A = \begin{bmatrix} 1 & x_{n_1} - x_m & 0 \\ \vdots & \vdots & \vdots \\ 1 & x_{n_m} - x_m & 0 \\ 1 & x_{f_1} - x_m & \left(x_{f_1} - x_m\right)^2 \\ \vdots & \vdots & \vdots \\ 1 & x_{f_n} - x_m & \left(x_{f_n} - x_m\right)^2 \end{bmatrix}, \; W = \begin{bmatrix} M_{n_1} & & & & \\ & \ddots & & & \\ & & M_{n_m} & & \\ & & & M_{f_1} & \\ & & & & \ddots \\ & & & & & M_{f_n} \end{bmatrix} \tag{11.12}$$

$$c = [a,b,c]^T \,, \ b = \left[y_{n_1}, \ldots, y_{n_m}, y_{f_1}, \ldots, y_{f_n} \right]^T \tag{11.13}$$

Under a *persistence-of-excitation* condition on the columns of A (namely A is full rank), the matrix $A^T W A$ results invertible and the sought solution is given by

$$c := \left(A^T W A \right)^{-1} A^T W b. \tag{11.14}$$

The procedure is applied to both stripes of the lane. Note also that linear fitting is a special case of LP fitting where the weighted quadratic error E_1 reduces to

$$E_1 = \sum_{i=1}^{m} M_{n_i} \left[y_{n_i} - f\left(x_{n_i} \right) \right]^2. \tag{11.15}$$

Also, one has

$$A = \begin{bmatrix} 1 & x_{n_1} - x_m \\ \vdots & \vdots \\ 1 & x_{n_m} - x_m \end{bmatrix}, \ W = \begin{bmatrix} M_{n_1} & & \\ & \ddots & \\ & & M_{n_m} \end{bmatrix}. \tag{11.16}$$

$$c = [a,b]^T \,, \ b = \left[y_{n_1}, \ldots, y_{n_m} \right]^T. \tag{11.17}$$

Again, under persistence of excitation, the sought solution is given by (11.14) with A, W, c, and b given in Equations 11.16 and 11.17.

11.3   LT Algorithms

The discussion thus far has focused on LD from a single image or video frame. This section turns to the problem of LT from video signals containing both spatial and temporal information. In general, the tracking of an object in a video is defined as the problem of establishing correspondences between the same object identified in subsequent video frames. In LT applications, these correspondences are used for recognizing the surrounding environment. There are various methods for addressing the tracking problem that are based on the consistency of motion, and the appearance and transformation of certain patterns. In general, object tracking is a fairly complex process. If an object moves within the scene, it changes its position and orientation with respect to the background and in multiple frames may look different.

Another challenge in object tracking, especially encountered in road environments, is occlusion, which refers to partial of full coverage of the target object by other objects, for example, a vehicle on the lane's stripes, or environmental characteristics such as shadows, snow, or leaves on the road. To maintain correspondence, it is typically necessary that at least a small part of the object is visible in successive frames. This may be possible if the segmentation is accurate and resolution sufficient. Otherwise, the tracking can be lost periodically. In this case, it is important that the algorithm is able to reset itself and reidentify the object as soon as possible.

In the following text, we present and contrast several algorithms well suited for lane stripe tracking. Note also that the tracking algorithms are synergistic with lane stripe reidentification and fitting algorithms, in that the knowledge of stripe position at the previous frame allows a more accurate identification and fitting of the stripe in the current frame.

The most commonly used fitting algorithms in the LDWS literature include recursive filters on subsequent N frames [14] and Kalman filters [15], described in Sections 11.3.1 and 11.3.2.

11.3.1 Recursive Filters on Subsequent *N* frames

The simplest filter that can be used for the tracking of a line within an image is a recursive filter that works on N consecutive frames. In practice, it performs an average of the position of the line on a sliding window of N previous frames.

The location of the line in the next frame is assumed to not deviate too much from the location in the current frame. In particular, if we define $\phi(k)$ as the parameter vector of the line identified in the frame k, we have

$$z(k+1) = z(k) + \phi(k) - \phi(k-N), \quad k \geq N+1 \tag{11.18}$$

with N arbitrarily selected. Then, the LDWS will use the filtered parameter vector $z(k)$ instead of $\phi(k)$.

For instance, in the case of a Hough transformed binary image with a lane boundary defined by ρ and θ, the parameter vector is $\phi(k) = [\rho(k), \theta(k)]$, and the filtered parameter vector $z(k) = [\rho_f(k), \theta_f(k)]$ is achieved by solving the recursion (11.18).

11.3.2 Kalman Filter

The Kalman filter is frequently employed for object tracking. It is usually assumed that the object moves with constant or linearly varying speed among successive frames. Thus, if we consider a dynamical linear model affected by random measurement noise, the Kalman filter is the optimal estimator if the noise is white, zero mean, and uncorrelated with the process variables. As new measures are achieved, the estimation of the position is improved at each iteration.

Consider the following linear time-invariant dynamical system:

$$\begin{aligned} x_k &= A x_{k-1} + w_{k-1} \\ z_k &= H x_k + v_k \end{aligned} \tag{11.19}$$

Here, x_k is the state of the system at time k; w_k and v_k are zero-mean noises with covariance matrices Q and R, respectively; A is the state transition matrix; H is the output matrix; and z_k are observed outputs (e.g., parametric estimates of lane stripes). At each sampling time, an estimate of the state is achieved as

$$\hat{x}_k = \bar{x}_k + K_k \left(z_k - H\bar{x}_k \right), \tag{11.20}$$

where

$$\bar{x} = A\hat{x}_{k-1}. \tag{11.21}$$

$$\bar{P}_k = A\hat{P}_{k-1}A^T + Q. \tag{11.22}$$

$$\hat{P}_k = \left(I - K_k H \right) \bar{P}_k. \tag{11.23}$$

$$K_k = \bar{P}_k H^T \left(H\bar{P}_k H^T + R \right)^{-1}. \tag{11.24}$$

This iteration is initialized with $\hat{x}_0 = 0$, and the covariance matrix $\hat{P}_0$ is suitably chosen.

It should be noted that the Kalman filter may give erroneous results in the presence of outliers, a common situation in many practical applications. One thus needs to test every prediction to determine if it deviates significantly from what would be nominally expected in light of recent estimates. If this happens, the object in question is probably partially or completely hidden. A simple solution is to assume that the object exhibits the same motion and to wait for it to reappear in a future frame.

It is also prudent to store in the computer memory multiple candidate locations for a future frame, so that the object can be easily reidentified in case it reappears in an unexpected location.

11.4 Implementation of an LD and LT Algorithm

This section presents an algorithmic implementation based on the previously described techniques for LD and LT that are similar to methods commonly present in today's commercial LDWSs. It uses images of the roadway directly acquired from a simulated road-facing camera. Preprocessing includes corrections for brightness and contrast. The lane departure warning is accomplished by the analysis of the suitably angled lines representing the lane stripes within the image and through appropriate thresholds determined experimentally. These methods do not provide an actual measurement in meters of the vehicle position within the lane. Rather, a warning (simulating the presence of "rumble stripes") is issued if the vehicle is excessively close to the lane stripes. The algorithm uses the Hough transform for LD and the Kalman filter for LT. The whole process is depicted in Figure 11.16.

Lane Detection

The identification of the lane stripes is obtained by extracting the contours of the scanned image with the Sobel operator and seeking two local maxima of the Hough transformed binary image that provide optimal estimates ρ and θ for the two lane boundaries.

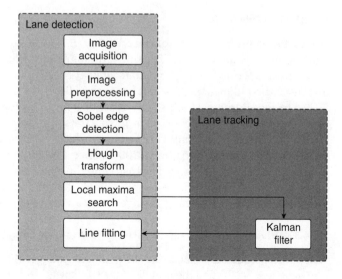

Figure 11.16 Algorithm implementation for lane detection and tracking.

Lane Tracking

A Kalman filter is used to predict and track the values of ρ and θ along consecutive frames. The details are

$$x_k = Ax_{k-1} + w_{k-1}$$
$$z_k = Hx_k + v_k$$

(11.25)

with

$$x = \begin{bmatrix} \rho & \theta & \Delta\rho & \Delta\theta \end{bmatrix}^T, \quad A = \begin{bmatrix} 1 & 0 & 1 & 0 \\ 0 & 1 & 0 & 1 \\ 0 & 0 & 1 & 0 \\ 0 & 0 & 0 & 1 \end{bmatrix}, \quad H = \begin{bmatrix} 1 & 0 & 0 & 0 \\ 0 & 1 & 0 & 0 \end{bmatrix}.$$

(11.26)

The terms w and v denote zero-mean white noise signals with covariance matrices Q and R, respectively.

The state estimates are given by Equations 11.20–11.24. The estimates $\hat{z} = [\hat{\rho}, \hat{\theta}]$ represent the current parameters of the lane stripe achieved by the algorithm.

11.4.1 Simulations

This section reports simulations undertaken for performance assessment. The simulations aim at evaluating the ability of the proposed method to identify the correct position of the lane stripes.

The algorithm has been coded in MATLAB/Simulink and tested in the CarSim simulation framework. The analysis of the algorithms has been performed on a relevant driving scenario generated by CarSim. For assessment purposes, the accuracy of the identification of the lane stripes position and their inclination θ within the images was analyzed. Each measure was compared with the true value provided by the software simulator.

11.4.2 Test Driving Scenario

To test the algorithm under realistic scenarios, it is necessary to know instant by instant the position of the vehicle within the lane. The software simulator CarSim 8, produced by Mechanic Simulation, is a software package for the dynamic simulation of vehicles. It allows the use of virtual cameras, produces three-dimensional animations, and allows the analysis of the behavior of a vehicle subjected to certain control inputs. The dynamical model of the cars consists of a number of nonlinear differential equations and more than 150 datasets from which to choose the model parameters to use in the current run, enabling very realistic vehicle behavior. The maneuvers described later were carried out on a straight roadway 1200 m long. The road is 4 m wide and consists of two lanes, one for each direction of traveling. The road centerline in Figure 11.17 (vehicle position) is the x-axis, and that figure depicts the vehicle position along the time in the two lanes, where the right lane has negative ordinate values while positive values are those corresponding to the left lane.

11.4.3 Driving Scenario: Lane Departures at Increasing Longitudinal Speed

The vehicle repeatedly executes a series of lane changes (Figure 11.17a), moving from the center of the right lane (−2 [m]) to the middle of the left lane (2 [m]) at increasing longitudinal speed as shown

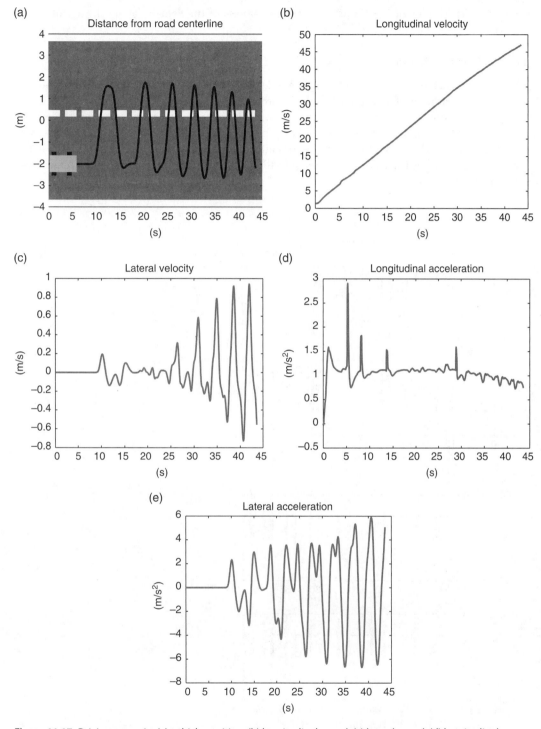

Figure 11.17 Driving scenario: (a) vehicle position, (b) longitudinal speed, (c) lateral speed, (d) longitudinal acceleration, and (e) lateral acceleration.

in Figure 11.17b. The longitudinal and lateral accelerations of the vehicle (along the x and y axes, respectively) are depicted in Figure 11.17d and e.

11.4.4 The Proposed Algorithm

Figure 11.18 depicts the orientation of the right and left lane stripes during the simulations. Note that the proposed algorithm performs the LT with a Kalman filter.

The tracking of the lines is indeed accurate. Between two consecutive lane changes, the error is <1%. On the contrary, in correspondence of lane changes, the Kalman filter is reset and the error reaches peaks of 100% because the stripes identification process restarts from scratch. The difficulty in reidentifying a stripe is increasingly greater with the increase in the longitudinal and lateral speeds of the vehicle.

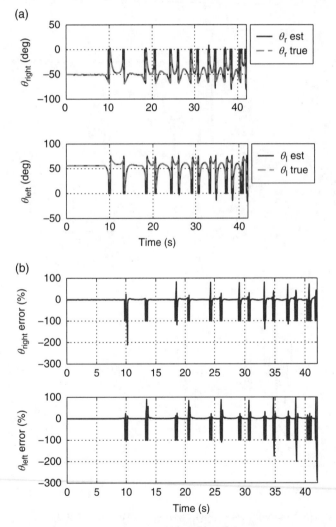

Figure 11.18 Orientation of the right and left lane stripes.

11.4.5 Conclusions

The simulations allow one to draw the following conclusions. In the driving scenario taken into consideration, the proposed algorithm shows a considerable capability of achieving good LD and LT performance, both in terms of estimation accuracy and robustness again noise and model imperfections. The use of the Kalman filter for LT is probably the main reason for this success. For this algorithm, the best results were achieved at constant driving speed (figures not reported here for brevity), whereas a slightly greater difficulty in tracking the lane stripes was observed in the scenario where driving speed was increasing. At constant driving speed (50 kmph) scenario (simulations not reported for brevity), the average error on the estimation of the stripes orientation θ is lower than 1 %, with a maximum of $100-150\%$. On the contrary, under the considered increasing driving speed scenario, an average relative error on θ of $1.2-1.8\%$ were observed in the simulations, with a maximum of $100-200\%$.

References

1 TRaffic Accident Causation in Europe (TRACE Project) 2007 Review of crash effectiveness of Intelligent Transport Systems' 2007—Lane Departure Warning and Control, http://www.trace-project.org/ (accessed on September 20, 2016).

2 McCall J.C. and Trivedi M. 2006 Video-based lane estimation and tracking for driver assistance: Survey, system and evaluation. *IEEE Transactions on Intelligent Transportation Systems*, 7(1), 20–37.

3 Lee J.W. 2002 A machine vision system for lane-departure detection. *Computer Vision and Image Understanding*, 86, 52–78.

4 Godthelp H., Milgram P., and Blaauw G.J. 1984 The development of a time-related measure to describe driver strategy. *Human Factors*, 26, 257–268.

5 Gonzalez R.C. and Woods R.E. 2007 *Digital Image Processing*, 3rd Ed. Upper Saddle River: Prentice Hall.

6 Muthukrishnan R. and Radha M. 2011 Edge detection techniques for image segmentation. *International Journal of Computer Science and Information Technology (IJCSIT)*, 3(6), 259–267.

7 Marr D. and Hildreth E. 1980 Theory of edge detection. *Proceedings of the Royal Society of London, Series B: Biological Sciences*, 207(1167), 187–217.

8 Marr D. 1982 *Vision: A Computational Investigation into the Human Representation and Processing of Visual Information*. New York: W.H. Freeman.

9 Gunn S.R. 1999 On the discrete representation of the Laplacian of Gaussian. *Pattern Recognition*, 32, 1463–1472.

10 Canny J.F. 1986 A computational approach to edge detection. *IEEE Transaction on Pattern Analysis and Machine Intelligence*, 8(6), 679–698.

11 Pratt W.K. 1978 *Digital Image Processing*. New York: Wiley-Interscience.

12 Duda, R.O. and Hart P.E. 1972 Use of the Hough transformation to detect lines and curves in pictures. *Communications of the ACM*, 15(1), 11–15.

13 Jung C.R. and Kelber C.R. 2005 A lane departure warning system based on a linear-parabolic lane model. *XVIII Brazilian Symposium on Computer Graphics and Image Processing*, Natal, Rio Grande do Norte, Brazil, October 9–12, 2005.

14 Lee J.W. and Yi U.K. 2005 A lane-departure identification based on LBPE, Hough transform, and linear regression. *Computer Vision and Image Understanding*, 99, 359–383.

15 Voisin V., Avila M., Emile B., Begot S., and Bardet J.C. 2005 Road markings detection and tracking using Hough transform and Kalman filter. *Advanced Concepts for Intelligent Vision Systems*, 3708, 76–83.

12

Vision-Based Integrated Techniques for Collision Avoidance Systems

Ravi Satzoda and Mohan Trivedi

University of California, San Diego, CA, USA

12.1   Introduction

In recent years, there has been a significant increase in the number of the embedded electronic processors in automobiles. For example, according to Chakraborty et al. [1], the prorated cost of incorporating these embedded systems has increased from 1% in 1980 to over 22% in 2007; this percentage is continuously increasing, which has also increased the complexity of the software interactions exponentially [2]. The presence of such systems directly impacts the overall power consumption, which is especially critical in present times when the automotive industry is moving toward battery-powered hybrid and electric vehicles.

Among the different embedded subsystems, vision-based advanced driver assistance systems (ADAS) are becoming more popular in recent times because of the availability of low-cost, high-resolution, and pervasive cameras [3]. Figure 12.1 shows some vision-based ADAS that are being actively researched. Such ADAS have two levels of processing. The first level of processing is pixel-level processing where visual data (pixels) is analyzed to extract features. Operations include filtering, edge detection, and enhancements. This can be accomplished by single instruction, multiple data (SIMD) type of data processing architectures. The complexity of such processing is directly proportional to the amount of image data that is being processed. The second level of processing is termed "higher order processing." This does not involve visual data directly. Instead, features that are extracted from the visual data are analyzed for different operations. Such features include edges, histograms of gradients, and gradient magnitudes. Considering that the features can be varied and not uniformly distributed, SIMD processing will no longer be applicable for this data. Therefore, multiple instruction, multiple data (MIMD) processing is employed. The computational complexity of such tasks is proportional to the complexity of the operations that are being employed to process the data.

In order to meet the increasing demand in vision-based ADAS, manufacturers such as Texas Instruments, etc., are releasing newer embedded platforms that are specifically catered for implementing vision algorithms in automobiles [4]. Although such platforms do provide an advantage over conventional processors and hardware systems, the efficiency (both power and real-time operation) are constrained by the implementation of the constituent vision algorithms [5, 6].

Computer Vision and Imaging in Intelligent Transportation Systems, First Edition.
Edited by Robert P. Loce, Raja Bala and Mohan Trivedi.
© 2017 John Wiley & Sons Ltd. Published 2017 by John Wiley & Sons Ltd.
Companion website: www.wiley.com/go/loce/ComputerVisionandImaginginITS

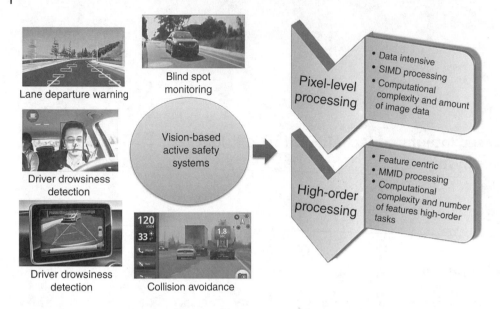

Figure 12.1 Vision-based driver assistance systems (images courtesy google.com).

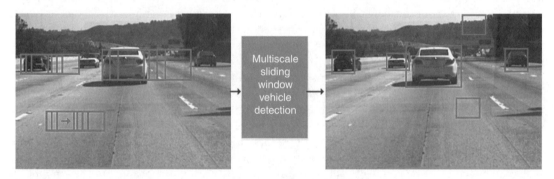

Figure 12.2 Lane and vehicle detection techniques when seen in isolation.

Various computer vision algorithms have been explored for ADAS tasks such as detection of lanes, vehicles, and pedestrians in Refs. [7, 8], wherein various feature extraction methods, classification techniques, and tracking methods have been proposed for improving accuracy. While robustness is critical in such vision-based active safety systems, exporting such data-intensive algorithms on embedded platforms is still a challenging task [5, 9]. Although implementations such as lane detection in Refs. [10, 11] demonstrate the realization of ADAS algorithms on embedded platforms, designing the algorithm itself to cater to more efficient embedded realizations is less explored in the current literature [6, 9, 11, 12]. For example, consider the illustration in Figure 12.2, which shows an example of multiscale sliding window approach for vehicle detection such as in Refs. [7, 13, 14]. There are a number of such techniques in the literature [15, 16]; however, false-positive windows such as those shown in Figure 12.2 are commonly found during the vehicle detection phase in Ref. [7]. Such techniques require additional computational steps such as tracking to further eliminate the false positives.

In this chapter, we propose an integrated approach called efficient lane and vehicle detection using integrated synergies (ELVIS) that incorporates the lane information to detect vehicles more efficiently in an informed manner using a novel two-part-based vehicle detection technique. The vehicle detection outputs are further used in ELVIS to improve the efficiency of the lane feature extraction process. We will also present detailed evaluations that show significant gains in computational efficiency using ELVIS, without compromising on accuracy of both lane and vehicle detection.

12.2 Related Work

In this section, we present a brief survey of some related works. A survey of existing literature shows a number of techniques that are separately described for lane and vehicle detection. A detailed survey of recent lane detection methods is presented in Refs. [16, 17]. Most of these lane detection techniques employ different kinds of filters such as Sobel edge detection kernels, steerable filters [17], and Gabor filters on the entire input frame (or below the horizon or vanishing line) to extract lane features. The lane features are further processed and tracked using techniques such as random sample consensus (RANSAC) [17] and Kalman filtering [18]. Computational efficiency of such estimation methods has been less addressed in most of the aforementioned works, but techniques such as those in Refs. [9, 11] and the hardware implementation in Ref. [10] present lane detection methods that are realizable on embedded platforms. The main challenge in such techniques is the robust extraction of lane features, and presence of features in the region of interest such as shadows and vehicles introduces incorrect lane features that affect the overall accuracy of lane estimation. Therefore, the focus is on extracting more accurate features, which increases the accuracy as well as the computational cost.

On-road vehicle detection has been extensively investigated in Ref. [19]. Various kinds of features such as histogram of oriented gradient (HOG) [8] and Haar-like features [20] have been explored in multiple studies. As shown in Figure 12.2, the features are computed using a sliding window approach that is applied on multiple scales in order to detect vehicles of varying sizes. The features are then sent for classification using support vector machines (SVMs) [21] and cascaded classifiers [20]. Different variants of such classification methods have been proposed, and the primary challenge such methods face is the variation in the road scene that leads to false detections as shown in Figure 12.2. In order to improve the accuracy, either the number of features is increased or changed, or the classifier is made more robust. However, most existing reliable vehicle detection techniques do not inherently address computational complexity at the algorithmic level, although hardware realizations such as [22] present hardware architectures for generic object detection including vehicles using Haar-like features and AdaBoost classifiers.

12.3 Context Definition for Integrated Approach

In this section, we will define the context under which the proposed integrated approach will function. Furthermore, we will show that the proposed context caters to most ADAS applications that involve detection of lanes and vehicles. First, we set the scope of the study presented in the chapter and limit to highways and marked urban roads, that is, roads that have lane markings, which form one of the major segment of existing techniques for ADAS. Now, let us look into the requirements of most commonly researched and deployed ADAS applications in the context of lane and vehicle detection. Based on commercial products that are available (or being developed), and academic literature such as Refs. [3, 23], we present a list of the most commonly researched ADAS applications involving detection of lanes and vehicles in Table 12.1. For each ADAS application, the

Table 12.1 Significance of lanes and vehicles in related ADAS applications.

ADAS application	Lanes	Vehicles
Lane departure warning	Host lane	—
Lane change assistance	Host and neighbor lanes	Nearest vehicles in host and neighbor lanes
Collision avoidance	Host lane	Nearest leading vehicle
Proximity estimation	Host and neighbor lanes	Nearest vehicles in host and neighbor lanes
Lane change intent prediction	Host and neighbor lanes	Nearest vehicles in host and neighbor lanes
Blind spot monitoring	Lanes in blind spots	Nearest vehicles from rear and blind spots
Merge assist	Neighbor lanes in front, rear, and blind spots	Nearest vehicles in front, rear, and blind spots
Lane-level micro navigation	Multiple lane detection	—
Leading vehicles intent prediction	Host and neighbor lanes	Nearest vehicles in host and neighbor lanes

specific requirements of the lane and vehicle detection modules are listed in separate columns in Table 12.1. In some applications, while both lanes and vehicles are required, some applications need lanes only.

In all the applications listed in Table 12.1, the following can be deduced:

1) The nearest vehicles to the ego-vehicle pose the highest risk, and hence they must be detected.
2) Detection of lanes till the nearest vehicles in the lanes is critical for most ADAS applications.

The first item is self-explanatory from the kind of ADAS application and its specification listed in Table 12.1. The second item needs further elaboration. In applications such as lane departure warning and lane-level micro navigation, lanes that are closest to the host vehicle are sufficient [24]. Other applications such as lane change assistance require the detection of lanes in deeper field of view of the host vehicle. However, the primary area of interest in such applications is limited by the leading vehicles in front of the ego-vehicle. Therefore, the nearest vehicles can define the region of interest in which the lane detection needs to be performed.

Based on these observations, we define the context as the detection of the nearest vehicles to the ego-vehicle, which are leading or approaching the ego-vehicle in the host and adjacent lanes. The listing in Table 12.1 shows that the proposed context of lane and vehicle detection caters to all the ADAS applications of interest. In Section 1.4, the proposed integrated approach exploits the context about the vehicles and lanes to detect them *robustly* in more *computationally efficient* manner.

12.4 ELVIS: Proposed Integrated Approach

In this section, we will first propose the vehicle detection component of ELVIS, followed by the lane feature extraction method that benefits from vehicle detection.

12.4.1 Vehicle Detection Using Lane Information

Let us consider a typical road scene as shown in Figure 12.3. Conventional classifier-based vehicle detection techniques apply a sliding window of multiple scales (or the image is scaled keeping the size of the window same) across the input frame as shown in Figure 12.3a to detect vehicles of multiple sizes. Although such techniques exhaustively look for vehicles, they also result in false positives and can be computationally overwhelming for embedded realization with a few hundreds of thousands of windows for processing. There is an additional computational overhead for eliminating false positives using tracking, the cost of which increases with higher number of false-positive windows.

In ELVIS, we consider the following observations about on-road vehicle detection. The on-road vehicles appear in an orderly manner along the road surface, with sizes that are changing in an orderly way, especially in highway and urban driving. Therefore, as shown in Figure 12.3b, if the image is scanned along the road surface, the vehicles that appear first in the direction of the road will appear bigger than the vehicles that appear later.

In the proposed method, we use the lane information and the contextual information described before to detect the vehicles in an informed manner. The first part of the proposed method is a one-time process to generate a lookup table (LUT) of the positions and sizes of the windows which should be used to detect the vehicles. In order to do this, we use the inverse perspective mapping (IPM) of the image, which can be readily derived based on the camera calibration [25]. Considering that IPM generation is an integral part of most lane detection methods such as in Refs. [9, 17, 25, 26], it will be reused for lane detection in ELVIS later.

Figure 12.4 illustrates the LUT generation process. Given an input image I, the IPM image I_W is generated using the homography matrix $\mathbf{H}$ [25]. Therefore, every point $P(x, y)$ in I is transformed to P_w in I_W using $\mathbf{H}$, that is,

$$\begin{bmatrix} x_w & y_w & 1 \end{bmatrix}^T = k\mathbf{H}\begin{bmatrix} x & y & 1 \end{bmatrix}^T, \tag{12.1}$$

where k is the calibration constant. Therefore, the four points P_1' to P_4' in Figure 12.4 in the image domain correspond to the minimas and maximas in the IPM domain along the $x_w - y_w$ coordinates. The following parameters for each row I_W are determined: (i) the start position of road surface in the

(a)

(b)

Figure 12.3 (a) Sliding window of multiple sizes used across the image and (b) relationship between sliding window and the position of the leading vehicles.

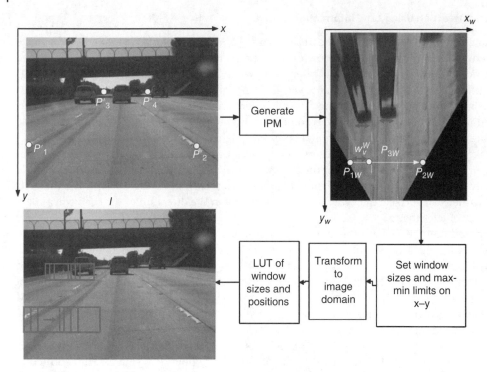

Figure 12.4 Generating the LUT of windows and positions using lane and IPM information.

row, that is, P_{1W} (Figure 12.4);, (ii) the end position of the road surface P_{2W}; and (iii) point P_{3W} such that $P_{3W} - P_{1W} = w_V^W$, where w_V^W is the width of the vehicle as can be seen from the top view. Considering most consumer vehicles usually have a standard axle length, w_V^W can be predetermined. Given P_{1W}, P_{2W}, and P_{3W}, we use the inverse of $\mathbf{H}$, that is, $\mathbf{H}^{-1}$ to determine the corresponding points P_1, P_2, and P_3 in the image domain. For each row index y in I, we now have the following:

$$x_{\min} = x_1, \quad x_{\max} = x_2, \quad w_V = x_3 - x_1. \tag{12.2}$$

Here, w_V is the width of the window that should be used for vehicle detection in the yth row of I, and $x_{\min}$ and $x_{\max}$ are the minimum and maximum indices along the x-axis in I where the windows will be processed. The height h_V of the window was set to $1/1.2$ times w_V, which was found suitable to detect most consumer vehicles. An LUT is generated with these variables for each row of the input image I. Therefore, unlike the conventional way of running the sliding window all over the image, we now have a defined search space and specific scales of the window for every row in the LUT to detect the vehicles.

In addition to the LUT generation, we will now propose a two-part-based vehicle detection scheme in ELVIS that incorporates the next contextual information about vehicles moving in the road scene. In an ADAS application that requires on-road vehicle detection such as collision avoidance or lane change assistance, it is evident that the vehicles are moving on the road surface. Therefore, the windows are applied in the direction of the decreasing y-coordinate (referring to the axes in Figure 12.4), that is, from the front of the host-vehicle toward the vanishing line. Next, the window $\mathbf{I}_V$ is divided into two parts $\mathbf{P}_1$ and $\mathbf{P}_2$ such that if the window is placed on the vehicle, they capture the two parts of the vehicle as shown in Figure 12.5. For an $M \times N$ (rows by columns) sized image patch $\mathbf{I}_V$, $\mathbf{P}_1$ and $\mathbf{P}_2$ are defined as follows:

$$\mathbf{P}_1 = \mathbf{I}_V\left(1 : \frac{2}{3}M, 1 : N\right) \quad \mathbf{P}_2 = \mathbf{I}_V\left(\frac{2}{3}M : M, 1 : N\right). \tag{12.3}$$

In ELVIS, we first detect the presence of lower part of the vehicle, that is, $\mathbf{P}_2$ because while traversing along the road, the lower part is expected to be seen first. The HOG feature [8] of part $\mathbf{P}_2$ in $\mathbf{I}_V$ denoted as h_2 is computed first, and it is classified using SVM classifier for $\mathbf{P}_2$. If the classification of the lower part is positive, then HOG feature h_1 for $\mathbf{P}_1$ is computed and classified using the SVM classifier for $\mathbf{P}_1$. The classifier scores from the SVMs of parts $\mathbf{P}_1$ and $\mathbf{P}_2$ are used in the following way to give the final classification result:

$$\mathbf{I}_V\left(x, y\right) = \text{vehicle if } p_1\left(x, y\right) \times p_2\left(x, y\right) > T_p. \tag{12.4}$$

Here, p_1 and p_2 are the probability scores of the SVMs for parts $\mathbf{P}_1$ and $\mathbf{P}_2$, respectively, and T_p is the threshold for overall classification score. Nonmaximal suppression is applied to remove overlapping windows.

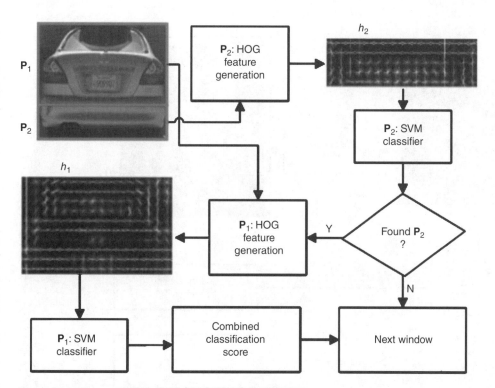

Figure 12.5 Two-part-based vehicle detection method in ELVIS.

12.4.2 Improving Lane Detection using On-Road Vehicle Information

In this section, the performance of lane detection is improved using the results obtained from the vehicle detection algorithm. For most ADAS applications, such as lane departure warning, lane change assistance, and collision avoidance, lanes are required to either localize the ego-vehicle in the lane or to maneuver the ego-vehicle between the host lane and the adjacent lanes. Therefore, if there is a leading vehicle in front of the ego-vehicle, detecting the lanes between the two vehicles is sufficient. Also, the presence of a leading vehicle obstructs the view in front of the ego-vehicle and lanes are not visible in the image plane resulting in false lane features that affect the accuracy of lane estimation [18]. We propose to use the positions of the vehicles that were determined by the proposed vehicle detection scheme in ELVIS to determine the regions in the input image where the lanes must be detected.

We demonstrate this using one of the recently proposed lane detection methods [9, 24] called lane analysis using selective regions (LASer). In this chapter, we limit the scope of the proposed lane detection approach to the host lane detection for the sake of explanation. Also, LASeR algorithm in Ref. [9] is designed for host lane detection, which will be used to demonstrate the proposed integrated method. In LASeR, N_B number of scan bands at positions $Y = \left\{ y_j^B \right\}$ along the y_w axis of the IPM image are processed where $1 \leq j \leq N_B$. The bands are shown in Figure 12.6 where the band closest to the ego-vehicle is indexed as 1. Given that we have the positions of the vehicles at $P^V(x^V, y^V)$ in the image domain I from ELVIS, we find their positions in the IPM domain using (12.1) to get $P_W^V \left(x_W^V, y_W^V \right)$, that is, $P_W^V = k\mathbf{H}P^V$. Therefore, with the leading vehicle at y_W^V, the index of the maximum band along y_W axis in the IPM image that must be processed in LASeR is computed using the following equation:

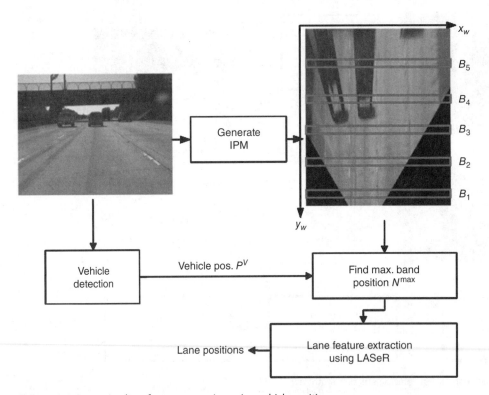

Figure 12.6 Improving lane feature extraction using vehicle positions.

$$N^{\max} = \arg\max_{j} y_W^V \leq y_j^B.$$ (12.5)

Referring to Figure 12.6, the conventional LASeR algorithm would have processed all the five scan bands shown in the IPM image in Figure 12.6. Due to the presence of vehicles in bands B_4 and B_5, false lane features could be generated. However, the integration of the vehicle detection method into the lane detection process allows choosing $N^{\max} = 3$ bands, which are processed by the next steps of LASeR algorithm. We will demonstrate the advantages of the proposed integration between lane and vehicle detection in terms of computational complexity and accuracy in Section 12.5.

12.5 Performance Evaluation

12.5.1 Vehicle Detection in ELVIS

12.5.1.1 Accuracy Analysis

The accuracy analysis setup involved a training set comprising 1700 positive samples and 2500 negative samples. Each subimage was divided into two parts as shown in Figure 12.5, and two different classifiers were trained using the two sets of training samples. The testing of the proposed vehicle detection component in ELVIS was performed using three different datasets that do not contain any of the training samples—Caltech 1999 (126 images), LISA dataset 2 (300 images), and LISA dataset 3 (300 images) [7]. We use the following metrics described in Ref. [7] for evaluating and comparing accuracy with results in Ref. [7]:

$$\text{True positive rate}\,(\text{TPR}) = \frac{\text{True detections}}{\text{Total number of vehicles}}$$

$$\text{False detection rate}\,(\text{FDR}) = \frac{\text{False positives}}{\text{True detection} + \text{False positives}}$$

$$\text{Average FP / frame} = \frac{\text{False positives}}{\text{Total number of images}}$$ (12.6)

$$\text{Average TP / frame} = \frac{\text{True positives}}{\text{Total number of images}}$$

Figure 12.7 plots the receiver operating curves (ROCs) showing the TPR versus FDR on applying ELVIS to the three datasets. It can be seen that the proposed two-part-based method performs well with the Caltech dataset giving over 90% TPR for <5% of false detection rate. Coming to the LISA 2 dataset, the proposed method detected the single vehicle in all cases giving 100% true positive rates for all data points considered. In LISA 3 dataset, ELVIS gives over 95% TPR for <10% FDR.

Table 12.2 compares the measures for the four different metrics and three different datasets between the two-part-based vehicle detection technique in ELVIS and the passive learning AdaBoost-based classifier in Ref. [7]. It can be seen that in most cases the proposed method gives at least 1–2% better TPRs with more than 15% lower FDRs as compared to Ref. [7]. The lower FDRs in the proposed method can be explained by the fact that the two-part-based method has a self-check mechanism in allowing positives. In other words, a subimage is considered positive if and only if both parts ($\mathbf{P}_1$ and $\mathbf{P}_2$) give a high classification score for the vehicle parts. This is however missing in conventional

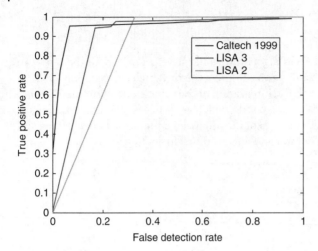

Figure 12.7 ROCs for detecting fully visible vehicles using the proposed two-part-based method.

Table 12.2 Comparison between ELVIS and Alvert.

	TPR (%)	FDR (%)	TP/frame	FP/frame
Caltech '99 [7]	85	5	NA	NA
LISA 2 [7]	83.5	79.7	1	4
LISA 3 [7]	98.1	45.8	3.16	2.7
ELVIS–Caltech '99	95	7	0.95	0.07
ELVIS–LISA 2	100	53.1	1	1.13
ELVIS–LISA 3	97.5	26.7	2.92	1.06

Source: Sivaraman and Trivedi [7]. Reproduced with permission of IEEE.

non-part-based method such as in Ref. [7]. Additionally, the use of IPM and the lane information to determine the region of interest for vehicle detection also removes false positives due to features outside the road area. It should also be noted that the figures in Table 12.2 for Ref. [7] involves tracking for LISA 2 and 3 datasets, whereas we have reported detection rates for ELVIS without tracking. We can see that FPs/frame have reduced by 2–3 times in ELVIS, which implies that tracking will be more effective in reducing the false positives further. Figure 12.8 shows the detection of fully visible vehicles on sample images from different datasets that were evaluated in Table 12.2.

12.5.1.2 Computational Efficiency

In order to compare the computational efficiency of the proposed vehicle detection method in ELVIS against variants of existing sliding window–based techniques, the following assumptions are made. First, the evaluation is performed for feature extraction and classification steps only, and the cost of peripheral steps such as nonmaximal suppression and tracking is assumed the same for all methods. Next, all techniques are assumed to use the same kinds of feature vectors and classifiers. It is to be noted that ELVIS can be implemented using different combinations of the types of feature vectors and classifiers. Therefore, the computational cost is evaluated based on the number of windows that are processed in each method. Also, it is to be noted, unlike computation time which is platform- and

Figure 12.8 Sample results of vehicle detection in ELVIS.

implementation-dependent itself, that computational cost is a more generic metric to compare different techniques. Computation cost can also be directly translated to computation time if required.

Let us first look at the computation cost of existing recent techniques such as in Refs. [7, 13]. In such methods, a sliding window of multiple scales is traversed all over the image to detect vehicles in every frame. Therefore, considering an $m \times n$ (rows by columns) sized image frame I, if the window slides over every s_m and s_n pixels along the rows and columns, respectively, the total number of windows that will be processed in existing methods is equal to

$$C_1 = n_s \frac{m \times n}{s_m \times s_n}, \tag{12.7}$$

where n_s is the number of scales that are used for detection.

In the proposed method, the scale of the window for each row in the image frame, and the start and end positions to slide the window in each row are determined by the formulations shown in Equations 12.1 and 12.2. Therefore, the total number of windows which is equivalent to the computational cost C_2 of the proposed method is equal to

$$C_2 = \sum_{j=0}^{\frac{y_{max}}{s_m}-1} \frac{x_{s_m j+1}^{max} - x_{s_m j+1}^{min}}{s_n}, \tag{12.8}$$

where x^{max} and x^{min} are the maximum and minimum x-coordinates in the image plane for the $s_m j+1$ th row. Equation 12.8 is the summation of the number of windows for each row in the image plane

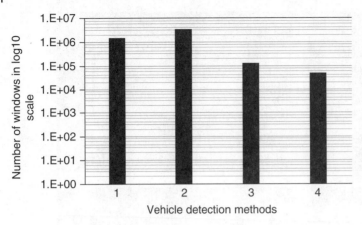

Figure 12.9 Comparison of computation costs in terms of number of windows (in log scale) on the *y*-axis and the following different techniques on the *x*-axis: (1) sliding window approach such as [7] bound by y_{max} and y_{min}, (2) sliding window approach on the entire image [7], (3) proposed method considering both parts of all windows are computed, and (4) proposed method considering both parts are computed for 10% of the total windows. Source: Sivaraman and Trivedi [7]. Reproduced with permission of IEEE.

and within y^{max}th and y^{max}th rows of the image where the step size of window traversal is s_m and s_n pixels along the rows and columns, respectively. It is to be noted that the number of scales n_s is not a factor in C_2. Additionally, Equation 12.8 gives the worst-case scenario for the proposed method because (12.8) assumes that the entire window is processed throughout the image. However, as discussed in Section 12.4, we process one-third of the window first, and the next two-third of the window is processed only if the lower part is detected.

Figure 12.9 shows the comparison of total computation cost in terms of the number of windows for four different types of methods. In order to compute the results in Figure 12.9, $m = 720$, $n = 1280$, $n_s = 10$, $s_m = 1$, and $s_n = 3$ are considered. The first two methods (methods 1 and 2 on the *x*-axis of Figure 12.9) refer to the conventional sliding window–based methods such as Ref. [7]. Method 2 refers to approaches where no knowledge of the region of interest is considered and the windows are applied on the entire image. In method 1, we assume that the vanishing line is available. We consider $y_{max} = 682$ and $y_{min} = 382$ for the results shown in Figure 12.9, which are derived from some of the test cases we considered during evaluation. Method 3 and 4 refer to the proposed method. As discussed previously, we are assuming the worst case in Method 3 by considering both parts $\mathbf{P_1}$ and $\mathbf{P_2}$ of all windows are computed. In Method 4, we consider 10% of the windows only for processing both parts, whereas the remaining 90% are processed for the lower part only (10% is significantly higher than the actual percentage we observed during our evaluation). Also, for Methods 3 and 4, we consider the same region of interest as in Method 1 in terms of *y*-axis. The *x*-axis limits are determined by the proposed LUT generation step. Figure 12.9 shows that the proposed method involves only 1/10th the number of computations as compared to 1 and 2. When the part-based computations are also accounted in Method 4, we see that the computations are further reduced. With such significant savings in computations, ELVIS is more suitable for embedded realization.

12.5.2 Lane Detection in ELVIS

In this section, we evaluate the performance of lane feature extraction step in ELVIS. Figure 12.10 shows sample results of a sequence on which the conventional LASeR algorithm and ELVIS are applied. From the first column of Figure 12.10, it can be seen that the LASeR algorithm tries to detect lane features beyond the obstructing lead vehicle in the ego-lane. In such scenarios, the leading vehicle tends to introduce features that are often mistaken for lane features. It can be seen that the lanes are estimated as curving at the far end of the field of view. Therefore, in the presence of an obstructing vehicle, the lane features that are detected by LASeR are not *credible* enough to extend the lanes

| Sequence from conventional LASeR | Sequence from ELVIS |

Figure 12.10 The first column shows lane detection on a sequence using the conventional LASeR algorithm. The second column shows the results of lane detection using the proposed integrated approach.

beyond the leading vehicle. Such observation about false lane features was also reported about other lane detection methods [18].

The second column in Figure 12.10 shows the lane detection results from ELVIS, wherein the lanes are detected till the position of the leading vehicle. It can be seen that the features of the vehicles do

Table 12.3 Comparison of lane position deviation (LPD) metric between ELVIS and LASeR.

Method	Mean absolute LPD (pixels)	Stan. dev. of absolute LPD (pixels)	Maximum absolute LPD (pixels)
ELVIS	8.3	3.3	18.7
LASeR [9]	9.3	3.3	19.1

Source: Satzoda and Trivedi [9]. Reproduced with permission of IEEE.

Table 12.4 Comparison of computational cost per frame for lane detection.

		This work 2014	
VioLET 2006 [17]	LASeR 2013 [9]	Average	Worst case
300,000	50,000	32,500	50,000

not interfere in the process of lane detection. It can be argued that the lanes are not being detected in the full field of view. It should be noted that in most ADAS applications, detecting the lanes between the ego-vehicle and the leading vehicle (if present) is more critical and necessary. If there is a requirement to detect the lanes beyond the obstructing vehicle, it is more accurate to extrapolate the lanes from existing lanes rather than depending on the lane features that are not visible due to obstructing vehicles. We compare the accuracy of lane estimation using lane position deviation (LPD) metric, which computes the average error between the positions of the lanes in ground truth and those estimated by a lane detection algorithm [24] for every frame. Table 12.3 shows the mean, standard deviation, and maximum absolute LPDs for the right lane from a set of 200 frames of the sequence shown in Figure 12.10. In the case of ground truth marking and ELVIS algorithm, if the lane markings are not visible due to the leading vehicle, the lanes are extrapolated as a tangent till the maximum y-coordinate that is used by conventional LASeR. It can be seen that the lanes estimated by ELVIS are more aligned with the ground truth. In a similar manner, the vehicles detected in the image scene can also be used to eliminate false lane features that are detected due to vehicles in neighboring lanes.

Let us now consider computational efficiency. If we consider ADAS applications that involve both lane and vehicle detection tasks, it can be seen that the proposed ELVIS will be computationally more efficient than LASeR. Considering that the number of bands changes dynamically based on the presence or absence of the leading vehicles in ELVIS, the computational cost can be compared based on a given test sequence. Using the formulation of computation cost for LASeR from Ref. [9], the cost in LASeR is equivalent to $C_{L1} = N_B w_B h_B$ where each band is w_B pixels wide and h_B pixels high. In other IPM-based lane detection algorithms such as in Ref. [17], the cost is given by $C_{L2} = w_B y_{max}^W$, where y_{max}^W is the height of the IPM image [9]. However, in ELVIS, the average computational cost per frame for detecting lanes is given by

$$C_{L3} = \frac{1}{N_f} \sum_{j=1}^{N_f} N_j^{max} w_B h_B \qquad (12.9)$$

For the test sequence that is evaluated in the chapter, Table 12.4 lists the average computational cost per frame. It is assumed that the lane detection is being performed for applications that employ vehicle detection also; therefore, the cost of vehicle detection is assumed to be the same for all methods. It can be seen that for the sequence considered, the ELVIS gives an average computation cost savings of nearly 35% as compared to LASeR and reduces the cost of the conventional lane detection method such as VioLET by at least 90%. The worst-case computation cost of the proposed approach is maximum when there are no leading vehicles and all bands in LASeR are processed. Therefore, the worst-case average computation cost of the integrated approach will never be more than conventional LASeR.

12.6   Concluding Remarks

In the chapter, a novel integrated approach called "ELVIS" is presented to detect on-road vehicles and lanes. We have shown that using the information about the lanes and the vehicle detection in a synergistic manner can help in detecting both vehicles and lanes in a computationally more efficient and robust manner. The proposed techniques have been shown to detect fully visible vehicles with lower false-positive rates and nearly 90% lesser computations as compared to existing methods. Similarly, the lane feature extraction method is also shown to benefit with significant cost savings by integrating the vehicle detection information. The proposed techniques need to be further extended for occluded vehicles, which is the next step in the proposed research.

References

1 Chakraborty, S., Lukasiewycz, M., Buckl, C., Fahmy, S., Leteinturier, P., Adlkofer, H. (2012) Embedded systems and software challenges in electric vehicles. *Proceedings of the 2012 15th Design, Automation and Test in Europe Conference and Exhibition (DATE)*, Dresden, March 12–16, 2012, 424–429, doi:10.1109/DATE.2012.6176508.

2 Venkatesh Prasad, K., Broy, M., Krueger, I. (2010) Scanning Advances in Aerospace and Automobile Software Technology. *Proceedings of the IEEE*, 98(4), 510–514, doi:10.1109/JPROC.2010.2041835.

3 Trivedi, M.M., Gandhi, T., McCall, J. (2007) Looking-In and Looking-Out of a Vehicle: Computer-Vision-Based Enhanced Vehicle Safety. *IEEE Transactions on Intelligent Transportation Systems*, 8(1), 108–120, doi:10.1109/TITS.2006.889442.

4 Lin, Z., Sankaran, J., Flanagan, T. (2013) Empowering automotive vision with TI's Vision AccelerationPac. *Texas Instruments White Papers*, 1–7. http://www.ti.com/lit/wp/spry251/spry251. pdf (accessed on September 20, 2016).

5 Stein, F. (2012) The challenge of putting vision algorithms into a car. *2012 IEEE Computer Society Conference on Computer Vision and Pattern Recognition Workshops*, Providence, RI, June 16–21, 2012, 89–94, doi:10.1109/CVPRW.2012.6238900.

6 Satzoda, R.K., Lee, S., Lu, F., Trivedi, M.M. (2015) Snap-DAS: A vision-based driver assistance system on a Snapdragon™ embedded platform. *IEEE Intelligent Vehicles Symposium (IV)*, Seoul, June 28 to July 1, 2015, 660–665.

7 Sivaraman, S. Trivedi, M.M. (2010) A General Active-Learning Framework for On-Road Vehicle Recognition and Tracking. *IEEE Transactions on Intelligent Transportation Systems*, 11(2), 267–276, doi:10.1109/TITS.2010.2040177.

8 Dalal, N. Triggs, W. (2005) Histograms of oriented gradients for human detection. *Proceedings of the 2005 IEEE Computer Society Conference on Computer Vision and Pattern Recognition (CVPR05)*, San Diego, CA, 1(3), 886–893.

9 Satzoda, R.K., Trivedi, M.M. (2013) Vision-based lane analysis: Exploration of issues and approaches for embedded realization. *Proceedings of the 2013 IEEE Conference on Computer Vision and Pattern Recognition Workshops on Embedded Vision*, Portland, June 23–28, 2013, 604–609, doi:10.1109/CVPRW.2013.91.

10 Marzotto, R., Zoratti, P., Bagni, D., Colombari, A., Murino, V. (2010) A Real-Time Versatile Roadway Path Extraction and Tracking on an FPGA Platform. *Computer Vision and Image Understanding*, 114(11), 1164–1179, doi:10.1016/j.cviu.2010.03.015.

11 Hsiao, P.Y., Yeh, C.W., Huang, S.S., Fu, L.C. (2009) A Portable Vision-Based Real-Time Lane Departure Warning System: Day and Night. *IEEE Transactions on Vehicular Technology*, 58(4), 2089–2094.

12 Satzoda, R.K. Trivedi, M.M. (2015) On Enhancing Lane Estimation Using Contextual Cues. *IEEE Transactions on Circuits and Systems for Video Technology*, 25(11), 1870–1881, doi:10.1109/TCSVT.2015.2406171.

13 Felzenszwalb, P.F., Girshick, R.B., McAllester, D., Ramanan, D. (2010) Object Detection with Discriminatively Trained Part-Based Models. *IEEE Transactions on Pattern Analysis and Machine Intelligence*, 32(9), 1627–45, doi:10.1109/TPAMI.2009.167.

14 Satzoda, R.K. Trivedi, M.M. (2016) Multi-part Vehicle Detection Using Symmetry Derived Analysis and Active Learning. *IEEE Transactions on Intelligent Transportation Systems*, 17, 926–937.

15 Sivaraman, S. Trivedi, M.M. (2013) Looking at Vehicles on the Road: A Survey of Vision-Based Vehicle Detection, Tracking, and Behavior Analysis. *IEEE Transactions on Intelligent Transportation Systems*, 14(4), 1773–1795, doi:10.1109/TITS.2013.2266661.

16 Bar Hillel, A., Lerner, R., Levi, D., Raz, G. (2012) Recent Progress in Road and Lane Detection: A Survey. *Machine Vision and Applications*, 25(3), 727–745, doi:10.1007/s00138-011-0404-2.

17 McCall, J. Trivedi, M. (2006) Video-Based Lane Estimation and Tracking for Driver Assistance: Survey, System, and Evaluation. *IEEE Transactions on Intelligent Transportation Systems*, 7(1), 20–37, doi:10.1109/TITS.2006.869595.

18 Sivaraman, S. Trivedi, M. (2013) Integrated Lane and Vehicle Detection, Localization, and Tracking: A Synergistic Approach. *IEEE Transactions on Intelligent Transportation Systems*, 14, 1–12.

19 Sun, Z., Bebis, G., Miller, R. (2006) On-Road Vehicle Detection: A Review. *IEEE Transactions on Pattern Analysis and Machine Intelligence*, 28(5), 694–711, doi:10.1109/TPAMI.2006.104.

20 Viola, P., Way, O.M., Jones, M.J. (2004) Robust Real-Time Face Detection. *International Journal of Computer Vision*, 57(2), 137–154.

21 Satzoda, R.K. Trivedi, M.M. (2014) Efficient lane and vehicle detection with integrated synergies (ELVIS). *Proceedings of the IEEE Conference on Computer Vision and Pattern Recognition Workshops on Embedded Vision*, Columbus, OH, June 23–28, 2014, 708–713.

22 Kyrkou, C. Theocharides, T. (2011) A Flexible Parallel Hardware Architecture for AdaBoost-Based Real-Time Object Detection. *IEEE Transactions on Very Large Scale Integration (VLSI) Systems*, 19(6), 1034–1047.

23 Enkelmann, W. (2001) Video-Based Driver Assistance: From Basic Functions to Applications. *International Journal of Computer Vision*, 45(3), 201–221.

24 Satzoda, R.K. Trivedi, M.M. (2013) Selective salient feature based lane analysis. *Proceedings of the 2013 IEEE Conference on Intelligent Transportation Systems*, The Hague, October 6–9, 2013, 1906–1911.

25 Bertozzi, M. Broggi, A. (1998) GOLD: A Parallel Real-Time Stereo Vision System for Generic Obstacle and Lane Detection. *IEEE Transactions on Image Processing*, 7(1), 62–81, doi:10.1109/83.650851.

26 Borkar, A., Hayes, M., Smith, M.T. (2012) A Novel Lane Detection System with Efficient Ground Truth Generation. *IEEE Transactions on Intelligent Transportation Systems*, 13(1), 365–374, doi:10.1109/TITS.2011.2173196.

13

Driver Monitoring

Raja Bala[1] and Edgar A. Bernal[2]

[1] *Samsung Research America, Richardson, TX, USA*
[2] *United Technologies Research Center, East Hartford, CT, USA*

13.1   Introduction

The US Department of Transportation's National Highway Traffic Safety Administration (NHTSA) reports that over 3000 fatalities from automobile accidents are caused by driver drowsiness or distraction [1]. In response, there is an emerging body of research on the use of in-vehicle cameras and sensors coupled with computer vision techniques to monitor driver behavior for enhanced safety [2–5]. As is often the case in the auto-industry, driver monitoring technologies enter the market initially in high-performance automobiles since the technologies rely upon sophisticated image capture and processing afforded by specialized hardware embedded in the vehicle. Over time, when the technologies mature and costs reduce due to scale of manufacturing, the same technologies make their way into the mainstream line of automobiles.

Computer vision is an attractive approach for driver monitoring due to ongoing advances in camera technology that can operate robustly on moving platforms, coupled with recent progress in the automatic analysis and recognition of facial pose and expression. A successful driver monitoring system is however not likely to depend solely on camera data, but will rather synergistically combine video with data from motion sensors, biosensors, and vehicle telematics to keep the driver alert and engaged with ongoing monitoring and feedback [6]. Such a scheme is shown schematically in Figure 13.1. A key goal is to look for *early* signs of driver fatigue so that feedback mechanisms can be activated ahead of an imminent mishap.

Given the emphasis of this text, we restrict the majority of the discussion on the camera as a sensor for monitoring driver behavior. A brief treatment on multisensor data fusion is given in Section 13.7. Broadly, driver monitoring addresses early detection of instances where attention is diverted, or about to be diverted from the task of driving. Two primary causes for inattention are fatigue and distraction. The former is induced by involuntary factors on the human condition, while the latter is usually voluntarily induced. As will be seen in the chapter, there is considerable overlap in techniques detecting fatigue and distraction as both require the ability to capture and localize the driver's head, face, and eyes under challenging conditions subject to motion, as well as widely varying lighting that can result in shadows and overexposure.

Computer Vision and Imaging in Intelligent Transportation Systems, First Edition.
Edited by Robert P. Loce, Raja Bala and Mohan Trivedi.
© 2017 John Wiley & Sons Ltd. Published 2017 by John Wiley & Sons Ltd.
Companion website: www.wiley.com/go/loce/ComputerVisionandImaginginITS

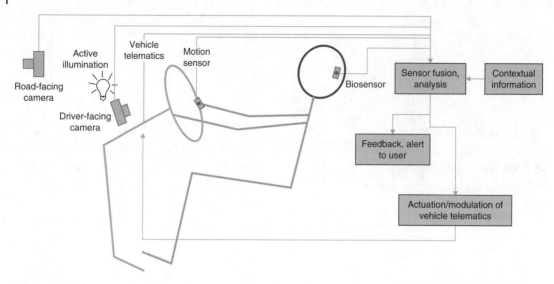

Figure 13.1 Schematic of a driver monitoring system utilizing cameras, sensors, and vehicle telematics.

The literature on the topic of fatigue monitoring can be discussed along two threads of investigation: methods that use a driver-facing camera to directly analyze driver state, and techniques that employ a road-facing camera to assess driver performance as an indirect indicator of fatigue. Methods in the second category focus on problems such as lane adherence monitoring and vehicle proximity estimation, and are covered in other chapters. The chapter is thus devoted primarily toward direct driver state monitoring with the inward-facing camera. The following sections describe the key functional elements required for such a system. Special attention is devoted in Section 13.5 to the fundamental problem of driver gaze estimation, required for measuring both fatigue and distraction.

13.2   Video Acquisition

The monitoring system begins at the acquisition step that captures video footage of the driver in real time. Factors to be considered include spatial and temporal resolution, spectral bands being captured, dynamic range, cost, form factor, lighting, and capture geometry. A variety of devices have been employed in the literature, ranging from smartphones [7–10] and webcams [11], to specially instrumented high-end cameras [12]. With the increasing abundance and declining cost of high-performance image sensors, many cameras today meet the requisite needs for spatial, temporal, and tonal resolution in a compact form factor. A principal remaining challenge is ensuring consistent and adequate lighting on the driver, especially during the night when fatigue is likely to occur. The common solution is to use near-infrared (NIR) active illumination, since this is effective in both daytime and nighttime conditions, and is less obtrusive to the driver than visible lighting. The method in Ref. [13], for example, uses an LED NIR illuminator with the LEDs distributed symmetrically in two concentric rings centered about the camera's focal axis. The two rings are used to generate bright pupil and dark pupil images to facilitate accurate eye detection and tracking, described in Section 13.4. Note that NIR imaging also requires that the sensor and filters in the camera be responsive in the NIR spectral regime. Fortunately, most standard silicon sensors are NIR-responsive, as are many common filters.

The geometrical setup of the camera is another important factor. The prime considerations are that the camera be on a stable surface, and that nominal driving behavior place the driver within the camera's field of view (FOV). A geometry that facilitates accurate estimation of head pose and gaze direction is desirable. A location on the dashboard is an obvious choice for a portable after-market solution. Automobile manufacturers have additional flexibility to embed the camera into the front panel or steering column in an unobtrusive manner. Multicamera configurations have also been proposed for providing enhanced FOV or stereo imaging [13].

Finally, motion stabilization is an important element needed to mitigate jitter due to vehicle movement. Optical stabilization is becoming increasingly prevalent in digital cameras; however, many digital algorithmic counterparts are available in the literature [14].

13.3 Face Detection and Alignment

Many fast and robust face detection technologies are available today [15–18] and are commonplace in almost all consumer digital cameras and smartphones. The classical method by Viola and Jones [15] is now presented. The algorithm parses the image in local windows of 24×24 pixels in size, and extracts binary Haar features, since these are efficient to compute and are discriminative of facial regions. The base Haar features are shown in Figure 13.2, wherein differences between sums of pixels in adjacent blocks are computed as the feature outputs. The base features are computed at multiple scales within the window resulting in a large number of over 180,000 features.

Next, a classification technique is required that maps the features to a binary class label (face vs. non-face). The Viola–Jones detector employs adaptive boosting (AdaBoost), a machine learning technique that learns both a small set of features and a binary classifier. Specifically, AdaBoost learns a strong classifier by successively learning and combining T weak classifiers $h_1, ..., h_T$, each of which need to only perform better than random chance. For simplicity, each weak classifier performs a binary test on only on one of the input feature dimensions. In detail, let x_i be the D-dimensional Haar feature vector for the ith training sample with corresponding binary class label y_i, $1 \leq i \leq N$, where $y_i = 0, 1$ for negative (i.e., non-face) and positive (i.e., face) samples, respectively. Let L and M denote the total number of positive and negative samples, respectively, in the training set, with $L + M = N$. The goal is to derive a classifier $h: x \rightarrow y$ that minimizes training error and generalizes well to independent test samples. AdaBoost learning proceeds as follows:

- Initialize sample weights $w_{1,i} = 1/2M, 1/2L$ respectively for $i = 0, 1$.
- For $t = 1, ..., T$:
  1) Normalize the weights to serve as a probability distribution.

$$w_{t,i} \leftarrow \frac{w_{t,i}}{\sum_{j=1}^{N} w_{t,j}}$$

  2) Derive optimal binary classifiers $h_j(\cdot)$, $1 \leq j \leq D$, that perform classification *only* along the jth scalar component of feature vectors x_i to minimize the classification error.

$$E_j = \sum_{i=1}^{N} w_{t,i} |h_j(x_i) - y_i|$$

  3) Choose the classifier $h_t(\cdot)$ with the lowest classification error denoted E_t.

4) Update the weights to assign smaller values to correctly classified samples. $w_{t+1,i} \leftarrow w_{t,i}\,\beta_t^{1-e_i}$, where $e_i = 0$ if x_i is classified correctly and 1 otherwise, and $\beta_t = E_t/1 - E_t$.

- The final strong classifier is given by

$$h(\mathbf{x}) = \begin{cases} 1 & \sum_{t=1}^{T}\alpha_t h_t(\mathbf{x}) \geq \dfrac{\sum_{t=1}^{T}\alpha_t}{2} \\ 0 & \text{otherwise} \end{cases}$$

where $\alpha_t = \log(1/\beta_t)$.

Figure 13.3 is an illustration of how 3-weak binary classifiers operating on a two-dimensional (2D) feature space are learned and combined to form a strong classifier with a complex decision boundary.

Two other crucial elements of the Viola–Jones algorithm are (i) the precomputation of an integral image, that is, an accumulating sum of pixel values over rectangular blocks that enables Haar features to be calculated in constant time; and (ii) a staged cascade of AdaBoost classifiers designed so that simpler classifiers (i.e., derived with small T) quickly eliminate a large number of false positives, while more complex classifiers (i.e., derived with large T) handle the fewer more difficult decisions.

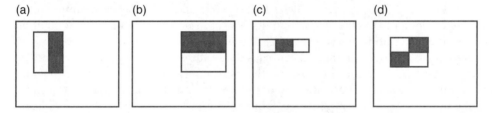

(a) (b) (c) (d)

Figure 13.2 Haar features used in the Viola–Jones face detector. The sum of pixel values within white rectangles is subtracted from the sum of pixels within gray rectangles: (a) horizontal two-rectangle feature, (b) vertical two-rectangle feature, (c) three-rectangle feature, and (d) four-rectangle feature.

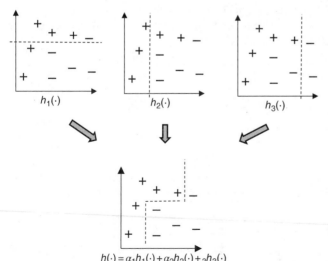

Figure 13.3 Schematic illustration of AdaBoost. Three weak classifiers $h_1(\cdot)$, $h_2(\cdot)$, and $h_3(\cdot)$ are trained on 2D training samples x_i, where + and − denote positive and negative classes, respectively. The first classifier operates only on the vertical dimension, while the second and third classifiers operate on the horizontal dimension. The final strong classifier formed as a weighted combination of the weak classifiers produces a more complex decision boundary, accurately separating the two classes. A desirable property of AdaBoost is that it is robust to overfitting even in the case when the input features are of very high dimensionality and the number of training samples is small.

Several enhancements and variants of this algorithm have been proposed, and the reader is referred to Refs. [16, 17] for details. Researchers have also treated face detection as an instance of the more general problem of object detection, and have adapted state-of-the-art object detection techniques such as the deformable parts model [19] for detecting faces. Yin et al. [20] employ a combination of Gabor filters, local binary patterns, and dynamic temporal histogramming to extract facial features for driver fatigue detection. A key requirement for the driver monitoring application is that face detection executes with speed at or close to the video frame rate.

After the face has been localized within the image or video frame, a useful next step is to identify features along important facial landmarks such as the eyes, nose, mouth, and chin. One approach is to train dedicated classifiers to detect each of these landmarks as object classes. Another approach is facial alignment, wherein a generic face shape with pre-identified landmarks is warped so that its landmarks align with those on the specific face within the image. Considerable progress has been made in the latter methodology in recent years [21–26]. In particular, iterative optimization on cascaded regression trees has yielded performance that is both computationally efficient and robust with respect to pose and illumination variations [21–25].

13.4 Eye Detection and Analysis

Eye behavior provides critical cues for driver fatigue. The first step is to reliably locate the eyes, a primary challenge being robustness with respect to varying lighting and occlusion from eyewear. Two categories of methods have been proposed. The first employs a passive methodology to locate eyes as features based on appearance or geometrical attributes within a detected face. Section 13.5 describes in detail an approach in this category. Face alignment techniques described in Section 13.3 can also be leveraged. The second flavor uses active IR illumination to identify the bright/dark pupil effect which helps localize the eyes [13, 27]. In the technique in Ref. [13], circuitry is devised to synchronize the inner and outer concentric rings of NIR LEDs with the even and odd fields of the interlaced video signal, respectively. Specifically, the inner LED ring produces the bright pupil image that is captured by even fields, and the outer LED ring produces the dark pupil image captured by odd fields. The difference between the bright and dark images produces a strong feature at the location of the feature. The even and odd fields are differenced, thresholded, and passed through various image processing operations to produce binary blobs denoting candidate eye regions. These regions in the original image are then classified into eye versus non-eye regions using a support vector machine (SVM) classifier. The flow diagram is shown in Figure 13.4.

The approach in Ref. [27] also produces a pair of bright/dark pupil images with a different optical configuration comprising two sets of NIR LED bars: one set placed immediately adjacent to the camera and the other placed in two clusters, 19.5 cm on either side of the camera. As with Ref. [13], camera capture is synchronized with the lighting, and the difference between the bright/dark pair of images is computed and thresholded. Binary blobs are mapped back to the original dark pupil image, and these image regions are sent as candidates to a binary classifier predicting the presence or absence of eye within the region. The step of bright/dark pupil imaging significantly speeds up eye detection by narrowing down the candidate regions to be processed through the classifier. Three classifiers are compared: SVM, artificial neural network (ANN), and AdaBoost. In their experiments, both AdaBoost and SVM performed comparably, reporting 97–98% classification accuracy, and both slightly outperformed ANN. AdaBoost is much faster to train than the other techniques; hence, it is recommended by the authors for practical eye detection applications. Once the eyes are detected, they must be tracked over time. In Ref. [13], the authors track the bright pupil

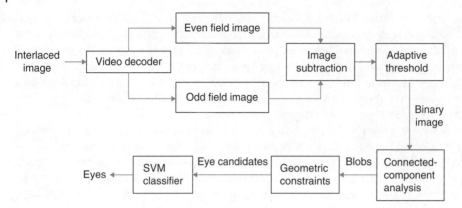

Figure 13.4 Block diagram of driver eye detection using dual active IR illumination. *Source*: Ji et al. [13]. Reproduced with permission of IEEE.

frames using a combination of Kalman filtering and mean-shift tracking. Many alternative tracking techniques may be used [28].

Following detection of the eyes, several metrics can be computed as indicators of fatigue. A popular measure is percent eye closure (PERCLOS) defined as the percentage of time within 1 min that the driver's eyelids are at least 80% closed, not counting rapid blinks [29]. If active IR illumination is used in the eye detection step, then PERCLOS can be measured by the presence and shape of the pupil since this is directly affected by eyelid occlusion. The authors in Ref. [13] observe that as the eyelid closes, the pupil shape becomes increasingly elliptical; hence, they track the eccentricity of the pupil ellipse over time. In a passive appearance-based framework, a standard classification pipeline can be used to distinguish "closed" versus "open" eye states over time. In Ref. [7] for example, SURF features are used in a linear SVM classifier to predict eye state on a per-frame basis, from which PERCLOS is easily calculated. Another indicator of driver fatigue is the rate of rapid blinks [7, 13], and can be obtained using the same techniques used to measure PERCLOS, provided the detection of open versus closed eye can be performed in real time.

13.5 Head Pose and Gaze Estimation

Head pose and eye gaze direction have been found to be the most important cues to infer driver attentiveness and intent [30]. There is an inherent trade-off between how informative a parameter is and how robustly and affordably it can be estimated. On the one hand, gaze estimates showcase slightly higher correlations with driving performance metrics than head-based measures. Gaze has been shown to be indicative of driving experience and to correlate well with awareness of surrounding vehicles [31]. On the other hand, research indicates that sensors required to measure head pose are more robust and affordable than those required to measure eye gaze [32].

13.5.1 Head Pose Estimation

Murphy-Chutorian and Trivedi [33] present an excellent overview of camera-based head pose estimation methods. The task of pose estimation consists of finding three Euler angles of rotation around three axes orthogonal to each other (i.e., yaw, pitch, and roll) best describing the orientation of a person's head. Depending on the type of imagery on which they rely, head pose estimation algorithms

Figure 13.5 Embedding of image poses in a manifold in two-dimensional feature space.

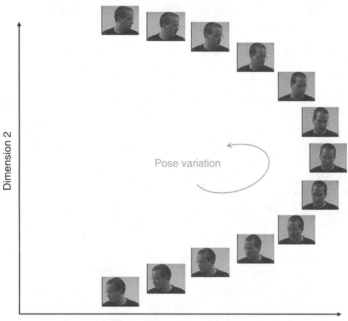

Dimension 2

Pose variation

Dimension 1

can be classified into 2D- and depth-based techniques. Approaches that rely on traditional 2D images can be classified into appearance- and feature-based methods. Appearance-based models typically discretize the head pose space and learn a separate classifier for each particular pose. For instance, the algorithm in Ref. [34] relies on a convolutional neural network (CNN) to achieve synergistic face detection and pose estimation. Feature-based models extract appearance features and train classifiers to learn the relationship between the different poses and the features. For example, the algorithm in Ref. [35] computes Haar-like features and estimates head pose with a neural network.

Many of these approaches rely on the assumption that the features describing the different poses of the head can be represented as a set of data points lying on a low-dimensional manifold, as illustrated in Figure 13.5 [36]. The majority of these techniques rely on learning, in a training stage, the parameters of the manifold that best models the statistical behavior of the feature representation of head poses in a training dataset. Once the manifold parameters are learned, the incoming test image of a previously unseen face and pose combination is projected or embedded into the manifold. The head pose corresponding to the incoming image is estimated based on the known pose of training images that lie in the neighborhood of the location where the test image was embedded. Manifold learning can be thought of as a nonlinear dimensionality reduction technique that preserves the coordinates capturing the intrinsic degrees of freedom of the training data samples. Two classes of approaches to manifold learning have been proposed: global and local models [36]. Isomap [37] is an example of a global modeling approach. It operates by constructing local neighborhoods of the training data in a feature representation space. The neighborhood of a given training sample in the manifold M may include its K-nearest neighbors or the set of all points within some fixed radius $\in$. A weighted graph G is constructed that represents the local neighborhoods, with the edge between sample feature i and j having a distance-based weight $d(i,j)$. Geodesic distances $d_M(i,j)$ between all pairs of points, measured along the surface of manifold M, are then computed by calculating shortest path distances $d_G(i,j)$ in the graph G. Embedding of the data in the lower-dimensional Euclidean space Y in which the manifold M resides is achieved by computing coordinate vectors y_i for each data point i that minimize the cost function

$$E = \| \tau(D_G) - \tau(D_Y) \|_{L^2} \tag{13.1}$$

where D_G denotes the matrix of graph distances with entries $d_G(i,j)$, D_Y denotes the matrix of Euclidean distances with entries $d_Y(i,j) = \| y_i - \mathbf{y}_j \|$, and $\| \cdot \|_{L^2}$ denotes the L^2 matrix norm $\sqrt{\Sigma_{ij} A_{ij}^2}$. The τ operator converts distances to inner products.

Local manifold modeling techniques rely on the assumption that the manifold is well represented by the training data so that the local geometry of the manifold can be well approximated by linear patches. If the assumption holds, then the error incurred in approximating a sample y_i with a weighted linear combination of its neighbors, that is, $\sum_j W_{ij} \mathbf{y}_j$, should be small. In order to minimize the reconstruction error, the optimal set of weights W_{ij} needs to be determined. Two constraints are enforced on W_{ij}: first, $W_{ij} = 0$ if $\mathbf{y}_j$ is not a neighbor of $\mathbf{y}_i$; second, the rows of the weight matrix W_{ij} have to sum to 1. The optimal set of weights is found by solving the least-squares problem that minimizes the reconstruction error across all training samples subject to the constraints listed earlier.

Another family of head pose estimation techniques rely on depth map acquisition and processing attempt to overcome the limitations imposed by traditional imagery including sensitivity to illumination, shadows, and imaging artifacts. The technique proposed in Ref. [38] computes a database of reference depth maps corresponding to different poses. When a new incoming depth map becomes available, signatures are extracted and compared with signatures extracted from the reference depth maps. The algorithm outputs the pose corresponding to the reference depth map that best matches the incoming depth map. The approach from Ref. [39] relies on offline training of a random forest regression model that learns the relationship between the nose position in 3D space and the gaze direction. As new depth maps become available, the location of the nose is determined and input to the model, which in turns output a head pose estimate. A third category of algorithms relies on both intensity images and depth maps in order to achieve head pose estimation. For instance, the algorithm in Ref. [40] employs random forests and tensor regression algorithms operating on fused depth and intensity data to achieve head pose estimation.

In the context of driver monitoring, head pose behavior across time is often more informative than instantaneous head pose [41]. For instance, frequent head tilts can be indicators of a fatigued driver. A driver who is looking frequently at directions other than the road may be distracted [13]. Studies have indicated that head motion is a good indicator of a driver's intention to turn [42] and change lanes [43]. The authors of Ref. [44] use continuous, nonrigid face tracking (i.e., determining the motion of mouth, eyebrows, cheeks, and facial expression) to estimate a driver's mental health.

13.5.2 Gaze Estimation

Several approaches have been proposed to track eye gaze in order to indirectly measure driver state [8, 45–47] or to guide attention toward regions of perceived increased risk [48], potentially missed traffic signs, and road events [49], proving to reduce accident rates. Eye gaze estimation has also been studied extensively in the domain of human–computer interaction [50, 51]. As mentioned earlier, gaze estimation approaches can be classified into passive methods which operate in ambient light conditions, or active methods that rely on external sources of illumination, usually in the NIR or IR band [52]. A primary challenge with active systems is the presence of interfering ambient light.

Passive gaze estimation systems can be classified into feature- and appearance-based [50]. Feature-based methods compute descriptors indicative of landmark locations including the iris and the eye corners [53] or shape primitives such as ellipses that indicate the location and orientation of the pupil and/or the iris [54]. The main downside of feature-based approaches is that high-resolution imagery is typically needed to accurately extract the relevant features. Appearance-based models, on the other hand, do not explicitly extract features, but rather learn a mapping from the image space onto gaze coordinate space. The learning is performed on large amounts of labeled gaze data [55, 56], and, as such, the resulting algorithms may not generalize well to conditions beyond those represented in the training dataset.

In order to address the intrinsic lack of generalizability of learning-based methods, the algorithm proposed in Ref. [8] introduces a novel *in situ* approach whereby a previously trained classifier is fine-tuned for a given driver/vehicle/camera setup in a driver monitoring application. This algorithm is now described in more detail. The primary motivation in Ref. [8] is to democratize real-time driver monitoring by foregoing reliance on specialized in-vehicle imaging and computing devices. To that end, the driver's cell phone camera and processing capabilities are used to carry out driver monitoring in real time. Thus computational efficiency and generalizability of the algorithm are critical. The algorithm implements a staged gaze estimation approach where an initial stage provides a coarse gaze direction estimate largely based on head pose, and a subsequent stage provides fine-grain gaze classification. The coarse estimate corresponds to one of three gaze directions, namely frontal and left- and right-side profiles, and is effected by a traditional three-class face detector with an AdaBoost classifier trained on Haar features of labeled images of the three face poses. The fine estimate operates only on frontal poses from the first stage, and classifies gaze into one of eight directions that are prevalent in a driving task, including left, top and right mirror, road, dashboard, traffic sign, music console, and phone, as illustrated in Figure 13.6.

The flow of the algorithm is illustrated in Figure 13.7. After the first stage of coarse head pose estimation, AdaBoost cascaded classifiers trained to detect the different facial landmarks, including the eyes, nose, and mouth, are run on the detected frontal face area, as illustrated in Figure 13.8. The initially identified facial regions are then tracked across subsequent frames as long the frontal pose is maintained, as determined by the coarse estimate. The location (x_i, y_i) and size (w_i, h_i) of each identified facial region i (where i corresponds to left eye, right eye, nose, and mouth) are used as input features to a multiclass linear SVM classifier trained to perform fine-grain gaze classification. The SVM is an eight-class classifier where each of the classes corresponds to one of the predetermined

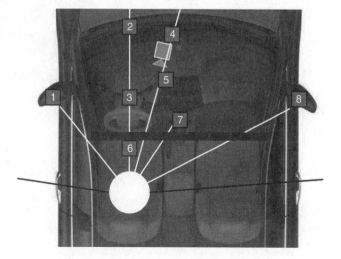

Figure 13.6 Fine-grain gaze locations supported by the algorithm in Ref. [8]. The fine estimate operates only on frontal poses from the first stage and classifies gaze into one of eight directions that are prevalent in a driving task. As shown in Figure 13.6, the selected directions are left rear mirror (1), road (2), dashboard (3), traffic sign (4), top mirror (5), phone/texting (6), music console (7), and right rear mirror (8). *Source*: Chuang et al. [8]. Reproduced with permission of IEEE.

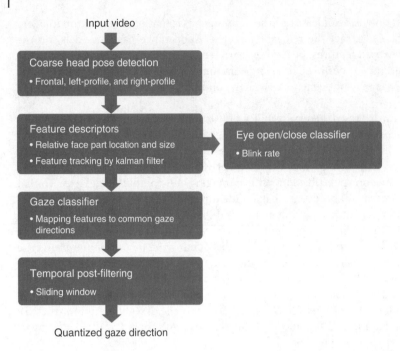

Figure 13.7 Flow diagram of the algorithm in Ref. [8]. *Source*: Chuang et al. [8]. Reproduced with permission of IEEE.

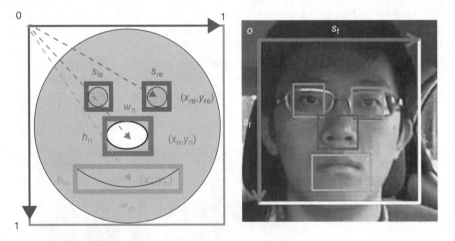

Figure 13.8 Features used for fine-grain gaze classification by the algorithm in Ref. [8]. *Source*: Chuang et al. [8]. Reproduced with permission of IEEE.

gaze directions. The space on which the SVM operates consists of a 14D feature space resulting from the concatenation of the location and size parameters for the different facial regions, namely

$$\left(x_{le}, y_{le}, s_{le}, x_{re}, y_{re}, s_{re}, x_n, y_n, w_n, h_n, x_m, y_m, w_m, h_m \right) \tag{13.2}$$

where subscripts le, re, n, and m denote left eye, right eye, nose, and mouth, respectively. Note that the eye regions are assumed to be square, and consequently only a single size parameter is used to describe their size attributes. Per-frame gaze direction decisions are buffered and a temporal smoothing filtering is applied to the decisions in order to reduce the effect of outliers and erroneous decisions. Additionally, analysis of the identified eye regions across time is used to estimate blink rate.

In order to improve generalizability across drivers, cameras, and vehicles, the method further proposes a short training procedure that personalizes the system to each particular driving configuration, whereby training data is gathered at system start-up, while the vehicle is stationary. To this end, the driver was asked to gaze in each of the eight supported directions for 4 s while video was acquired. While this step adapts the system to a specific driver, vehicle and camera, generalization from stationary to actual driving conditions was still limited. To address this issue, the stationary training data was augmented with data gathered during a few minutes of driving, and clustered in an unsupervised manner. The underlying assumption is that the dominant gaze direction when driving is "road facing," and so data samples belonging to the largest resulting cluster are assigned to the "road" class. Specifically, let X denote a random variable representing the behavior of the 14D feature vectors corresponding to the different gaze positions. Using the expectation–maximization (EM) algorithm [57], the authors estimated the parameters of a Gaussian mixture model that best describes the distribution of the features extracted during the driving stage of the training procedure. This model takes the following form:

$$P(X) = \sum_{i=1}^{K} w_i \Phi\left(X, \mu_i, \Sigma_i\right) \tag{13.3}$$

In Equation 13.3, K is the number of clusters (i.e., $K = 8$), w_i, μ_i, and Σ_i are estimates of the weight, the mean value, and the covariance matrix of the ith Gaussian component in the mixture, respectively; $\Phi(\cdot)$ is the multivariate Gaussian probability density function, namely

$$\Phi\left(X, u_i, \Sigma_i\right) = \frac{1}{(2\pi)^{D/2}|\Sigma_i|^{1/2}} \exp\left\{-\frac{1}{2}(X - u_i)^T \Sigma_i^{-1}(X - u_i)\right\} \tag{13.4}$$

where D is the dimensionality of the feature space (i.e., $D = 14$). Once a set of distribution parameters has been estimated, the probability that the feature vector extracted from the nth frame belongs to the jth cluster, $(m_n)_j$, can be computed. The dominant or "road" class can be identified by first determining the index j_{dom} of the class with the largest number of samples assigned to it by the model, and then by finding the set of samples that satisfies

$$\left\{X_i \mid \operatorname{argmax}_{j=1,\ldots,K}(m_n)_j = j_{dom}\right\} \tag{13.5}$$

The newly formed "road-facing" class during driving was combined with the training data acquired in the stationary vehicle to form an augmented training dataset based on which a new SVM classifier was trained. The obtained results are illustrated in Figure 13.9, which compares the performance of four different classifiers across four phone/vehicle configurations. The four classifiers considered include "Self" (i.e., when a classifier is used in the same environment it was trained), "Static to Moving" (i.e., when a classifier is trained in static conditions and tested in normal driving conditions), "Moving to Moving" (i.e., when a classifier is trained and tested in normal driving conditions), and "Proposed Method" which implements the described augmented dataset approach. It can be seen that the classifier based on the augmented dataset generalizes better than other classifiers studied, and its performance was shown to approach the upper bound case where a classifier is used for inference in the same environment in which it was trained.

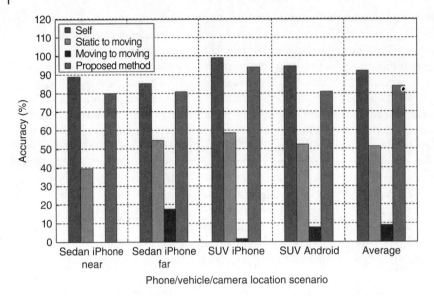

Figure 13.9 Generalizability of different classifiers proposed in Ref. [8]. *Source:* Chuang et al. [8]. Reproduced with permission of IEEE.

This approach of personalizing a computer vision pipeline with a human in the loop can be very powerful in improving generalizability from a controlled setting to a realistic "in the wild" environment.

13.6 Facial Expression Analysis

Facial expression can provide important cues about driver fatigue. The general problem of expression analysis is an active area of research within the computer vision community, and thus there is a rich body of literature to draw from to address the driver monitoring application. Many facial expression techniques draw upon the Facial Action Coding System (FACS) [58]. FACS describes every observable facial expression in terms of 46 basic action units (AUs) and their combinations. AUs correspond to localized muscular movements and formations in different regions of the face, as shown in Figure 13.10. A typical FACS- based facial expression pipeline comprises three stages: (i) face detection, (ii) detection of AUs, and (iii) mapping of AU activations to facial expression. A variety of machine learning techniques can be employed for each of these stages [60, 61]. This pipeline is typically carried out on a frame-by-frame basis.

Gu and Ji [62] apply the FACS framework to predict three classes of expressions pertaining to fatigue: inattention, yawning, and falling asleep. They identify AUs and AU combinations for each class; for example, AUs activated for the "falling asleep" category are head nods, blinks, nose wrinkle, and eyelid tighteners. Their novel contribution is that rather than simply applying static expression analysis on a frame-by-frame basis, they model spatiotemporal dependencies with a dynamic Bayesian network. Experiments show promising results for posed fatigue expressions.

Vural et al. [60] simulate driver drowsiness by asking subjects to participate in a video game for a 3 h session starting at midnight. The game simulates driving via the use of a steering wheel and a display, and randomly applies a "wind effect" that drags the vehicle to one side or the other, forcing

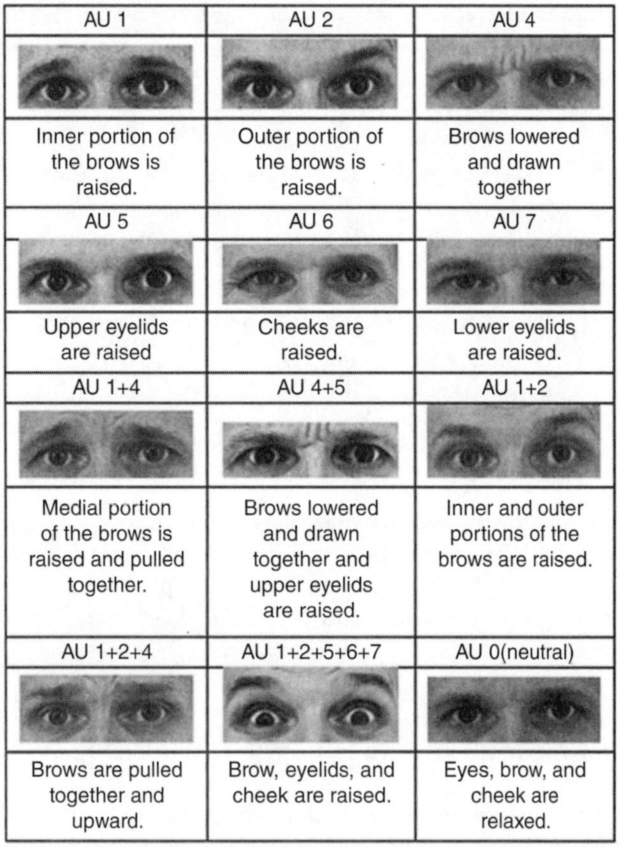

Figure 13.10 Examples of action units in the Facial Action Coding System. *Source:* Tian et al. [59]. Reproduced with permission of IEEE.

AU 1	AU 2	AU 4
Inner portion of the brows is raised.	Outer portion of the brows is raised.	Brows lowered and drawn together
AU 5	**AU 6**	**AU 7**
Upper eyelids are raised	Cheeks are raised.	Lower eyelids are raised.
AU 1+4	**AU 4+5**	**AU 1+2**
Medial portion of the brows is raised and pulled together.	Brows lowered and drawn together and upper eyelids are raised.	Inner and outer portions of the brows are raised.
AU 1+2+4	**AU 1+2+5+6+7**	**AU 0(neutral)**
Brows are pulled together and upward.	Brow, eyelids, and cheek are raised.	Eyes, brow, and cheek are relaxed.

the driver to take corrective action to stay on the road. The driver's facial expression is recorded during the entire session, and episodes during crashes are tagged. This procedure provides a dataset of facial expressions labeled as belonging to either the "alert" or "non-alert" class.

The computer vision pipeline begins with face detection, followed by preprocessing to align and crop the facial region to a fixed 96×96 block. Next, a Gabor filter bank is applied to the facial image to obtain 72 filter responses at nine spatial scales and eight orientations. The Gabor features are used to detect the presence of each of 31 facial AUs shown in Table 13.1 via binary SVM classifiers trained on several datasets of both posed and spontaneous facial expressions. For each SVM, the distance to the separating hyperplane is computed as a continuous measure of both the decision on presence of an AU, and confidence in the decision. The distance measures are then analyzed to assess which AUs are most discriminative of the alert versus non-alert classes from the driving simulation. Based on the principles of signal detection theory, the area under the receiver operating characteristic (ROC) curve for each binary detector is obtained as a measure of class separation afforded by a given AU. AU 45 (eye blink) was observed to be the most reliable predictor of drowsiness, confirming earlier studies on this subject. Interestingly, AU 2 (outer brow raise) was also found to be a discriminating feature; the authors' interpretation is that subjects often raised their eyebrows in an effort to stay awake. Equally interesting is that AU 26 (jaw drop) associated with yawning did not occur frequently in the moments prior to a crash episode. This suggests that while yawning is an indicator of overall tiredness, it is not a common occurrence in the final moments prior to falling asleep.

Table 13.1 Facial action units used for driver drowsiness analysis by Vural et al.

AU	Name	AU	Name	AU	Name	AU	Name
1	Inner brow raise	10	Upper lip raiser	18	Lip pucker	27	Mouth stretch
2	Outer brow raise	11	Nasolabial furrow deepener	19	Tongue show	28	Lips suck
4	Brow lowerer	12	Lip corner puller	20	Lip stretch	30	Jaw sideways
5	Upper lid raise	13	Sharp lip puller	22	Lip funneller	32	Bite
6	Cheek raise	14	Dimpler	23	Lip tightener	38	Nostril dilate
7	Lids tight	15	Lip corner depressor	24	Lip presser	39	Nostril compress
8	Lip toward	16	Lower lip depress	25	Lips part	45	Blink
9	Nose wrinkle	17	Chin raise	26	Jaw drop		

Source: Vural et al. [60]. Reproduced with permission of Springer.

The aforementioned analysis was carried out for single video frames. The authors additionally examine the effect of analyzing longer temporal windows, and found that performance improves as a function of time duration, reaching a plateau of 99% detection accuracy for 30 s intervals, and achieving diminishing returns beyond. The reader is referred to Ref. [60] for further details on the analysis.

13.7 Multimodal Sensing and Fusion

As mentioned at the beginning of the chapter, a driver monitoring system is more likely to achieve robust performance by acquiring and analyzing data from multiple sensors, both visual and nonvisual. Visual sensors offer the benefits of nonintrusiveness and richness of interpretation. Nonvisual sensors can often provide a more direct and accurate way to measure driver behavior and state; however, some devices are intrusive and involve contact with and cooperation from the driver. The subsequent discussion focuses on driver-facing technologies. Note that there is an analogous set of nonvisual sensors that complement the road-facing camera, such as radar and ultrasonic devices for determining proximity to other vehicles and objects. As such, multimodal sensor fusion is useful for combining complementary information offered by the different data modalities as well as reducing the effects of noise and disturbances in a moving vehicle.

One class of nonvisual sensors comprises inertial devices such as accelerometers and gyroscopes, as well as vehicle telematics that measure driving motion patterns indicative of driver inattention. Vural et al. [60] use accelerometers to observe motion of the driver's head and the steering wheel, and make the qualitative observation that both held motion and the correlation between head and steering wheel motion are greater in the non-alert than in the alert state. In a similar vein, Takei and Furukawa [63] perform simulated driving experiments, measure the angle of steering wheel motion, and suggest that Chaos theory could be used to detect fatigue from a wavelet domain representation of this signal. One general consideration is that data from motion sensors is often noisy, corrupt or missing, and thus require suitable preprocessing such as low-pass filtering, Kalman filtering, or transformation to an alternate representation such as the Fourier or wavelet domain.

Since drowsiness is a physiological condition, biological sensors are also commonly used to monitor driver behavior. Lee and Chung [11] use a pulse oximeter to measure the photo-plethysmographic

(PPG) signal representing blood flow under the skin of the finger, and from this compute heart rate and heart rate variability. Miyaji et al. [64] analyze electrocardiogram (ECG) signals to extract heart rate as an indicator of driver drowsiness. Wang and Gong [65] use contact sensors on the driver's left foot to measure skin conductance, respiration, blood pressure, and temperature to predict a number of emotional states including fatigue. Shin et al. [66] place ECG and PPG sensors on the steering wheel and estimate heart rate variability (HRV), defined as the variation in beat-to-beat interval in the temporal heart waveform. They compute the power spectral density of the HRV signal, and compute the ratio of the mean low-pass to mean high-pass power as a feature to distinguish alert versus drowsy states. The signal processing required to extract HRV from sensor data and infer driver state is carried out on a remote server. Lee and Chung [11] calculate four bio-features from PPG sensor data: (i) HRV, (ii) root mean square of the PPG signal over a specific time window, (iii) average signal gradient over a time window, and (iv) power spectral density. Additionally, they capture video of the driver's face; perform face detection via an active shape model [67]; and locate the eyes and estimate average eye size for each frame, yielding the following four features: (i) blink frequency, that is, the number of blinks per unit time period; (ii) speed of blinking; (iii) PERCLOS; and (iv) average eye closure speed. Electro-encephalographic (EEG) signals have also been analyzed for detecting driver drowsiness, although these sensors have not been generally favored in practical applications due to the difficulty in obtaining recordings in realistic driving scenarios [68].

Data that have been acquired from the various sensors, synchronized, and preprocessed must then be combined to facilitate inference on driver attention. Numerous multimodal fusion techniques exist in the literature. An excellent survey is presented by Shivappa et al. [69] who broadly categorize information fusion approaches into five categories as shown in Figure 13.11.

1) *Signal-level fusion*: Data streams are combined from sensors of the same type, such as a network of cameras or multiple microphones. A dual-camera configuration, for example, can help recover depth, thereby providing accurate information about driver pose and posture.
2) *Feature-level fusion*: This is a form of early fusion where features are extracted from each modality and combined into a single multimodal representation before being processed through a classifier. Combination techniques can include direct concatenation, weighted concatenation, and dimensionality reduction techniques such as principal components analysis and linear discriminant analysis.
3) *Classifier-level fusion*: This is a form of intermediate fusion wherein a statistical model such as a hidden Markov model (HMM) or dynamic Bayesian network (DBN) models each individual data stream. After each feature vector is separately processed, a composite classifier combines probability scores using a reliability weighting scheme on a frame-by-frame basis.
4) *Decision-level fusion*: Late fusion that processes features within a given modality through its own inference engine and pools class probability scores or likelihood values across all modalities and all video frames into a single decision via a reliability-based weighting scheme.
5) *Semantic-level fusion*: At this final level, other contextual information such as time of day, the driver's calendar and schedule, geo-location, and traffic conditions can be incorporated to aid the inference of driver state [70].

Hybrid architectures that fuse data at more than one of these levels are also conceivable [69]. With any of these strategies, it is important that inference take into account the temporally evolving state of the driver. As just mentioned, temporal Bayesian inference methods such as HMMs and DBNs have been frequently used for driver monitoring. Details can be found in Refs. [11, 65, 71].

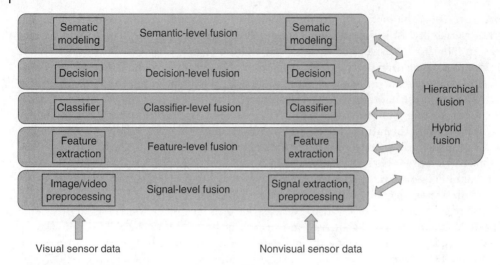

Figure 13.11 Hierarchy of data fusion at various levels of signal abstraction, described by Shivappa et al. *Source*: Shivappa et al. [69]. Reproduced with permission of IEEE.

13.8 Conclusions and Future Directions

This chapter has surveyed computer vision–based methods for driver monitoring with the end goal of safety and accident prevention. A sequence of modules have been described for analyzing eye behavior, head pose and movement, facial expression, and physiological state. The common aim for all these modules is early detection that the driver's attention is compromised due to fatigue or other distractions. One challenge that pervades all these functions is having to operate in a moving vehicle with widely varying illumination conditions. Active illumination and motion stabilization are key technologies for addressing these issues. Another challenge is the high level of inter-driver variability, prompting a need to personalize the inference and interpretation of driver behavior. This is exemplified in the driver gaze estimation problem in Section 13.5.

In the overall system, driver-facing analytics would be combined with environment-facing analytics such as lane departure and object proximity detection. Also additional relevant contextual information such as distance traveled and traffic conditions can be used to set thresholds for safe versus unsafe driving. Note that much of the same methodology used to monitor safety can also be used to monitor driver comfort and well being. For example 3D imaging has been used to monitor driver posture [72].

Once it has been detected that the driver's attention has been diverted, a suitable feedback mechanism must be triggered. A common proposal is tactile feedback such as vibration in the steering wheel [7], seat, or seatbelt [68]. Audio feedback is also a logical choice, ranging from simple alarms being proposed today [7] to sophisticated systems of the future wherein audio dialogs are auto-synthesized to engage the driver much like a companion passenger would. Also envisioned in the near future is automatic modulation of vehicle telematics as a more proactive form of feedback. Examples include automatic force feedback of the steering wheel during lane departure [68] or automatic braking if the vehicle is in close proximity with another object. Since such actions take over control of the vehicle, they require extremely accurate inferences on driver behavior and must be designed in such a fashion that the driver can easily override control.

Since driver monitoring must take place in real time, suitable computational architectures must be designed to accommodate fast execution of sophisticated inference algorithms. One approach is to design special-purpose processors embedded into the vehicle's computational platform and dedicated entirely to monitoring driver behavior [73]. Another alternative is to perform just the sensing, preprocessing, and feature extraction steps in the vehicle, and to execute computationally intensive analysis on a remote server [66]. While each has its pros and cons, it is likely that over time, dedicated in-vehicle computational engines will dominate as vendors of system-on-chip (SoC) architectures play an increasingly active role in smart automobile applications.

References

1 "Blueprint for ending distracted driving," NHTSA Report, 2012. Available in: http://www.distraction.gov (accessed on September 13, 2016).

2 M. Bertozzi, A. Broggi, A. Fascioli, "Vision-based intelligent vehicles: state of the art and perspectives," *Robotics and Autonomous Systems*, vol. 32, no. 1, pp. 1–16, July 2000.

3 M. Bertozzi, A. Broggi, M. Cellario, A. Fascioli, P. Lombardi, M. Porta, "Artificial vision in road vehicles," *Proceedings of the IEEE*, vol. 90, no. 7, pp. 1258–1271, July 2002.

4 L. M. Bergasa, J. Nuevo, M. A. Sotelo, R. Barea, M. E. Lopez, "Real-time system for monitoring driver vigilance," *IEEE Transactions on Intelligent Transportation Systems*, vol. 7, no. 1, pp. 63–77, March 2006.

5 M. M. Trivedi, T. Gandhi, J. McCall, "Looking-in and looking-out of a vehicle: computer-vision based enhanced vehicle safety," *IEEE Transactions on Intelligent Transportation Systems*, vol. 8, no. 1, pp. 108–120, 2007.

6 H. Kang, "Various approaches for driver and driver behavior monitoring: a review," In *Proceedings of the IEEE International Conference on Computer Vision Workshop*, Sydney Convention and Exhibition Centre, Sydney, December 1–8, 2013, IEEE Computer Society, Washington, DC, pp. 616–623, 2013.

7 C. You, N. D. Lane, F. Chen, "CarSafe App: alerting drowsy and distracted drivers using dual cameras on smartphones," In *Proceeding of the 11th Annual International Conference on Mobile Systems, Applications, and Services*, Taipei, Taiwan, June 25–28, 2013, ACM, New York, pp. 461–462, 2013.

8 M. Chuang, R. Bala, E. Bernal, P. Paul, A. Burry, "Estimating gaze direction of vehicle drivers using a smartphone camera," In *Proceedings of the IEEE CVPR Mobile Vision Workshop*, IEEE Computer Society, Washington, DC, pp. 165–170, 2014.

9 S. Makinist, E. Akin, A. Yilmaz, "Estimating driver behavior by a smartphone," In *IEEE Intelligent Vehicles Symposium*, Alcalá de Henares, June 3–7, 2012.

10 iOnRoad mobile app, Available in: http://www.ionroad.com (accessed on September 13, 2016).

11 B. Lee, W. Y. Chung, "Driver alertness monitoring using fusion of facial features and bio-signals," *IEEE Sensors Journal*, vol. 12, no. 7, pp. 2416–2422, 2012.

12 E. Murphy-Chutorian, A. Doshi, M. M. Trivedi, "Head pose estimation for driver assistance systems: a robust algorithm and experimental evaluation," In *IEEE Intelligent Transportation Systems Conference*, September 30–October 3, 2007, Bellevue, WA.

13 Q. Ji, Z. Zhu, P. Lan, "Real-time nonintrusive monitoring and prediction of driver fatigue," *IEEE Transactions on Vehicular Technology*, vol. 5, no. 4, pp. 1052–1068, 2004.

14 J. Yang, D. Schonfeld, M. Mohamed, "Robust video stabilization based on particle filter tracking of projected camera motion," *IEEE Transactions on Circuits and Systems for Video Technology*, vol. 19, no. 7, pp. 945–954, 2009.

15 P. Viola, M. Jones, "Rapid object detection using a boosted cascade of simple features," In *IEEE Computer Society Conference on Computer Vision and Pattern Recognition*, Kauai, HI, December 8–14, 2001.

16 J. Ren, N. Kehtarnavaz, L. Estevez, "Real-time optimization of Viola–Jones face detection for mobile platforms," In *IEEE CAS Workshop: SoC—Design, Applications, Integration, and Software*, 2008 IEEE Dallas Circuits and Systems Workshop on System-on-Chip, SoC: Design, Applications, Integration, and Software, DCAS 2008, Richardson, TX, October 19–20, 2008.

17 G. Gualdi, A. Prati, "Multistage particle windows for fast and accurate object detection," *IEEE Transactions on Pattern Analysis and Machine Intelligence*, vol. 34, no. 8, pp. 1589–1604, 2012.

18 J. Cheney, B. Klein, A. K. Jain, B. F. Klare, "Unconstrained face detection: state of the art baseline and challenges," In *Eighth International Conference on Biometrics, ICB 2015*, Phuket, May 19–22, 2015, Institute of Electrical and Electronics Engineers Inc., pp. 229–236, 2015.

19 P. F. Felzenszwalb, R. B. Girshick, D. McAllester, D. Ramanan, "Object detection with discriminatively trained part based models," *IEEE Transactions on Pattern Analysis and Machine Intelligence*, vol. 32, no. 9, pp. 1627–1645, 2010.

20 B. Yin, X. Fan, Y. Sun, "Multiscale dynamic features based driver fatigue detection," *International Journal of Pattern Recognition and Artificial Intelligence*, vol. 23, pp. 575–589, 2009

21 S. Ren, X. Cao, Y. Wei, J. Sun, "Face alignment at 3000 fps via regressing local binary features," In *Proceedings of the IEEE Conference on Computer Vision and Pattern Recognition (CVPR)*, Columbus, OH, June 24–27, 2014, pp. 1685–1692, 2014.

22 V. Kazemi and J. Sullivan, "One millisecond face alignment with an ensemble of regression trees," In *Proceedings of the IEEE Conference on Computer Vision and Pattern Recognition (CVPR)*, Columbus, OH, June 24–27, 2014, pp. 1867–1874, 2014.

23 D. Lee, H. Park, C. D. Yoo, "Face alignment using cascade Gaussian process regression trees," In *Proceedings of the IEEE Conference on Computer Vision and Pattern Recognition (CVPR)*, Boston, MA, June 7–12, 2015, pp. 4204–4212, 2015.

24 S. Zhu, L. Cheng, C.C. Loy, X. Tang, "Face alignment by coarse-to-fine shape searching," In *Proceedings of the IEEE Conference on Computer Vision and Pattern Recognition (CVPR)*, Boston, MA, June 7–12, 2015, pp. 4998–5006, 2015.

25 G. Tzimiropoulos, "Project-out cascaded regression with an application to face alignment," In *Proceedings of the IEEE Conference on Computer Vision and Pattern Recognition (CVPR)*, Boston, MA, June 7–12, 2015, pp. 3659–3667, 2015.

26 M.A. Haque, K. Nasrollahi, T. B. Moeslund, "Quality aware estimation of facial landmarks in video sequences," In *Proceedings of the IEEE Conference on Winter Applied Computer Vision*, Waikoloa Beach, HI, January 5–9, 2015, pp. 678–685, 2015.

27 R. C. Coetzer, G. P. Hancke, "Eye detection for a real-time vehicle driver fatigue monitoring system," In IEEE Intelligent Vehicles Symposium, Baden-Baden, June 5–9, 2011, pp. 66–71, 2011.

28 A. Yilmaz, O. Javed, M. Shah, "Object tracking: a survey," *ACM Computing Surveys*, vol. 38, no. 4, 109–153, 2006.

29 D. F. Dinges, R. Grace, "PERCLOS: A valid psychophysiological measure of alertness as assessed by psychomotor vigilance," US Department of Transportation, Federal highway Administration. Publication Number FHWA-MCRT-98-006.

30 A. Doshi, M. M. Trivedi, "A comparative exploration of eye gaze and head motion cues for lane change intent prediction," In *IEEE Intelligent Vehicles Symposium*, Eindhoven, June 4–6, 2008, pp. 49–54, 2008.

31 M. Mori, C. Miyajima, P. Angkititrakul, T. Hirayama, L. Yiyan, N. Kitaoka, K. Takeda, "Measuring driver awareness based on correlation between gaze behavior and risks of surrounding vehicles,"

In *Proceedings of the IEEE ITSC*, 15th International IEEE Conference on IEEE, Anchorage, AK, September 16–19, 2012, pp. 644–647, 2012.

32 H. Zhang, M. Smith, and R. Dufour. A final report of safety vehicles using adaptive interface technology: visual distraction (online). Available in: https://www.volpe.dot.gov/safety-management-and-human-factors/surface-transportation-human-factors/save-it-reports-and (accessed on September 13, 2016).

33 E. Murphy-Chutorian, M. Trivedi, "Head pose estimation in computer vision: a survey," *IEEE Transactions on Pattern Analysis and Machine Intelligence*, vol. 31, no. 4, pp. 607–626, 2009.

34 M. Osadchy, Y. Cun, M. Miller, "Synergistic face detection and pose estimation with energy-based models." *Journal of Machine Learning Research*, vol. 8, p. 1197–1215, 2007.

35 T. Vatahska, M. Bennewitz, S. Behnke, "Feature-based head pose estimation from images," In *Seventh IEEE-RAS International Conference on Humanoid Robots*, Omni William Penn Hotel, Pittsburgh, PA, November 29–December 1, 2007, pp. 330–335, 2007.

36 V. Balasubramanian, J. Ye, S. Panchanathan, "Biased manifold embedding: a framework for person-independent head pose estimation," In *IEEE Conference on Computer Vision and Pattern Recognition*, Hyatt Regency, Minneapolis, MN, USA, June 17–22, 2007, IEEE Computer Society, Washington, DC, pp. 1–7, 2007.

37 J. B. Tenenbaum, V. D. Silva, J. C. Langford. "A global geometric framework for nonlinear dimensionality reduction," *Science*, vol. 290, no. 5500, pp. 2319–2323, 2000.

38 M. Breitenstein, D. Kuettel, T. Weise, I. Van Gool, H. Pfister, "Real-time face pose estimation from single range images," In *IEEE Conference on Computer Vision and Pattern Recognition, CVPR 2008*, Anchorage, AK, June 23–28, 2008, pp. 1–8, 2008.

39 G. Fanelli, J. Gall, I. Van Gool. "Real time head pose estimation with random regression forests," In *Computer Vision and Pattern Recognition, CVPR*, Colorado Springs, CO, June 20–25, 2011, pp. 617–624, 2011.

40 S. Kaymak, I. Patras, "Exploiting depth and intensity information for head pose estimation with random forests and tensor models," In *Proceedings of the Asian Conference on Computer Vision Workshops*, Daejeon, November 5–9, 2012, pp. 160–170, 2012.

41 A. Tawari, S. Martin, M. M. Trivedi, "Continuous head movement estimator for driver assistance: issues, algorithms and on-road evaluations," *IEEE Transactions on Intelligent Transportation Systems*, vol. 15, no. 2, pp. 818–830, 2014.

42 S. Y. Cheng, M. M. Trivedi, "Turn-intent analysis using body pose for intelligent driver assistance," *IEEE Pervasive Computing*, vol. 5, no. 4, pp. 28–37, October–December 2006.

43 J. C. McCall, D. P. Wipf, M. M. Trivedi, B. D. Rao, "Lane change intent analysis using robust operators and sparse bayesian learning," *IEEE Transactions on Intelligent Transportation Systems*, vol. 8, no. 3, pp. 431–440, September 2007.

44 S. Baker, I. Matthews, J. Xiao, R. Gross, T. Kanade, T. Ishikawa, "Real-time non-rigid driver head tracking for driver mental state estimation," Carnegie Mellon University, Technical Report, CMU-RI-TR-04-10, Robotics Institute, Pittsburgh, PA, February, 2004.

45 T. Ishikawa, S. Baker, I. Matthews, T. Kanade, "Passive driver gaze tracking with active appearance models", In *Proceedings of the World Congress on Intelligent Transportation Systems*, Nagoya, October 17–22, 2004.

46 E. Wahlstrom, O. Masoud, N. Papanikolopoulos, "Vision-based methods for driver monitoring," In *IEEE Proceedings on Intelligent Transportation Systems*, vol. 2, 2003 IEEE International Conference on Intelligent Transportation Systems, ITSC, Shanghai, October 12–15, 2003, pp. 903–908, 2003.

47 Q. Ji, X. Yang, "Real-time eye, gaze, and face pose tracking for monitoring driver vigilance," *Real-Time Imaging*, vol. 8, pp. 357–377, 2002.

48 L. Pomarjanschi, M. Dorr, C. Rasche, E. Barth, "Safer driving with gaze guidance," In *Proceedings of Bionetics 2010—Fifth International ICST Conference on Bio-Inspired Models of Network, Information, and Computing Systems*, Boston, MA, December 1–3, 2010.

49 L. Fletcher, A. Zelinsky, "Driver inattention detection based on eye gaze—road event correlation," *International Journal of Robotics Research*, vol. 28, no. 6, pp. 774–801, 2009.

50 D. Hansen, Q. Ji, "In the eye of the beholder: a survey of models for eyes and gaze," *IEEE Transactions on Pattern Analysis and Machine Intelligence*, vol. 32, no. 3, pp. 478–500, 2010.

51 M. Reale, T. Hung, L. Yin, "Pointing with the eyes: gaze estimation using a static/active camera system and 3D iris disk model," In *IEEE International Conference on Multimedia and Expo (ICME)*, Singapore, July 19–23, 2010.

52 A. Tawari, K. H. Chen, M. M. Trivedi, "Where is the driver looking: analysis of head, eye and iris for robust gaze zone estimation," In *Proceedings of the IEEE Conference on Intelligent Transportation Systems*, Qingdao, October 8–11, 2014, pp. 988–994, 2014.

53 J. Chen, Q. Ji, "3D gaze estimation with a single camera without IR illumination," In *Proceedings of the International Conference on Pattern Recognition (ICPR)*, Rensselaer Polytechnic Institute, Troy, NY, December 8–11, 2008.

54 S. Kohlbecher, S. Bardinst, K. Bartl, E. Schneider, T. Poitschke, M. Ablassmeier, "Calibration-free eye tracking by reconstruction of the pupil ellipse in 3D space," In *Symposium on Eye Tracking Research and Applications (ETRA)*, March 26–28, 2008, Savannah, GA, 2008.

55 Y. Sugano, Y. Matsushita, Y. Sato, "Learning-by-synthesis for appearance-based 3D gaze estimation," In *Proceedings of the IEEE Conference on Computer Vision and Pattern (CVPR)*, Greater Columbus Convention Center, Columbus, OH, June 24–27, 2014, pp. 1821–1828, 2014.

56 T. Schneider, B. Schauerte, R. Stiefelhagen. "Manifold alignment for person independent appearance-based gaze estimation," In *Proceedings of the International Conference on Pattern Recognition (ICPR)*, IEEE, Stockholm, August 24–28, 2014.

57 A. P. Dempster, N. M. Laird, D. B. Rubin, "Maximum likelihood from incomplete data via the EM algorithm," *Journal of the Royal Statistical Society*, vol. 39, no. 1, pp. 1–38, 1977.

58 P. Ekman, W. Friesen, *Facial Action Coding System: A Technique for the Measurement of Facial Movement*, Consulting Psychologists Press, Palo Alto, CA, 1978.

59 Y. Tian, T. Kanade, J. F. Cohn, "Recognizing upper face action units for facial expression analysis," In *Proceedings of the IEEE Conference on Computer Vision Pattern Recognition*, Hilton Head, SC, June 13–15, 2000, pp. 294–301, 2000.

60 E. Vural, M. Cetin, A. Ercil, G. Littlewort, M. Bartlett, J. Movellan, "Drowsy driver detection through facial movement analysis," In M. Lew, N. Sebe, T. S. Huang, E. M. Bakker (eds.) *Human Computer Interaction*, Lecture Notes in Computer Science, vol. 4796, pp. 6–18, Springer-Verlag, Heidelberg, 2007.

61 Y. Tian, T. Kanade, J. F. Cohn, "Recognizing action units for facial expression analysis," *IEEE Transactions on Pattern Analysis and Machine Intelligence*, vol. 23, no. 2, pp. 97–115, 2001.

62 H. Gu, Q. Ji, "An automated face reader for fatigue detection," In *Proceedings of the IEEE Conference on Automatic Face and Gesture Recognition*, IEEE Computer Society, Seoul, May 17–19, 2004, pp. 111–116, 2004.

63 Y. Takei, Y. Furukawa, "Estimate of driver's fatigue through steering motion", In *Man and Cybernetics, 2005 IEEE International Conference*, vol. 2, October 10–12, 2005, Hilton Hotel Waikoloa, HI, pp. 1765–1770, 2005.

64 M. Miyaji, H. Kawanaka, K. Oguri, "Driver's cognitive distraction detection using physiological features by the Adaboost," In *Proceedings of the 12th International IEEE Conference on Intelligent*

Transportation Systems, Marriott St. Louis Union Station, St. Louis, MO, October 4–7, 2009, pp. 1–6, 2009.

65 J. Wang, Y. Gong, "Recognition of multiple drivers' emotional state," In *Proceedings of the 19th International Conference on Pattern Recognition*, Tampa, FL, December 8–11, 2008, pp. 1–4, 2008.

66 H. S. Shin, S. J. Jung, J. J. Kim, W. Y. Chung, "Real time car driver's condition monitoring system," In *Proceedings of IEEE Sensors*, Waikoloa, HI, November 1–4, 2010, pp. 951–954, 2010.

67 Z. L. Zheng, F. Yang, "Enhanced active shape model for facial feature localization," In *Proceedings of the International Conference on Machine Learning and Cybernetics*, Kunming, July 12–15, 2008, pp. 2841–2845, 2008.

68 Q. Wang, J. Yang, M. Ren, Y. Zheng, "Driver fatigue detection: a survey," In *Proceedings of the IEEE Sixth World Congress on Intelligent Control & Automation*, Dalian University of Technology, Dalian, June 21–23, 2006, pp. 8587–8591, 2006.

69 T. S. Shivappa, M. M. Trivedi, B. D. Rao, "Audiovisual information fusion in human–computer interfaces and intelligent environments: a survey," *Proceedings of the IEEE*, vol. 98, no. 10, pp. 1692–1715, 2010.

70 A. Tawari, M. M. Trivedi, "Contextual framework for speech based emotion recognition in driver assistance system," In *Proceedings of the IEEE Conference on Intelligent Vehicles Symposium*, University of California, San Diego, CA, June 21–24, 2010, pp. 174–178, 2010.

71 Y. Guosheng, L. Yingzi, B. Prabir, "A driver fatigue recognition model based on information fusion and dynamic Bayesian network," *Information Sciences*, vol. 180, no. 10, pp. 1942–1954, 2010.

72 C. Tran, M. M. Trivedi, "Towards a vision-based system exploring 3D driver posture dynamics for driver assistance: issues and possibilities," In *Proceedings of the IEEE Conference on Intelligent Vehicles Symposium*, University of California, San Diego, CA, June 21–24, 2010, pp. 179–184, 2010.

73 J. Kim, H. Shin (eds.), *Algorithm & SoC Design for Automotive Vision Systems*, Springer-Verlag, Dordrecht, 2015.

14

Traffic Sign Detection and Recognition

Hasan Fleyeh

Dalarna University, Falun, Sweden

14.1   Introduction

Traffic signs define a visual language that can be interpreted by drivers. They represent the current traffic situation on the road, show danger and difficulties around the drivers, give them warnings and help them with their navigation by providing useful information that makes driving safe and convenient [1, 2].

The human visual perception abilities depend on the individual's physical and mental conditions. In certain circumstances, these abilities can be affected by many factors such as fatigue and observatory skills. Giving this information in a good time to drivers can prevent accidents, save lives, increase driving performance and reduce the pollution caused by vehicles [3–5].

Traffic sign recognition is a field which is concerned with the detection and recognition of traffic signs in traffic scenes acquired by a camera. It is a technique which uses computer vision and artificial intelligence to extract the traffic signs from outdoor images taken in uncontrolled lighting conditions where these signs may be occluded by other objects, and may suffer from different problems such as colour fading, disorientation and variations in shape and size. It is the field of study that can be used to aid the development of an inventory system (for which real-time recognition is not required) or to aid the development of an in-car advisory system (when real-time recognition is necessary). Both road sign inventory and road sign recognition are concerned with traffic signs, face similar challenges and use automatic detection and recognition.

A traffic sign recognition system could in principle be developed as part of intelligent transport systems (ITS) that continuously monitors the driver, the vehicle and the road, for example, to inform the driver in time about upcoming decision points regarding navigation and potentially risky traffic situations. Detection of these signs in outdoor images from a moving vehicle will help the driver to take the right decision at the right time, which means fewer accidents, less pollution and better safety.

Computer Vision and Imaging in Intelligent Transportation Systems, First Edition.
Edited by Robert P. Loce, Raja Bala and Mohan Trivedi.
© 2017 John Wiley & Sons Ltd. Published 2017 by John Wiley & Sons Ltd.
Companion website: www.wiley.com/go/loce/ComputerVisionandImaginginITS

14.2 Traffic Signs

Traffic signs are those that use a visual/symbolic language about the road(s) ahead that can be interpreted by drivers. They provide the driver with pieces of information that make driving safe and convenient. Road signs are designed, manufactured and installed according to strict regulations [6]. However, they can appear in different conditions, including partly occluded, distorted, damaged and clustered in a group of more than one sign [7, 8].

Traffic signs are characterised by a number of features which ideally make them recognisable with respect to the environment, but certain factors can also affect a driver's perception. They are, but not limited to, as follows:

- They are designed in fixed two-dimensional (2D) shapes such as triangles, circles, octagons or rectangles [7, 9].
- The colours of the signs are chosen to contrast with the surroundings, which make them easily recognisable by drivers [10].
- The colours are regulated by the sign category [11].
- The information on the sign has one colour and the rest of the sign has another colour.
- The tint of the paint which covers the sign should correspond to a specific wavelength in the visible spectrum [6, 8].
- The signs are located in well-defined locations with respect to the road, so that the driver can, more or less, anticipate the location of these signs [11].
- They may contain a pictogram, a string of characters or both [8].
- In every country the traffic signs are characterised by using fixed text fonts and character heights.

14.2.1 The European Road and Traffic Signs

According to 1968 Vienna Convention on Road Signs and Signals, the European traffic signs are categorised into four groups:

- Warning signs: This group of traffic signs indicates a hazard ahead on the road. It is characterised by an equilateral triangle with a thick red rim and a white or yellow interior. A pictogram is used to specify different warnings. Other signs such as the Yield sign, the distance to level crossing signs and track level crossing also belong to this class.
- Prohibitory signs: They are used to prohibit certain types of manoeuvres for some types of traffic. The no entry, no parking and speed limit signs belong to this category. Normally, they are designed in a circular shape with a thick red rim and a white or yellow interior. There are few exceptions; the Stop sign is an octagon with a red background and white rim, the No Parking and No Standing signs have a blue background. The end-of-restriction signs are marked with black bars.
- Mandatory signs: They are characterised by a complete blue circle and a white arrow or pictogram. They control the actions of drivers and road users. Signs ending obligation have a diagonal red slash.
- Indicatory and supplementary signs: These signs are characterised by using rectangles with different background colours such as yellow, green or blue. The pictograms are either white or black. This category includes the diamond-shaped rectangle and the signs which give information about road priority.

The colours used on road signs have specific wavelengths in the visible spectrum. They are selected to be distinguishable from the natural and man-made surroundings so that they can be easily recognisable by road users. The Swedish traffic signs are illustrated in Figure 14.1.

(a)

(b)

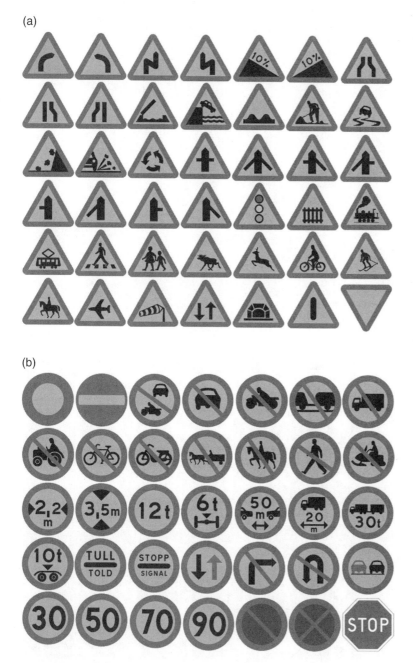

Figure 14.1 The Swedish traffic signs. (a) Warning signs, (b) prohibitory signs, (c) mandatory signs and (d–f) indicatory and supplementary signs.

(c)

(d)

(e)

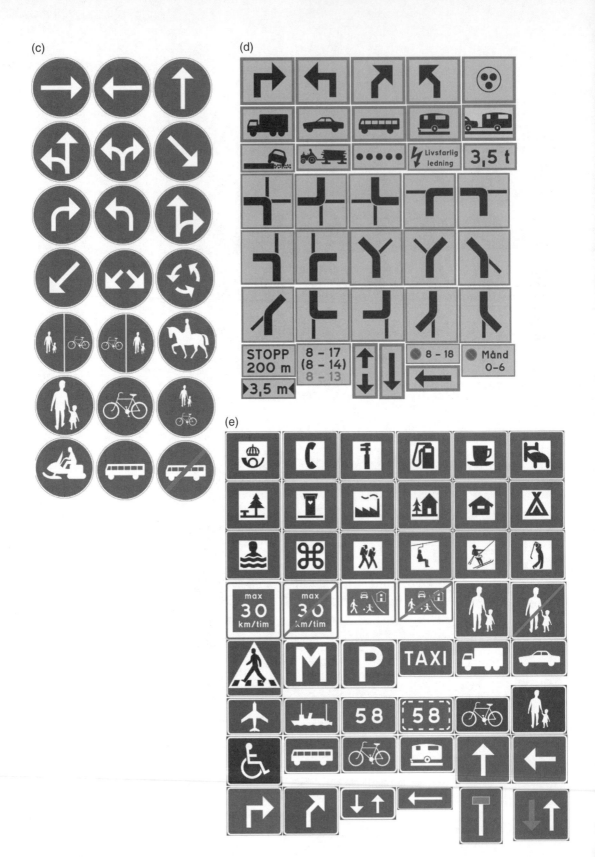

Figure 14.1 (Continued)

(f)

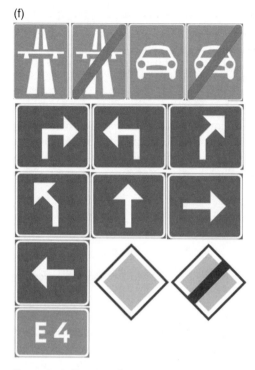

Figure 14.1 (Continued)

14.2.2 The American Road and Traffic Signs

American traffic signs follow the *Manual on Uniform Traffic Control Devices* and its companion 'Standard Highway Signs'. They did not follow Vienna Convention on Traffic Signs and Signals, and they are divided into eight categories: Regulator, Schools, Warning, Guide, Toll Road Signs, Hospital, Non-Compliant to MUTCD Signs and References. Regulatory and Warning signs are further divided into several subgroups called Series. These groups are completely different than that of European traffic signs. Yield and Stop signs are grouped in R1 Series, for instance, and speed limit signs are characterised by white background and black text [12].

14.3 Traffic Sign Recognition

The first paper on the subject was published in Japan in 1984 [13]. The aim was to try various computer vision methods for the detection of road signs in outdoor scenes. Since that time many research groups and companies have shown interest, conducted research in the field and generated an enormous amount of work. Different techniques have been used to cover different application areas, and vast improvements have been achieved during the past decade.

The identification of the traffic signs is achieved through two main stages: detection and recognition. In the detection phase, the image is preprocessed, enhanced and segmented according to the sign properties such as colour or shape or both. The output is a segmented image containing potential regions which could be recognised as possible road signs. The efficiency and speed of the detection are important factors because they reduce the search space and indicate only potential regions. In the recognition stage, each of the candidates is tested against a certain set of features (patterns) to decide whether it is in the group of traffic signs or not, and then according to these features they are classified into different groups. These features are chosen so as to emphasise the differences among the classes. The shape of the sign plays a central role in this stage and the signs are classified into different classes such as triangles, circles and octagons. Pictogram analysis allows a further stage of classification. By analysing pictogram shapes together with the text available in the interior of the sign, it is easy to decide the individual class of the sign under consideration. The system can be implemented by either colour information or shape information or both. Combining colour and shape may give better results if the two features are available, but many studies have shown that detection and recognition can be achieved even if one component, either colour or shape, is missing.

14.4 Traffic Sign Recognition Applications

Techniques for traffic sign detection and recognition have been developed in a range of application areas. These include the following:

- Driver support system (DSS): This system can detect and recognise traffic signs in real time. The system can also help to improve traffic flow and safety [14, 15], and avoid hazardous driving conditions, such as collisions. Until the past decade, there were very small number of studies about traffic sign detection and classification. Research groups focused on other aspects of sign detection, more related to the development of an automatic pilot, such as the detection of the road borders and/or the recognition of obstacles in the vehicle's path, for example, other vehicles or pedestrians. Intelligent vehicles such as Volvo S, V and XC series introduced systems which could take decisions about their speed, trajectory, etc., depending on the signs detected [16, 17]. Although, in the future, it can be part of a fully automated vehicle, now it can be a support to automatically limit the speed of the vehicle, send a warning signal indicating excess speed, warn or limit illegal manoeuvres or indicate earlier the presence of a sign to the driver. The general idea is to support the driver in some tasks, allowing him or her to concentrate on driving.
- Highway maintenance: This is used to check the presence and condition of the signs. Instead of an operator watching a video, which is a tedious work because the signs appear from time to time and the operator should pay a great attention to find the damaged ones, the traffic sign detection and recognition system can do this job automatically for the signs with good conditions and alerts the operator when the sign is located but not classified which can be interpreted as a sign in poor condition.
- Sign inventory: The many millions of roadway signs necessary to keep roadways safe and traffic flowing present a particular logistical challenge for those responsible for the installation and maintenance of those signs. Traffic signs must be properly installed in the necessary locations and an inventory of those signs must be maintained for future reference.
- Mobile robots: Road and traffic signs can be used to automatically mobilise robots (unmanned vehicles) depending on the detection and recognition of these landmarks by the robot [14]. Mei et al. developed an unmanned vehicle which has the ability to navigate in urban environments without GPS or any other satellite navigation system. The vehicle navigates solely by perceiving traffic signs [18].

14.5   Potential Challenges

In addition to the complex environment of the roads and the scenes around them, traffic signs can be found in different conditions such as aged, damaged and disoriented as depicted in Figure 14.2. Hence, the detection and recognition of these signs may face one or more of the following difficulties:

- The colour of the sign fades with time as a result of long exposure to sun light, and the reaction of the paint with the air [2, 5].
- Visibility is affected by the weather conditions such as the fog, rain, clouds and snow [2].
- Visibility can be affected by local light variations such as the direction of the light, the strength of the light depending on the time of the day and the season and the shadows generated by other objects [8, 19, 20].
- Colour information is very sensitive to the variations of the light conditions such as shadows, clouds and the sun [2, 5, 21]. It can be affected by illuminant colour (daylight), illumination geometry and viewing geometry [22].
- The presence of obstacles in the scene, such trees, buildings, vehicles and pedestrians or even signs which occlude other signs [19, 21].
- The presence of objects similar in colour and/or shape to the road signs in the scene under consideration, such as buildings or vehicles [5, 19]. They could be similar to the road sign in colour or shape or both.
- Signs may be found disoriented, damaged or occluded by any kind of obstacles, even by some other signs.
- The size of the sign depends on the distance between the camera and the sign itself. Traffic signs may appear rotated due to the imaging orientation [23].
- The acquired image often suffers from motion blur and car vibration [24]. This motion blur cannot be predicted above a certain level, because the car movements are not known to the recognition process. It is possible to make an assertion about the movements of objects in the future if the motion is continuous and unchanged.
- Sign boards can appear to have bright white or near-white spots due to first surface reflection from the light sources. In first surface reflection the light is reflected prior to penetrating to a depth where certain wavelengths are absorbed, thereby imparting a colour associated with the sign. This is called 'highlight'.
- Vandalism of sign boards by people who put stickers or write on them or damage the pictograms by changing the pictogram shape.
- Different countries use different colours (Figure 14.3) and different pictograms (Figure 14.4).
- The absence of a standard database for evaluation of the existent classification methods [25].

It is extremely important for the algorithms to be developed for the detection and recognition of road and traffic signs to have high robustness of colour segmentation, high insensitivity to noise and brightness variations, and should be invariant to geometrical effects such as translation, in-plane and out-plane rotations and scaling changes in the image [26, 27].

14.6   Traffic Sign Recognition System Design

A system to detect and recognise traffic signs should be able to work in two modes: the training mode in which a database can be built by collecting a set of traffic signs for training and validation, and the prediction mode in which the system can recognise a traffic sign which has not been seen before. Figure 14.5 illustrates the main stages to recognise a traffic sign.

Figure 14.2 Potential challenges when working with traffic signs. (a) Faded sign, (b) bad weather condition, (c) bad lighting geometry, (d) obstacles in the scene, (e) similar background colour, (f) damaged sign, (g) distance related size, (h) motion blur, (i) reflection from sign board and (j) stickers.

(g)

(h)

(i)

(j)

Figure 14.2 (Continued)

(a)

(b)

Figure 14.3 Different countries have different colour standard. (a) The Netherlands and (b) Sweden.

(a) 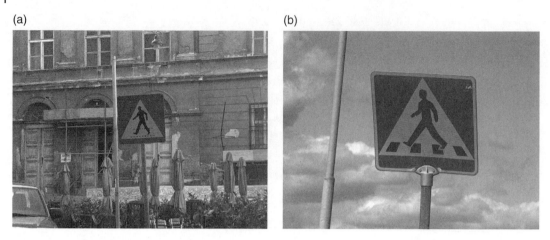 (b)

Figure 14.4 Traffic sign pictograms are different in different countries. (a) Austria and (b) Sweden.

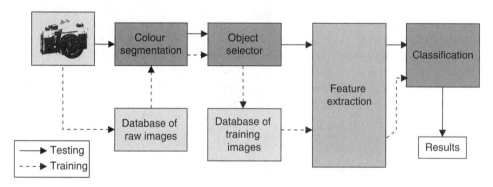

Figure 14.5 Block diagram of traffic sign recognition system.

Recognition and classification of traffic signs can be achieved by combining the two main traffic sign features: colour and shape. This method helps the recognition algorithm to perform in a better way and to reduce the number of false alarms generated by this algorithm. Therefore, the detection and recognition of different signs requires testing the presence of different colour combinations in the image together with the presence of the specific shape. Therefore, recognition and classification is carried out by two stages. In the first stage colour segmentation is applied. Two rim colours exist for traffic signs in Sweden: red and blue. A traffic sign shape tree is built according to these two colours as depicted in Figure 14.6. Based on the colour of the traffic sign's rim, the type of the traffic sign is specified by combing the shape of the rim with its pictogram. This requires two different shape analysis stages: the shape of the rim and the analysis of the pictogram. Very often, this shape analysis is achieved by training two classifiers to classify the rim and the pictogram.

14.6.1 Traffic Signs Datasets

There are three main public traffic sign datasets in Europe; among them are two Swedish datasets and one German. The datasets were collected for research purposes and create a standard benchmark for developing and testing algorithms for traffic sign recognition.

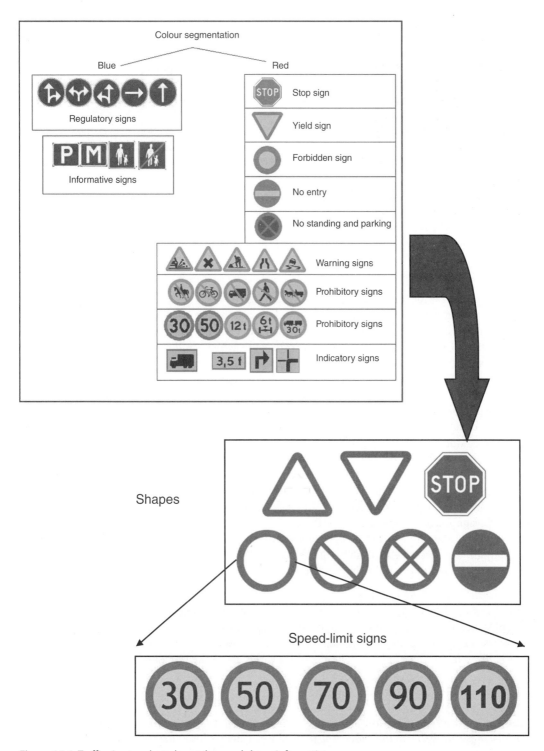

Figure 14.6 Traffic sign tree based on colour and shape information.

Dalarna University in Sweden released a traffic sign dataset which consists of 4338 image collected in Sweden and 330 images collected in other counties [28]. All still images were taken manually when traffic signs were seen by the camera operator. They were collected in different light conditions, in different weather conditions and in different road conditions including different speeds. For all images and without any exception, the camera was set to 640 × 480 pixels. Images in this database are classified into 30 categories depending on weather conditions, type of the sign, sign condition, image condition and light geometry.

Ruhr-Universität Bochum in Germany released the German Traffic Sign Benchmark dataset which is a multiclass, single-image in conjunction with the International Joint Conference on Neural Networks (IJCNN) 2011. The benchmark traffic sign dataset contains 40 traffic sign classes. It consists of more than 50,000 images [29].

Linköping University in Sweden released a public traffic signs dataset in conjunction with Scandinavian Conference on Image Analysis (SCIA) 2011 conference. The dataset contains 20,000 images with 20% of the images are labelled. It contains 3488 traffic signs which were collected from highways and cities from more than 350 km of Swedish roads [30].

14.6.2 Colour Segmentation

Colours represent an important part of the information provided to the driver to ensure the objectives of the traffic sign. Therefore, traffic signs and their colours are selected to be different from the nature or from the surrounding in order to be distinguishable. One of the important steps in Traffic Sign Recognition (TSR) is segmentation. It is the process by which candidate objects are specified for further analysis according to certain properties such as colour or shape. Colour segmentation has always been considered a strong tool for image segmentation because it is computationally inexpensive. Due to the fact that colour gives more information than grey it is used in segmentation algorithms instead of edge-based and luminance histogram-based techniques.

In order to develop a robust colour segmentation algorithm, it is necessary to understand the nature of the colour and the circumstances under which the traffic scene image is collected. The first problem for any colour segmentation algorithm is that the apparent colour of the object varies because of the chromatic variation of daylight. The second problem is that the colours of traffic sign boards change with age, which means newly installed traffic signs have different colours than older traffic signs. In addition to this, different countries use different standard colours for traffic signs. Finally, traffic sign images may suffer from the effect of shadows and highlights. The effect of shadows occurs when different parts of the object are exposed to different illumination levels, while in the case of highlight the object reflects some of the light of the illuminant directly to the viewer. These problems are discussed in the following text in detail.

1) *Colour variations in outdoor images*
   One of the most difficult problems in using colours in outdoor images is the chromatic variation of daylight which causes the apparent colour of the object to vary as daylight changes. The irradiance of any object in a colour image depends on the following three parameters:
   The colour of the incident light: Daylight's colour varies along the CIE curve. It is given by

$$y = 2.87x - 3.0x^2 - 0.275 \text{ for } 0.25 \leq x \leq 0.38 \tag{14.1}$$

   The variation of daylight's colour is a single variable which is independent of the intensity.
   The reflectance properties of the object: The reflectance of an object $s(\lambda)$ is a function of the wavelength λ of the incident light. It is given by

$$s(\lambda) = e(\lambda)\phi(\lambda) \tag{14.2}$$

where $e(\lambda)$ is the intensity of the light at wavelength λ and $\phi(\lambda)$ is the object's albedo at each wavelength.

The camera properties: The observed intensities depend on the lens diameter d, its focal length f and the image position of the object measured as angle a off the optical axis. This is given by

$$E(\lambda) = L(\lambda) \cdot \left(\frac{\pi}{4}\right)\left(\frac{d}{f}\right)^2 \cos(4a) \tag{14.3}$$

According to Equation 14.3, the radiance $L(\lambda)$ is multiplied by a constant which will not affect an object's observed colour. By cancelling the camera's lens chromatic aberration, only the density of the observed light will be affected.

As a result, the colour of the light reflected by an object located outdoors is a function of the temperature of the daylight and the object's albedo [22, 31].

2) *Aging of traffic signs*

Traffic signs may be mounted on a pole for a long time without any kind of maintenance or replacement. Over time the properties of the material used to reflect the light in these traffic signs changes its properties because of environmental reactions and so its colour fades. Figure 14.7 depicts two traffic signs; the sign on the left is a new one, while that on the right is an old one.

To investigate the effect of aging, a set of equal number of old and new traffic signs was selected. Colour located in the red part of each traffic sign was extracted and converted from RGB (red, green, blue) into HSV (hue, saturation, value). HSV was selected because it is invariant to the variations in light conditions as it is multiplicative/scale invariant, additive/shift invariant, and it is invariant under saturation changes. In addition, it has been proven by Gevers and Smeulders [32] that Hue is invariant against shadow and highlights.

Vitabile et al. [8] defined three different areas in the HSV colour space as follows:

1) The *achromatic* area, characterised by $s \leq 0.25$ or $v \leq 0.2$ or $v \geq 0.9$.
2) The *unstable chromatic* area, characterised by $0.25 \leq s \leq 0.5$ and $0.2 \leq v \leq 0.9$.
3) The *chromatic* area, characterised by $s \geq 0.5$ and $0.2 \leq v \leq 0.9$.

(a)

(b)

Figure 14.7 Effect of aging. (a) A new traffic sign. (b) An old traffic sign.

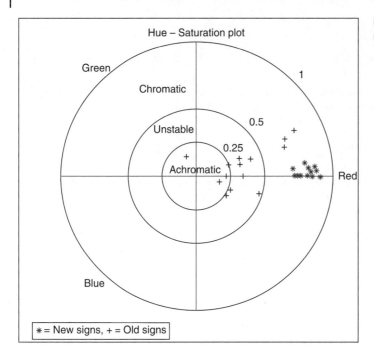

Figure 14.8 Hue–Saturation plot of new and old traffic signs.

Plotting the Hue and Saturation of each traffic sign in the Hue–Saturation plot is shown in Figure 14.8. The new traffic signs were grouped in the chromatic region, while old traffic signs shifted either towards the yellow part of the Hue or towards the unstable area or even towards the achromatic area of this plot. Old traffic signs which moved toward the yellow area but still in the chromatic area can be colour segmented, while the colour of the remaining traffic signs cannot be segmented because of the loss of Hue.

3) *Traffic signs from different countries*

   Different countries use different colours for their traffic signs. Figure 14.3 shows two yield traffic signs from the Netherlands (left) and Sweden (right). To understand the distribution of traffic signs from different countries on the Hue–Saturation plot, a number of traffic signs were collected from a number of countries in Europe, the United States and Japan, Figure 14.9. Plotting the red colour on the Hue–Saturation plot clearly indicates that the red colour of these traffic signs is not similar and those countries do not follow any standard colour.

4) *Robustness of colour spaces against shadows and highlights*

   Shadows and highlights represent a big challenge to computer vision researchers. In the first case different parts of the object are exposed to different illumination levels, and in the second case the object reflects some of the light of the illuminant directly to the viewer.

   Consider an image of a certain surface patch which is illuminated by an incident light with a certain spectral power density (SPD) denoted $e(\lambda)$. This image is taken by a camera with RGB sensors characterised by their spectral sensitivities $f_C(\lambda)$ for $C = \{R,G,B\}$. The Cth sensor response of the camera is given by

$$C = m_b\left(\mathbf{n,s}\right)\int_\lambda f_C\left(\lambda\right)e\left(\lambda\right)c_b\left(\lambda\right)d\lambda + m_s\left(\mathbf{n,s,v}\right)\int_\lambda f_C\left(\lambda\right)e\left(\lambda\right)c_s\left(\lambda\right)d\lambda \qquad (14.4)$$

Figure 14.9 Hue–Saturation of traffic sign colours of different countries.

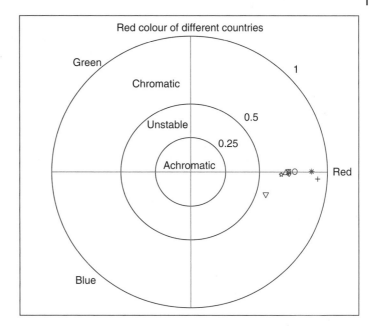

Figure 14.10 Traffic sign reflection model.

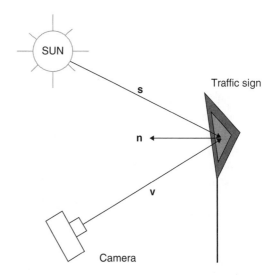

for $C = \{R, G, B\}$ where $c_b(\lambda)$ and $c_s(\lambda)$ are the body and surface albedo, respectively; λ is the wavelength at which the sensor responds; and **n**, **s**, **v** are unit vectors that represent the direction of the normal vector to the surface patch, direction of the source of illumination and direction of the viewer, respectively, Figure 14.10.

Furthermore, the terms m_b and m_s denote the geometric dependencies on the body and surface reflection component, respectively [32].

Assuming that surface albedo $c_s(\lambda)$ is constant and independent of the wavelength, and white illumination is used (white illumination means equal energy for all wavelengths within the visible spectrum), then $e(\lambda) = e$ and $c_s(\lambda) = c_s$ which are constants. The sensors responses can be modified as follows:

$$C_w = em_b\left(\mathbf{n},\mathbf{s}\right)k_C + em_s\left(\mathbf{n},\mathbf{s},\mathbf{v}\right)c_s \int_\lambda f_C\left(\lambda\right)d\lambda \ \text{ for } \ C_w = \left\{R_w, G_w, B_w\right\} \tag{14.5}$$

In Equation 14.5, C_w is the response of the RGB sensors under the assumption of white light source, and k_C is given by

$$k_C = \int_\lambda f_C\left(\lambda\right)c_b\,d\lambda \tag{14.6}$$

where k_C is the compact formulation depending on the sensors and the surface albedo only. If the assumption of white illumination holds, then

$$\int_\lambda f_R\left(\lambda\right)d\lambda = \int_\lambda f_G\left(\lambda\right)d\lambda = \int_\lambda f_B\left(\lambda\right)d\lambda = f \tag{14.7}$$

and the reflection of the surface can be given by

$$C_w = em_b\left(\mathbf{n},\mathbf{s}\right)k_C + em_s\left(\mathbf{n},\mathbf{s},\mathbf{v}\right)c_s\,f \tag{14.8}$$

The first term of Equation 14.6 represents the effect of shadows on colour invariance, while the second term is effect of highlights on colour invariance. Fleyeh [33] has tested different colour spaces and showed that Hue is the only colour feature which is invariant against viewing direction, surface orientation, highlight, illumination direction and illumination intensity, as depicted in Table 14.1.

Colour segmentation can be designed and implemented in different ways. Simple rule-based colour segmentation algorithms were developed in the literature [34]. However, these algorithms cannot deal with all of the aforementioned problems. Therefore, more advanced and adaptive algorithm which can tackle these problems is needed. Fleyeh and Mumtaz [35] suggested a colour segmentation algorithm for traffic signs based on self-organising maps (SOM). SOM is a powerful unsupervised clustering technique which can be utilised to solve the problem of colour segmentation. Colour segmentation is based on reducing the number of colours in an image. It is achieved by selecting the most representative colours from all colours present in the original image, making a colour palette, and then mapping each colour in the image to the nearest colour in the palette. Since each colour in the segmented image represents an object, only a few colours are desired in the colour palette.

Table 14.1 Invariance of colour models to imaging conditions.

Colour feature	Viewing direction	Surface orientation	Highlight	Illumination direction	Illumination intensity
I	N	N	N	N	N
RGB	N	N	N	N	N
Nrgb	Y	Y	N	N	N
H	Y	Y	Y	Y	Y
S	Y	Y	N	Y	Y

'Y' denotes invariance and 'N' denotes sensitivity of colour models to imaging conditions.

Figure 14.11 Dynamic search boundary by the proposed algorithm for the red colour.

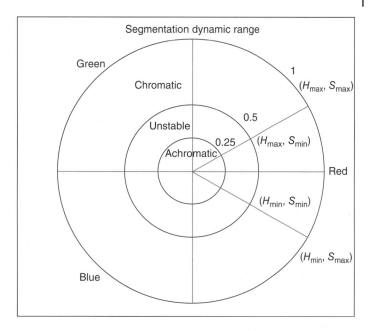

An important feature on which the quality of segmentation depends is the search space boundaries. RGB images were converted into HSV colour space and the *H* and *S* values of the traffic signs were observed to assign a reasonable search boundary in the *H* and *S* space. Since the values of *H* and *S* differ from one traffic sign to another, it is not wise to choose static boundaries. Instead, boundaries of a search subspace in the *H–S* plot are specified dynamically by the SOM which is trained to find the upper and lower limits of *H* and *S* through the best matching neurons. It is, therefore, believed that the algorithm is adaptive and can be used in a wide range of environmental conditions and in different countries.

The values H_{min}, H_{max}, S_{min} and S_{max} in Figure 14.11 are specified dynamically by the SOM. This means that the position and the size of the search space dynamically depends on the desired colour to be segmented.

The training process divides the grid of nodes into most dominant colours depending on the size of the grid. For example if the grid size is 3, a maximum of nine dominant colours present in the test image are obtained. These nine colours can then be used to assign the boundary values for the colour under consideration through its neurons obtained by the proposed approach. The best matching neurons have their own associated weight vectors which can be used to segment the image. Therefore the larger the size of the grid the greater the number of representative colours which will be considered and thus better segmentation can be achieved. Results of colour segmentation are depicted in Figures 14.12 and 14.13, respectively.

14.6.3 Traffic Sign's Rim Analysis

The shape of the traffic sign's rim can be determined by different methods. One method is by exploiting Hough transform. Hough transform in its original form is not suitable for complex object detection such as triangles, circles and rectangles due to noise and shape imperfection. Circular Hough transform can instead be invoked to detect these shapes. It can be described as a transformation of centre point of a circle in *x–y* plane to the parameter space. The equation of a circle in *x–y* plane is given by

Figure 14.12 Colour segmentation results of traffic sign images in different light conditions.

$$(x-a)^2 + (y-b)^2 = r^2 \tag{14.9}$$

where a and b are the centre point of the circle in the x and y direction and r is the radius of the circle. The parametric representation of the circle is given by

$$x = a + r\cos(\theta) \tag{14.10}$$

$$y = b + r\sin(\theta) \tag{14.11}$$

Figure 14.13 Traffic sign images from different countries. (Top left to bottom right) Austria, France, Germany, Japan, The Netherlands, and Poland.

To determine the presence of any traffic sign, it is necessary to accumulate votes in the 3D parameter space (a, b, r). The objective is to find the coordinates (a, b) of the object's centroid and the locus of the objects circumference. A voting mechanism which aims to find the distribution of the votes in the Hough space is illustrated in Figure 14.14.

This voting mechanism specifies the number of votes given to any object in the image and its location. In this mechanism, each white pixel of the object is considered as the centre of a set of concentric circles with different radii. For each radius of the set of circles, a vote is given to each intersection of

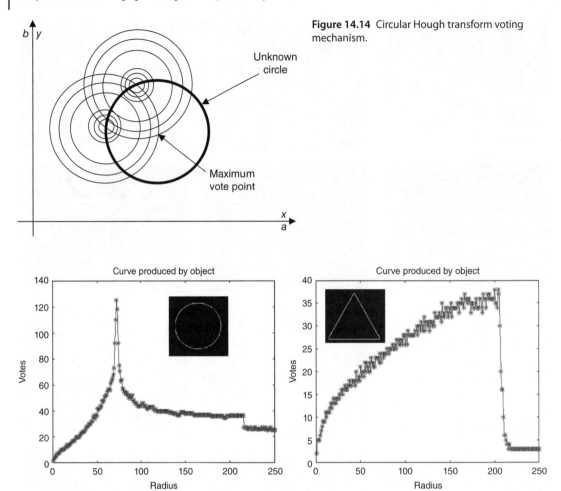

Figure 14.14 Circular Hough transform voting mechanism.

Figure 14.15 Property curves of a circle (a) and a triangle (b).

the circles, and the maximum number of votes is computed. This means that for each radius of the concentric circles there is one point in the Hough space (h, k) which represents the maximum voting.

A circle is detected by its peak generated by the voting mechanism, while a triangle is characterised by a smooth curve and the absence of the peak in the voting curve, as depicted in Figure 14.15. Property curves generated by the voting mechanism were smoothed by LOWESS regression [36] and normalised in both axes to [0, 1] as depicted in Figure 14.16. A training set of curves can be collected and an SVM classifier can be trained with these curves to recognise the different rims of traffic signs. In order to reduce the amount of computations in the recognition phase, edge detection algorithm such as Canny edge detector can be invoked to produce the edges of each of the candidate objects. These objects will sequentially be separated from the image and a voting curve will be created, smoothed and exposed to the SVM classifier for classification. Usually, Canny edge detector generates two edges which represent the outer and inner edges of the object under consideration, as depicted in Figure 14.17. To consider one edge only, the following set of rules were proposed [37, 38]:

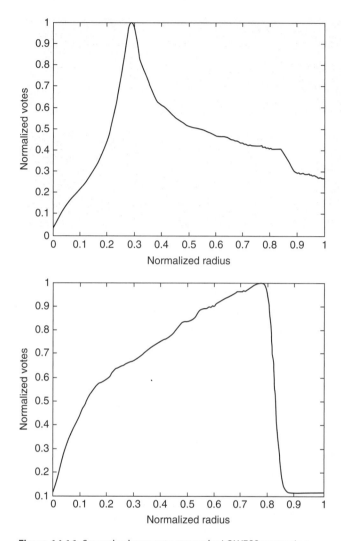

Figure 14.16 Smoothed property curves by LOWESS regression.

- If two concentrated edges are detected, the two edges of the traffic sign are healthy and one traffic sign is detected. The outer edge of the traffic sign is considered.
- If one edge, which is located inside another object, is detected, the outer edge of the traffic sign is destroyed, while the inner edge is healthy. The inner edge of the traffic sign is considered.
- If one edge, which is not located inside another object, is detected, the inner edge of the traffic sign is destroyed, while the outer edge is healthy. The outer edge of the traffic sign is considered.

Another simple method to detect traffic sign rim is by training a classifier by a set of features extracted from the segmented rim. Invariant image features based on integration over a transformation group which were introduced by Schulz-Mirbach [39] can be invoked for this purpose. In many cases during image retrieval, the exact position and orientation of objects in an image are only of secondary value. Thus, it is desirable to have features which are invariant to certain transformations, say translation and rotation.

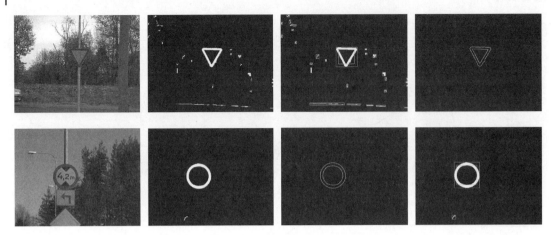

Figure 14.17 Results of rim detection.

Let $I = \{I(i,j)\}$, $0 \le i < N$, $0 \le j < M$ be an image, with $I(i,j)$ representing the grey value at the pixel coordinate (i,j). Let G be the transformation group of translations and rotations with elements $g \in G$ acting on the image, such that the transformed image is gI. An invariant feature must satisfy $F(gI) = F(I)$, $\forall g \in G$. Such invariant features can be constructed by integrating $f(gI)$ over the transformation group G.

$$F(I) = \frac{1}{|G|} \int_G f(gI) dg \tag{14.12}$$

For a segmented binary traffic sign image, transformations will be restricted to a certain group of translations. As a kernel function, binary operations among neighbour pixels are proposed. For example, a two-point kernel evaluated at a point (x,y) would be

$$k(x,y) = I(x,y)\,\mathrm{XOR}\,I(x+\Delta_x,\ y+\Delta_y) \tag{14.13}$$

where the pair (Δ_x, Δ_y) determines the local support.

The ith invariant feature is then given by

$$F_i = \frac{1}{MN} \sum_{x=0}^{N-1}\sum_{y=0}^{M-1} \mathrm{XOR}\!\left(I(x,y),\ I\!\left(x+\Delta_x^i,\ y+\Delta_y^i\right)\right) \tag{14.14}$$

where the pair $\left(\Delta_x^i, \Delta_y^i\right)$ are the translation parameters for the ith kernel. The boundary pixels may optionally be discarded if the corresponding translated point falls outside the image.

Theoretically, the values of $\left(\Delta_x^i, \Delta_y^i\right)$ are $0 \le \left(\Delta_x^i, \Delta_y^i\right) < \infty$; but practically, they should not exceed the size of the image. Figure 14.18 illustrates the features which are essentially discriminative for this classification task. The first column is the original image. The other columns depict the result of XORing of the original image with its translated version by the amount mentioned in the first row.

14.6.4 Pictogram Extraction

Once the rim of the traffic sign is specified, the next step is to extract its pictogram. All binary objects in the segmented image are labelled using connected components labelling, and all objects with red rims, yellow or white interiors, and appropriate dimensions will be selected as candidates for further

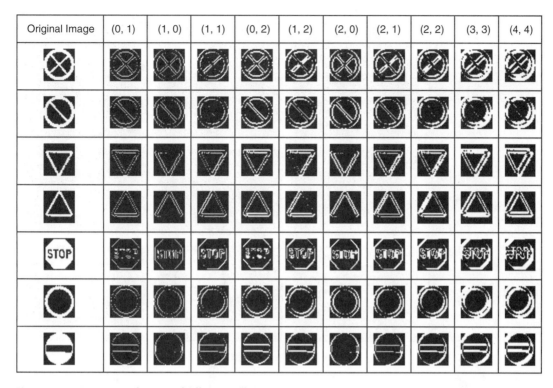

Original Image	(0, 1)	(1, 0)	(1, 1)	(0, 2)	(1, 2)	(2, 0)	(2, 1)	(2, 2)	(3, 3)	(4, 4)

Figure 14.18 Binary Haar features of different traffic sign rims.

investigation. As illustrated in Figure 14.19, the process of extracting the pictogram from the image is described in the following steps [40]:

1) Fill the red region which represents a candidate sign (sub-image A) with white pixels in order to produce (sub-image B).
2) Extract the area corresponding to the pictogram of the sign by XOR operator between sub-image A and sub-image B, that is (sub-image C = sub-image A XOR sub-image B). The resulting area will correspond to the pictogram of the traffic sign.
3) Extract the corresponding pictogram and convert it into grey level (sub-image D).

14.6.5 Pictogram Classification Using Features

In order to classify the pictogram, a set of features which describe it is required. A classifier, which is trained by this set of features, is the tool by which this pictogram is specified. There are many features and descriptors in the literature, but histogram of oriented gradients (HOGs), which was developed by Dalal and Triggs [41] to detect pedestrian, has become a source of attention for many researchers.

To compute the HOG descriptors of any image containing the extracted pictogram, this image is divided into a number of cells and a number of orientation bins as depicted in Figure 14.20. For each cell a local 1D histogram of the gradient directions of edge orientations over the pixels of the cell is collected. For better invariance to illumination such as shadows, the local histogram is accumulated over a larger area called 'blocks'. To improve the contribution of the cells in the final image descriptor, overlapping between these cells is invoked.

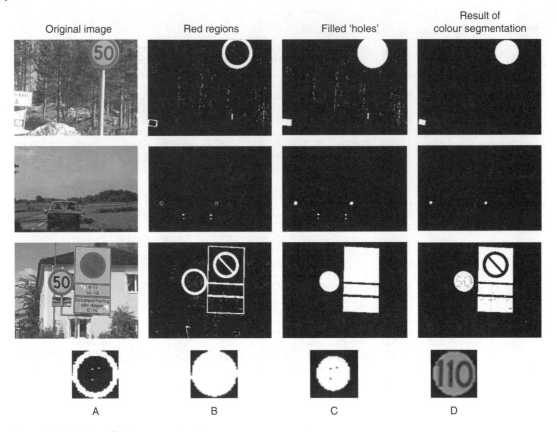

Figure 14.19 Steps of pictogram extraction.

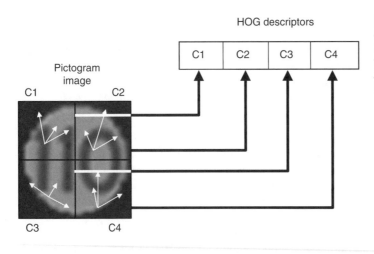

HOG descriptors

Figure 14.20 Computing the HOG descriptors of a pictogram. Source: Fleyeh and Roch [42]. Reproduced with permission of International Journal for Traffic and Transport Engineering.

Edge orientations are divided into a number of bins. These bins are equally spaced over the interval 0–180° for unsigned gradient and 0–360° for signed gradients. Edge orientations should fit into one of these bins. The histograms collected for the different cells in the bins in the same

Figure 14.21 The training dataset includes non-traffic sign objects and pictograms.

block are concatenated to make the final set of features of the object under consideration as illustrated in Figure 14.20.

The result of the HOG algorithm is a discrete amount of features which describe the input image. The number of features depends on the number of cells and orientation bins.

As an example of this stage, HOG features are invoked to classify speed limit traffic signs [42]. The dataset comprises 1727 images from which a total of 1710 speed limit signs and 1025 nontraffic sign objects were extracted by the pictogram extractor. A number of images that comprise the dataset are shown in Figure 14.21. Gentle AdaBoost, a boosting algorithm which was introduced by Friedman [43], was trained by the HOG descriptors of the pictograms. This Gentle AdaBoost is a robust and stable version of the Real AdaBoost and performs slightly better than the latter on regular data and considerably better on noisy data. The algorithm uses adaptive Newton steps rather than exact optimization at each step to minimise the exponential criterion in order to stabilise the learning processing. The performance of the HOG descriptors for traffic sign recognition is evaluated in the following sections.

14.6.5.1 Effect of Number of Features
A Gentle AdaBoost classifier was trained and tested with different numbers of features. The number of features was based on different numbers of cells per block and a constant number of orientation bins. The number of orientation bins was nine, while the number of cells varied from 2×2 to 6×6 with an increment of 1 in each direction, that is, $2 \times 2, 3 \times 3, ..., 6 \times 6$. This gives 4, 9, 16, 25 and 36 cells in each block. A plot of the classification error versus the number of features indicates clearly that increasing the number of features deduced per block decreases the classification error. Figure 14.22 depicts the relationship between the classification error and the number of features. Since the extracted traffic signs which were exploited in this experiment were of different scales, it is obvious that HOG descriptors are scale invariant. This property is essential to traffic sign recognition because images or footage may contain traffic signs with different sizes depending on the distance between the vehicle and the traffic sign. To have a set of descriptors which performs with scale invariance means that time required for normalisation can be saved for real-time applications.

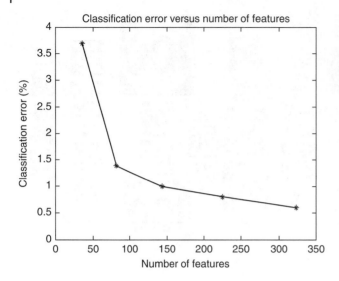

Classification error versus number of features

Figure 14.22 Classification errors versus number of features. Source: Fleyeh and Roch [42]. Reproduced with permission of International Journal for Traffic and Transport Engineering.

14.6.5.2 Classifying Disoriented Traffic Signs

Traffic signs are usually installed on poles which are always vertical with respect to the ground level. However, for many reasons such as the nature of the ground or environmental effects, these traffic signs depart from the vertical situation. The Gentle AdaBoost classifier was trained with 360 HOG features (6 × 6 cells and 10 bins). All of these features were derived from vertically oriented speed limit signs. The classifier was tested with HOG features derived from speed limit images which were rotated by different angles in clockwise and counter clockwise directions. Figure 14.23 depicts a plot of the classification rate versus the angle of rotation.

The plot shows a large drop in the classification rate when disoriented traffic signs were classified with the trained classifier, which means that HOG descriptors are rotation variant and cannot be utilised in these situations. As this is a crucial issue as far as traffic signs are concerned, a proper solution is essential in this case to avoid this kind of invariance.

The Gentle AdaBoost was trained again with a set of HOG descriptors which was derived from speed limit sign rotated by different angles between −90° and 90°. These HOG descriptors were derived in the same manner as described before. The classification rate did not drop as traffic signs rotate. Although there was a slight variation in the classification rate, the average classification rate was 92%. The plot of the classification rate versus the angle of rotation when Gentle AdaBoost was trained with rotated signs is depicted in Figure 14.24.

14.6.5.3 Training and Testing Time

Timings of training the Gentle AdaBoost classifier with 2735 descriptors together with the classification time of a speed limit sign using different numbers of HOG descriptors are depicted in Figure 14.25, showing a plot of training and testing times versus number of features. While training time increases with respect to the number of features, testing time is almost constant regardless of the number of features. The increment in the testing time between a low number of features and a high number of features is not crucial.

This system was tested on real traffic scenes collected on different environmental conditions when the vehicle was driven in the same speed of the traffic sign to be recognised. A number of speed limit results are shown in Figure 14.26.

Figure 14.23 Classification rate of the Gentle AdaBoost trained with vertically aligned signs versus angle of rotation of traffic signs. Source: Fleyeh and Roch [42]. Reproduced with permission of International Journal for Traffic and Transport Engineering.

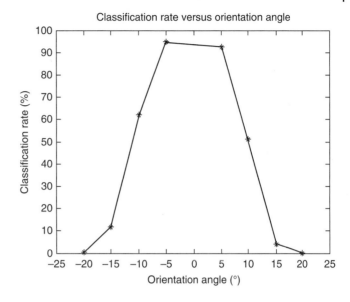

Figure 14.24 Classification rate of the Gentle AdaBoost trained with rotated signs versus angle of rotation of traffic signs. Source: Fleyeh and Roch [42]. Reproduced with permission of International Journal for Traffic and Transport Engineering.

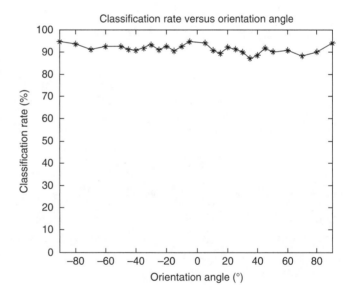

14.7 Working Systems

Traffic sign recognition systems are adopted by many car manufacturers. Traffic recognition systems for speed limit applications similar to the one described earlier was introduced by BMW in 2008 and then followed by Mercedes-Benz. A second generation which can detect overtaking restrictions was introduced by many can manufacturers such as Volkswagen. Volvo used a system called road sign information (RSI) [44]. The system is designed to lower the risk that the driver misses any traffic sign. The vehicle is equipped with a camera and a traffic sign recognition system which looks for the traffic

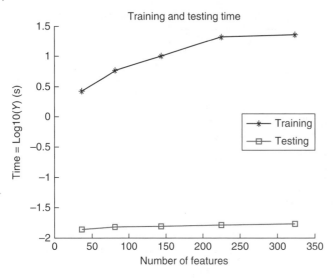

Figure 14.25 Training and testing time of the classifier with HOG descriptors. Source: Fleyeh and Roch [42]. Reproduced with permission of International Journal for Traffic and Transport Engineering.

Figure 14.26 Correctly classified speed limit traffic signs.

sign in the scene in front of the vehicle in real time; once the traffic sign is detected, it will be displayed as symbols on the speedometer of the vehicle. This technique, together with other new technologies such as adaptive cruising system, collision warning, automatic queue assistant system and automatic lane detection system, increases traffic safety which is one of the most important goals to achieve in the intelligent vehicles.

References

1 Fang, C., Chen, S. & Fuh, C. (2003) Road-sign detection and tracking. *IEEE Transactions on Vehicular Technology*, 52, 1329–1341.

2 Fang, C., Fuh, C., Chen, S. & Yen, P. (2003) A road sign recognition system based on dynamic visual model. *The 2003 IEEE Computer Society Conf. Computer Vision and Pattern Recognition*, Madison, WI, 18–20 June 2003, pp. 750–755.

3 de la Escalera, A., Moreno, L., Puente, E. & Salichs, M. (1994) Neural traffic sign recognition for autonomous vehicles. *20th International Conference on Industrial Electronics Control and Instrumentation, 1994*, IEEE, Bologna, Italy, 5 September 194, pp. 841–846.

4 Estevez, L. & Kehtarnavaz, N. (1996) A real-time histographic approach to road sign recognition. *IEEE Southwest Symposium on Image Analysis and Interpretation*, San Antonio, TX, 8–9 April 1996, pp. 95–100.

5 Miura, J., Kanda, T. & Shirai, Y. (2000) An active vision system for real-time traffic sign recognition. *2000 IEEE Intelligent Transportation Systems*, Dearborn, MI, 1–3 October 2000, pp. 52–57.

6 Vitabile, S. & Sorbello, F. (1998) Pictogram road signs detection and understanding in outdoor scenes. *Conference on Enhanced and Synthetic Vision*, Orlando, FL, April 1998, pp. 359–370.

7 Parodi, P. & Piccioli, G. (1995) A feature-based recognition scheme for traffic scenes. *Intelligent Vehicles '95 Symposium*, Detroit, MI, 25–26 September 1995, pp. 229–234.

8 Vitabile, S., Gentile, A. & Sorbello, F. (2002) A neural network based automatic road sign recognizer. *The 2002 International Joint Conference on Neural Networks*, Honolulu, HI, 12–17 May 2002, pp. 2315–2320.

9 Plane, J. (1992) *Traffic Engineering Handbook*. Englewood Cliffs, NJ: Prentice-Hall.

10 Jiang, G. & Choi, T. (1998) Robust detection of landmarks in color image based on fuzzy set theory. *Fourth International Conference on Signal Processing*, Beijing, China, 12–16 October 1998, pp. 968–971.

11 Lalonde, M. & Li, Y. (1995) Road sign recognition – survey of the state of art. Technical Report for Sub-Project 2.4, Center de recherche informatique de Montrèal, CRIM-IIT-95/09-35.

12 Wikipedia. (2014) Road Signs in the United States (online). Available: https://en.wikipedia.org/wiki/Road_signs_in_the_United_States (accessed 20 July 2014).

13 Paclik, P. (1984) ITS, Intelligent Transport System (online). Available: http://euler.fd.cvut.cz/research/rs2/articles/itsp.html (accessed 10 September 2016).

14 de la Escalera, A., Armingol, J. & Mata, M. (2003) Traffic sign recognition and analysis for intelligent vehicles. *Image and Vision Computing*, 21, 247–258.

15 Kehtarnavaz, N. & Kang, D. (1993) Stop-sign recognition based on color/shape processing. *Machine Vision and Applications*, 6, 206–208.

16 Volvo. (2011) Samlar den bästa säkerhetstekniken. *Aftonbladet* (online). Available: http://www.aftonbladet.se/kampanj/arkiv/volvo/innovation/article13635643.ab (accessed 10 September 2016).

17 Volvo. (2015) Säkerhet på en ny nivå (online). Available: http://www.volvocars.com/se/om-volvo/innovationer/intellisafe/driver-support (accessed 10 September 2016).

18 Mei, T., Liang, H., Kong, B., Yang, J., Zhu, H., Li, B., Chen, J., Zhao, P., Xu, T., Tao, X., Zhang, W., Song, Y., Wei, H. & Wang, J. (2012) Development of "Intelligent Pioneer" unmanned vehicle. *IEEE 2012 Intelligent Vehicles Symposium*, Alcalá de Henares, Spain, 3–7 June, pp. 938–943.

19 Blancard, M. (1992) Road sign recognition: a study of vision-based decision making for road environment recognition. In: Masaki, I. (ed.) *Vision-based Vehicle Guidance*. Berlin, Germany: Springer-Verlag.

20 Vitabile, S., Gentile, A., Dammone, G. & Sorbello, F. (2002) Multi-layer perceptron mapping on a SIMD architecture. *12th IEEE Workshop on Neural Networks for Signal Processing*, Martigny, Switzerland, 4–6 September 2002, pp. 667–675.

21 Vitabile, S., Pollaccia, G., Pilato, G. & Sorbello, F. (2001) Road sign recognition using a dynamic pixel aggregation technique in the HSV color space. *11th International Conference on Image Analysis and Processing*, Palermo, Italy, 26–28 September 2001, pp. 572–577.

22 Buluswar, S. & Draper, B. (1998) Color recognition in outdoor images. *International Conference on Computer Vision*, Bombay, India, 4–7 January 1998, pp. 171–177.

23 Luo, R., Potlapalli, H. & Hislop, D. (1993) Outdoor landmark recognition using fractal based vision and neural networks. *1999 IEEE/RSJ International Conference on Intelligent Robots and Systems*, Yokohama, Japan, 26–30 July 1993, pp. 612–618.

24 Paclik, P. & Novovicova, J. (2000) Road sign classification without color information. *Sixth Annual Conference of the Advanced School for Computing and Imaging*, Lommel, Belgium, 14–16 June 2000.

25 Paclik, P., Novovicova, J., Pudil, P. & Somol, P. (2000) Road sign classification using Laplace kernel classifier. *Pattern Recognition Letters*, 21, 1165–1173.

26 Kang, D., Griswold, N. & Kehtarnavaz, N. (1994) An invariant traffic sign recognition system based on sequential color processing and geometrical transformation. *IEEE Southwest Symposium on Image Analysis and Interpretation*, Dallas, TX, 21–24 April 1994, pp. 88–93.

27 Perez, E. & Javidi, B. (2000) Composite filter bank for road sign recognition. *13th Annual Meeting IEEE Lasers and Electro-Optics Society*, Rio Grande, Puerto Rico, 13–16 November 2000, pp. 754–755.

28 Fleyeh, H. (2009) Traffic Sign Database (Online). Available: http://users.du.se/~hfl/traffic_signs/ (accessed 10 September 2016).

29 Stallkamp, J., Schlipsing, M., Salmen, J. & Igel, C. (2011) The German traffic sign recognition benchmark: a multi-class classification competition. *IEEE International Joint Conference on Neural Networks*, San Jose, CA, 31 July–5 August.

30 Larsson, F. & Felsberg, M. (2011) Using Fourier descriptors and spatial models for traffic sign recognition. In: Heyden, A. & Kahl, F. (eds.) *Image Analysis*. 17th Scandinavian Conference on Image Analysis, SCIA 2011. Berlin/Heidelberg: Springer, pp. 238–249

31 Buluswar, S. & Draper, B. (1994) Non-parametric classification of pixels under varying outdoor illumination. *Intelligent Robots and Computer Vision XIII: Algorithms and Computer Vision*, October 1994, Vol. 2353, pp. 529–536.

32 Gevers, T. & Smeulders, A. (1999) Color-based object recognition. *Pattern Recognition*, 32, 453–464.

33 Fleyeh, H. (2006) Shadow and highlight invariant colour segmentation algorithm for traffic signs. *2006 IEEE Conference on Cybernetics and Intelligent Systems*, Bangkok, Thailand, June 2006, pp. 108–114.

34 Fleyeh, H. (2004) Color detection and segmentation for road and traffic signs. *2004 IEEE Conference on Cybernetics and Intelligent Systems*, Singapore, 1–3 December 2004, p. 808–813.

35 Fleyeh, H. & Mumtaz, A. (2011) Adaptive shadow and highlight invariant colour segmentation for traffic sign recognition based on Kohonen SOM. *Journal of Intelligent Systems*, 20, 15–31.

36 Cleveland, W. & Devlin, S. (1988) Locally weighted regression: an approach to regression analysis by local fitting. *Journal of the American Statistical Association*, 83, 596–610.

37 Fleyeh, H., Bhuiyan, N. & Biswas, R. (2011) Prohibitory traffic signs detection using LVQ and windowed Hough transform. *Fifth Indian International Conference on Arificial Intelligence*, Tumkur, India, 14–16 December.

38 Fleyeh, H., Biswas, R. & Davami, E. (2013) Traffic sign detection based on AdaBoost based color segmentation and SVM classification. *IEEE EuroCon 2013*, Zagreb, Croatia, 1–4 July 2013.

39 Schulz-Mirbach, H. (1995) Invariant features for gray scale images. *17 DAGM – Symposium "Mustererkennung"*, Bielefeld. Reihe Informatik aktuell, Springer, pp. 1–14.

40 Fleyeh, H. & Davami, E. (2011) Eigen-based traffic sign recognition. *IET Intelligent Transport Systems*, 5, 190–196.

41 Dalal, N. & Triggs, B. (2005) Histograms of oriented gradients for human detection. *IEEE Computer Society Conference on Computer Vision and Pattern Recognition (CVPR)*, IEEE Computer Society Washington, DC, 20 June 2005, pp. 886–893.

42 Fleyeh, H. & Roch, J. (2013) Benchmark evaluation of HOG descriptors as features for classification of traffic signs. *International Journal for Traffic andTransport Engineering*, 3, 448–464.

43 Friedman, J., Hastie, T. & Tibshirani, R. (2000) Additive logistic regression: a statistical view of boosting. *Annals of Statistics*, 28, 337–407.

44 Wikipedia. (2014) Traffic Sign Recognition (online). Available: https://en.wikipedia.org/wiki/Traffic_sign_recognition (accessed 20 July 2014).

15

Road Condition Monitoring

Matti Kutila[1], Pasi Pyykönen[1], Johan Casselgren[2] and Patrik Jonsson[3]

[1] *VTT Technical Research Centre of Finland Ltd., Tampere, Finland*
[2] *Luleå University of Technology, Luleå, Sweden*
[3] *Combitech AB, Östersund, Sweden*

15.1   Introduction

Statistics regarding road accidents indicate that slippery road conditions are often the primary cause of severe vehicular crashes or road departures [1]. Hence, a good estimation of traction is of great importance for safety and handling applications in vehicles; therefore, this has been the subject of intense research for many years. Incorporating a device that estimates the traction in front of the vehicle could enable more effective interventions aiming to avoid or mitigate the negative consequences of accidents. The reduced traction caused by icy or snowy road is estimated to be associated in approximately 75% of all accidents overall over each winter season in Finland. Moreover, the figures in the CARE database [2, 3] indicate that adverse road conditions are involved in 3800 annual fatalities in Europe overall. The primary challenge for the vehicle driver is what is known as "black ice," something which is very difficult to see from road surface. A study in Finland [4] arrived at the conclusion that almost one-half of the drivers misjudged the available traction on a road. This is a ground for motivating the development of roadside traction monitoring applications in order to prevent aquaplaning and warning drivers of slippery road conditions when the degree of traction has degraded.

Today, there are many prototypes of optical devices using optical characteristics to make a classification of various road conditions. Yamada et al. [5] pioneered the work in the area of polarization-based detection using standard closed-circuit television (CCTV) cameras. Their work includes texture and coarseness analysis in order to improve the classification performance. Gustafsson [6] has reviewed the existing automotive safety systems where tire-based traction monitoring technology has been used. The obvious conclusion is that tire-monitoring technology is not sufficient on its own to obtain a reliable state of the road conditions estimation; hence, sensor fusion with alternative sensors such optical devices should be more exhaustively investigated, not only roadside, but also in vehicle onboard units.

The modern cooperative traffic safety systems have been developed to make their surveys in adverse and locally varying road conditions. Moreover, the information is useful for road operators who plan the necessary maintenance services such as salting or snow plowing. There many examples of prior European-level projects which have engaged with development of cooperative traffic safety

Computer Vision and Imaging in Intelligent Transportation Systems, First Edition.
Edited by Robert P. Loce, Raja Bala and Mohan Trivedi.
© 2017 John Wiley & Sons Ltd. Published 2017 by John Wiley & Sons Ltd.
Companion website: www.wiley.com/go/loce/ComputerVisionandImaginginITS

systems related to road conditions [7–11]. The projects have been the driving force in the development of camera-based systems for in-vehicle and roadside units to support driver decision making. The existing road condition measurement products use a spectrometric measuring principle [12].

Another somewhat different sensor technique employs laser diodes of two wavelengths and a photo detector [13]. Two wavelengths are chosen because of the differences in spectral responses between water, ice, and snow [14, 15]; moreover, they are inexpensive off-the-shelf laser diodes. Yusheng and Higgins-Luthman have introduced an opportunity to use polarization differences for black ice detection [16].

In this chapter, we will not only focus on camera-based technology and classification but also make a review considering the system called "Road eye." Casselgren et al. [17] have performed exhaustive investigations concerning the changes of light intensity with varying incident angle and the spectrum of light when asphalt is covered with water, ice, or snow with using wavelengths from 1100 to 1700 nm. They observed that the snow reflection ratio drops consistently at wavelengths above 1400 nm, and as a conclusion they proposed two different bands in the NIR region (below and above 1400 nm) to classify road conditions. The benefit of this approach is that the equipment is robust and relatively cheap, but the drawback is the measuring area limited to one small spot on the road.

The chapter will first review the proposed measurement principles in the road traction monitoring area and then provide examples of the sensor solutions which are feasible for vehicle onboard and road sensing. Then, we will review opportunities to improve performance with the use of sensor data fusion, and finally discuss future opportunities. One important aspect is also the classification algorithms which create the backbone for signal processing algorithms.

15.2 Measurement Principles

The first road-embedded sensor that was installed for retrieving information about the road surface condition was the road temperature sensor, which was a crucial sensor in the very first Road Weather Information System (RWIS) installed in the beginning of the 1980s [18]. Thereafter, common measurement principles involve *in situ* measurements, where sensors are embedded in the road surface, and remote sensing technologies using visual cameras and IR sensors or cameras. It has been found that only knowing the meteorological conditions and the temperature of the road surface does not provide sufficient information for accurately estimating the slipperiness of the road. Therefore, additional sensors that could detect the presence of water and ice on the road surface were developed. In addition, sensors were developed that could also determine the freezing point of the water on road surfaces. These *in situ* sensors have some drawbacks, however, such as difficulties in terms of installation as the lanes need to be closed during the installation plus the sensors are also exposed to continual wear and degradation from the traffic. Another drawback is that a given sensor only monitors a very small area of the road surface, which to some extent is solved by the traffic itself as it will drag new fluid over the sensor area as the vehicles pass [19].

When digital cameras were first available in the 1990s, they were widely installed at the RWIS installation locations to be used for manual observations of road conditions and the results from winter road maintenance tasks. However, these camera systems needed a manual operator to be of sufficient value, and the acquired images were used solely for observations by the operators. It was more desirable to have camera systems that could automatically determine road conditions. As computational power increased, some initial tests with image analysis and learning-based computer models were performed [20]. From these tests, it was realized that it is possible to classify road conditions based on camera image data. In the research that followed, meteorological parameters and camera images were used in conjunction together for classifying road conditions. By using this additional information with image data, it was hypothesized that the classification performance could be improved compared to only using image data. Even though it was found that the combined RWIS and image feature dataset could be used for developing accurate road status classifiers, it was found that the most crucial situations, such as black ice occurring at temperatures in the neighborhood of 0°C, could be missed by using RWIS data and visual camera images.

Some studies found that it would be possible to detect ice and classify road conditions by using NIR light [21]. This principle was used to develop remote sensing, also called "contactless sensing." The typical sensors utilize NIR detectors operating in the wavelength region 1–2 μm and usually together with a light source. The light source should have an illumination power that is greater than the light from solar sources at the measurement area during daytime to get reliable measurements. Several new products based on this principle have recently been released on these markets too. One drawback of using the aforementioned NIR sensors is that they measure a relatively small spatial area, approximately a circle with 30 cm diameter. This means that differences in and between wheel tracks can be missed. It is thus desired to get broad lane coverage for remote sensing equipment. Although NIR-sensitive camera chips are more expensive than single-point sensors, research has been performed to find a cost-effective camera system operating in the NIR wavelength region that has the capability to classify road conditions [22].

15.3 Sensor Solutions

15.3.1 Camera-Based Friction Estimation Systems

Cameras are widely used in road condition measurement units in roadside installations. Due to the area they are cover, they need to have an appropriate resolution in order to depict the situation accurately. This follows the typical method of using textural analysis for road monitoring, for example, Kutila et al. [10]. A roadside camera should also have the ability to process images at nighttime, which means preferences to operate in the 0.7–0.9 μm wavelength band. It is also desired to have the road illuminated during nighttime, and at remote locations without any street lighting there is a need to provide IR lighting. Many cameras have internal light-emitting diodes that provide IR illumination, but they are usually only intended for short-range illumination. Thus, typical built-in illuminators will not provide satisfactorily illumination of the road surface during nighttime. Therefore, separate independent illuminators should be used, such as the powerful NIR illuminators used by common surveillance systems.

One example of a road condition monitoring system, called IcOR, is based on a stereo camera arrangement. The reason for use of a stereo camera system (see Figure 15.1) is not actually acquiring 3D measurements, but this is optimal solution for accurate spatial and temporal synchronization between captured images. This is the key requirement for comparing polarization differences between two images. For measuring illumination difference between cameras, horizontal and vertical polarization filters are mounted in front of the optics. The IR filter has been removed from the cameras to improve performance during dark time. On the other hand, the polarization filters allow only a narrow spectral band to pass to the detectors, which eliminates most of the detrimental ambient light. One camera with a horizontal field of view of 65° acquires monochrome images with a maximum resolution of 640 × 480 pixels.

Figure 15.2 shows the image processing part of the camera-based friction estimation system. There are two major steps which bases on filtering out the polarization planes and then estimating the

Figure 15.1 IR light source and stereo camera inside housing.

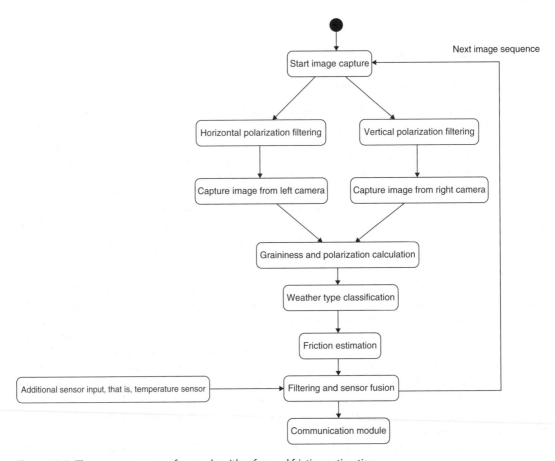

Figure 15.2 The camera system software algorithm for road friction estimation.

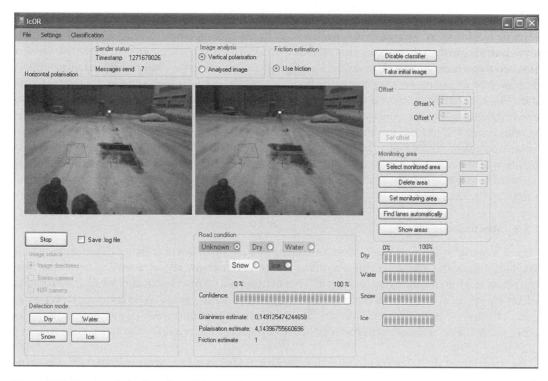

Figure 15.3 User interface of road condition monitoring system.

difference index. Furthermore, graininess value is estimated using frequency analysis in the image. These two factors (polarization and graininess) are finally classified into the following categories: dry, wet, snowy, and icy asphalt.

The user interface of the system enables the option to select, for example, lanes to be monitored from an intersection area. This is important when no automatic calibration and region-of-interest selection exists. In Figure 15.3, three different areas have been selected for classification. Lane selection enables monitoring of individual lanes and more precise detection of different road conditions. Graininess and polarization values from the specified areas are classified into the following road condition categories: dry, ice, snow, wet, and unknown.

The road condition monitoring system data contains a time stamp in seconds, sensor location, classified road condition (dry, ice, snow, wet, or unknown), and confidence level of the classified road condition. Classification of the road conditions is a relatively big operation, and even when the cameras are operated at 25 Hz, the system output frequency remains 1 Hz.

15.3.2 Pavement Sensors

The pavement sensors are installed as close to the road surface as possible without exposing the sensor to undue risk of being broken or otherwise damaged by traffic. The sensors provide information for prognosis models of when the road will reach a frozen condition. Knowledge about when the road will freeze is important information for initiating preventive anti-icing activities. A consequence of such temperature measurements and associated anti-icing efforts is that the presence and amount of anti-icing materials present on the road should be known in order to be able to determine when water on the surface will form ice. Several pavement-mounted sensors have been developed during the years, and they are primarily of two different types: active sensors and passive sensors. The passive sensors are by far the most common as their construction is simpler than the active sensors that need to perform additional tasks for heating and cooling as well as additional measurements. A common type of passive sensor detects the freezing point by measuring the conductivity of the road fluid. This conductivity is then translated to a freezing point value when the anti-icing substance is known. Additional passive sensors are needed for detecting the presence of water and snow by utilizing reflections of light from water and snow on the road surface [19]. Active sensors utilize a Peltier element for heating and cooling the fluid, ice, or snow on the road surface in order to perform a phase transition of the material on the road surface from fluid to solid. It is possible to detect this phase transition, and the temperature at this phase transition indicates the freezing point of the ice, snow, or fluid on the road surface.

15.3.3 Spectroscopy

When a material is illuminated by electromagnetic waves, the atoms can react in one of two ways depending upon the energy of the incoming photon. Generally, the atom will scatter the light redirecting it without otherwise altering it. This nonresonant scattering occurs when the incoming radiant energy is far away from the resonance frequencies of the atom (see Figure 15.4). When an atom in the lowest state interacts with a photon whose energy is too small to cause a transition to a higher excited state, the electromagnetic field drives the electronic cloud of the atom into oscillation without any transition.

The vibration of the electron cloud is the same as the frequency of the incident light. Once the electron cloud starts to vibrate with respect to the positive nucleus, the system constitutes an oscillating dipole and will immediately begin to radiate at the same frequency. The scattered light will consist of a photon with the same energy as the incident photon but possibly redirected in another direction.

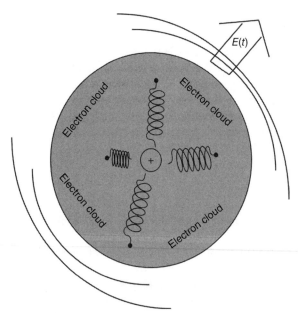

Figure 15.4 A mechanical representation of an oscillator in an isotropic material where the negatively charged shell is fastened to a stationary positive nucleus by identical springs.

The other atomic reaction occurs when the photon's energy matches that of one of the excitation states of the atom. The atom will then absorb the light and make a quantum jump to that higher energy level. In solids and liquids, which are used in this application, the atomic landscape is dense. For such an atomic landscape, it is likely that the excitation energy will rapidly be transferred, via collision, to random atomic motion (thermal energy) before a photon can be emitted. This process is referred to as "dissipative absorption." For a dense material, the closer the frequency of the incident light to an atomic resonance, the stronger the interaction. This results in more energy absorption. It is this mechanism that creates much of the visual appearance of matter, and it is the mechanism that makes it possible to characterize different materials using the absorbing spectra. These mechanisms also state that the complex refractive index is dependent on frequency (wavelength) [23].

In the NIR spectrum, different phases of water (water, ice, snow, and slush) have a distinctive absorption for specific wavelengths as shown in Figure 15.5. This physical property is one of the main characteristics that enable road condition classification using NIR light [13]. In Casselgren et al. [17], the optimal wavelengths (980, 1310, and 1550 nm) to classify dry, wet, icy, and snow asphalt within the NIR spectrum were calculated. These wavelengths are utilized in several road condition–monitoring sensors where Road eye is one example [24]. In Figure 15.5a, the spectrum between 1100 and 1700 nm are shown for dry asphalt and asphalt covered with water, ice, and snow. It can be seen that the two road conditions that are closest to each other are water and ice. These are also the conditions that have the most potential hazard, because a driver can think that a road is simply wet when in fact it is frozen and slippery. In Figure 15.5b, the spectral reflectance for five different water depths is shown, moist and 1–4 mm. From the figure, we can see that absorption makes it possible to distinguish the thickness of water films.

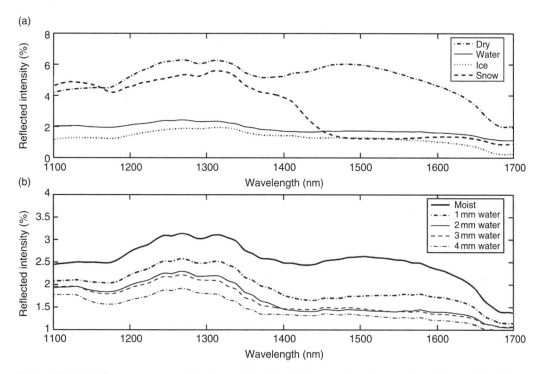

Figure 15.5 (a) Reflectance spectrum for dry asphalt and asphalt covered with water, ice, and snow. (b) Reflectance spectrum for five different depths of water on asphalt.

15.3.4 Roadside Fog Sensing

The fog detection system, which is based on a laser scanner (Figure 15.6) on 950 nm band, was implemented in a ROAD SIDE UNIT after the preliminary laboratory tests and results (see Figure 15.7). Based on the laboratory tests, it was noted that the laser beam does not sufficiently penetrate fog when fog density is increased. The distance to the reflection point is measured continuously, and the mean reflectance value over the last 10 s is calculated. This mean value is then compared to a reference value without fog to get a visibility estimate.

Figure 15.8 illustrates the basic laser scanner fog measurement system method. The scanner is installed over a road section, attached to a pole bridge structure. The laser scanner beams are orientated to the road surface to obtain a reference point and distance for clear visibility. When fog density is increasing, measured mean distance for the laser beams is decreasing and showing that fog is limiting visibility as a ratio of decreased distance to road surface.

The ROAD SIDE UNIT has been installed on a motorway ramp for testing and the visibility was continuously monitored. In addition, the previously mentioned IcOR module was installed onto the ROAD SIDE UNIT for monitoring the road surface friction at the same time. Thus, the system was versatile monitoring unit processing both fog and friction with the same laser- and camera-based unit.

In Figure 15.8, the algorithm for camera-based fog vision system is depicted. The system bases on assessing laser light reflectance from different distance for estimating visibility index. However, rain also tends to influence laser scanners, and these situations are classified separately to minimize influence for fog index calculation.

Figure 15.9 show benchmarking results of the fog measurement unit. The average distance of the laser reflection is measured over the motorway in meters. The red box on the left indicates a period when the snowfall was very dense. During dense snowfall, the average distance decreased significantly from 30 to 10 m, while the maximum visibility value decrease was proportionally smaller, from about 200 to 160 m. This result can be explained partially by the operation principle of the scanning

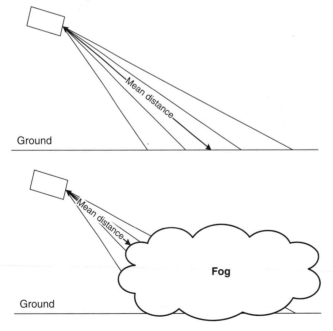

Figure 15.6 Basic measuring principle of fog detection system with a laser scanner.

Figure 15.7 Targets as seen in various fog densities in the fog chamber.

laser, in which the laser sends multiple beams to measure the environment. During snowfall, individual laser beams have a higher possibility to pass through snow crystals than to pass through fog since the fog is denser and more efficiently prevents the penetration of the laser scanner beams. Hence, in the snowfall condition, the maximum distance over all beams can be measured by a beam that misses most falling snow crystals and results in a distance closer to the clear visibility condition.

15.3.5 In-Vehicle Sensors

In-vehicle systems are used to adapt electronic stability control (ESC) skidding prevention system according to the available friction level. Current in-vehicle friction estimation units are based on lateral vehicle dynamics and rotation speed errors of tires [9]. Some tire-based sensors have been developed that utilize tire deformation. These sensors are installed inside the vehicle tire to detect possible shape changes due to reduced friction.

In addition, in-vehicle systems utilize inertia sensing of the vehicle to calculate lateral accelerations, which are indicators of skidding. The acceleration and speed variations are available in the vehicle controller area network (CAN) bus from the standard sensors which typically exists in cars.

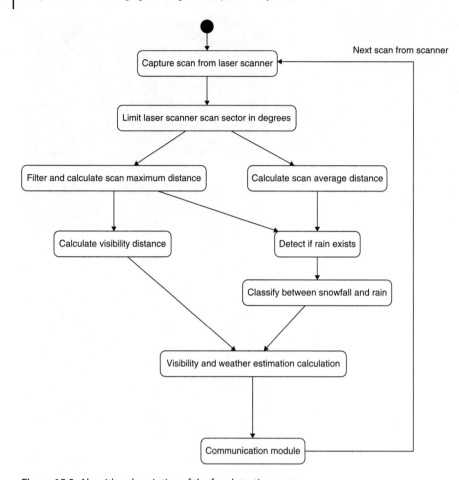

Figure 15.8 Algorithm description of the fog detection system.

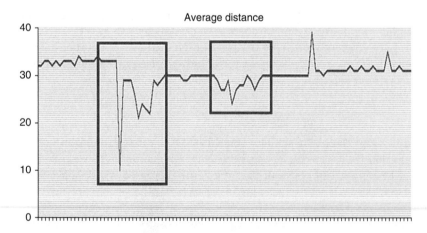

Figure 15.9 Average visibility distance measured by the laser scanner in meters. Two areas are marked during which there was dense snowfall.

Thus standard sensors are used to avoid an added cost in the manufacturing of the car, therefore are favored. However, the problem is that these systems do not predict friction in front of the car as the optical sensors do. Transition to automated vehicles needs this type of prediction where friction in front of the vehicle is calculated for adapting vehicle systems accordingly similar to a driver viewing the proceeding road condition.

The environmental system IcOR provides a classification of road conditions over the whole road surface at distance of 0–40 m in front of sensor, in this case mounted on a vehicle. The IcOR system was first developed as an onboard driver support system. The visible and NIR spectrum, with the help of the car's front lights, were utilized to avoid additional illumination installations for keeping the system price reasonable. As a detector, the IcOR system uses a stereo camera pair. Each camera acquires monochrome images with a resolution of 640×480 pixels. They have a global shutter that exposes all the pixels simultaneously. This prevents some artifacts that arise with a rolling shutter (e.g., skew). The system uses external synchronization, that is, all DSCG devices on the same FireWire bus capture the images at the same time. This is convenient because there is no need to synchronize the images afterward. A polarization filter is mounted on each camera in such a way that one camera measures the incoming horizontally polarized light, while the other measures the vertically polarized light.

The camera pair of the IcOR system was tested positioned inside the cab/interior of the vehicle behind the windshield, avoiding the windshield wipers and having a clear optical path. The installation corresponds with the typical lane tracker cameras since they are often mounted behind the interior rearview mirror together with the rain detector. Another benefit concerning the installation location is that expensive heated optics is not needed during wintertime. The evaluation of the IcOR system was first done in a passenger car (see Figure 15.10). To make results fully comparable between the IcOR and the Road eye systems, both systems were also installed into a specially instrumented vehicle.

The environmental sensor Road eye provides a classification of road conditions at a short distance of 0.5–1.5 m and has a short-wave IR (SWIR)-active illumination system consisting of three laser diodes emitting at wavelengths $\lambda 1 = 980$ nm, $\lambda 2 = 1310$ nm, and $\lambda 3 = 1550$ nm. As laser diodes emits linearly polarized light, it is possible to control the polarization of the illuminated light by mounting the diodes in a specific manner. The polarization is important because as shown in Figure 15.11 the

Figure 15.10 IcOR's stereo camera located inside the test vehicle.

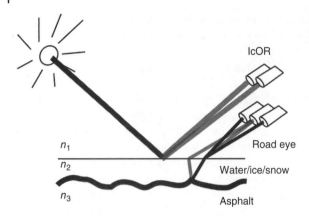

Figure 15.11 Road eye and IcOR measuring principles.

light measured by the Road eye needs to propagate into the medium to give a good classification. The laser diodes were mounted so that the emitted light is vertically polarized in order to allow the strongest possible signal back to the sensor. Hence, for angles around 50° all vertical polarized light will propagate down into the medium [25, 26], resulting in a reflection that is as strong as possible. The Road eye's focusing optics gives an illuminated spot with a radius of 10 mm on the road surface at a distance of 0.8 m. In order to acquire data from the reflected light, a lens focuses the reflected light onto a photodiode. The amplitude-modulated signals are sampled at 20 Hz. The output signal consists of three voltages (mV) representing the reflected intensity of the three wavelengths, respectively. The active amplitude-modulated illumination ensures insensitivity to disturbances, such as other vehicle's headlights or daylight.

In one study, the Road eye sensor was tested in a tube in front of the right front wheel of a truck. The tube was used to keep the sensor clear from splash and pollutions and worked without any problems. The mounting angle of the tube was around 45°, with a height of 0.8 m from the ground, resulting in a measuring distance in front of the wheel of 0.8 m. Due to the sampling rate of the Road eye sensor and the simple classification algorithm, the response time of the system ensures a preview measurement 0.8 m in front of the wheel.

In the INTEREG IVA Nord Intelligent Road initiative, the two in-vehicle optical sensors from Optical sensors Road Eye and Teconer RCM 411 were evaluated against a continuous road/tire friction apparatus to evaluate the friction estimation. The results for each road condition are shown in Figure 15.12. It can be seen that the optical sensors can classify certain slippery road conditions from nonslippery conditions. What is notable is the difference between the two sensors for slush, and this is because the definition of slush can alter between different manufacturers of the sensors. Also notable are the classifications that are erroneous; for example, the road condition is dry but the friction is below 0.6. This is caused by erroneous classifications, such as an icy road that is sanded.

15.4 Classification and Sensor Fusion

The classification of road conditions is not as easy as it first may seem, partly due to different road users and road maintenance operators having their own unique way of describing road conditions. It is possible to develop classifiers with well-defined functionality by limiting the road conditions to those that human eyes easily recognize (dry, wet, icy, and snowy). However, the separation of ice and snow is not obvious if wavelengths between 1 and 2 μm are used for classification. In this wavelength

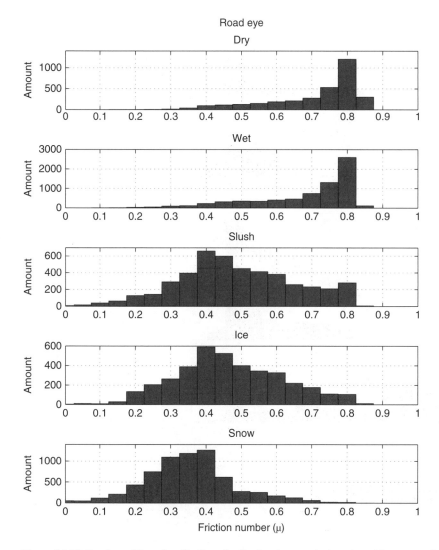

Figure 15.12 Road condition classifications for the Road eye and a benchmarking sensor for different road conditions.

band, ice and snow could look similar even though snow is clearly white and ice is gray in the visual spectral range. This phenomenon indicates that manual classification of ground truth road conditions needs to take this into consideration when the data is used for developing road condition classifiers. The development of classifiers should be done using supervised learning, because it is desired to map human observations to classified road conditions (see Figure 15.13). Using unsupervised learning for developing classifiers is not desired in this case, because unsupervised learning may find groups of classes that cannot be seen in the visual spectra; thus, it will be difficult for users to understand the classification models.

The initial step in the classification phase is to perform a principal component analysis (PCA) to find the internal relations of the dataset. By using PCA, it is possible to plot the data against the

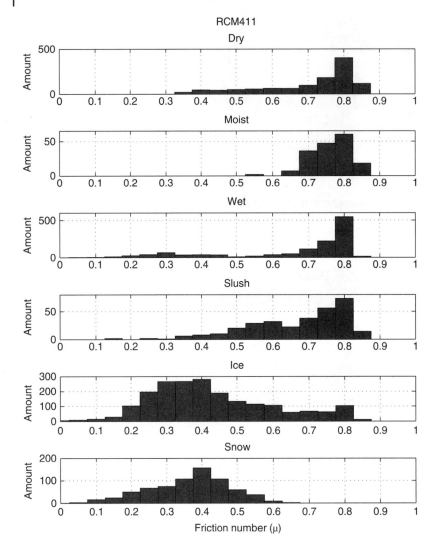

Figure 15.12 (Continued)

orthogonal axes that explains the data with highest variance. If groupings are found using PCA, it is possible to continue to develop classification models [27].

There are many classification methods, for example, K-nearest neighbor, neural networks (NNs), support vector machines, partial least squares (PLS), and discriminant analysis. The choice of a specific type depends on many things, for example, the success rate of the developed classifier, how much computational power is available, and how much memory is available in the processing unit, how the classifier is intended to be used. The performance of the developed classifier cannot be evaluated until training and validation of a dataset is done. Therefore, it is necessary to test a few different classification methods in order to find the most suitable one. K-nearest neighbor is an old and well-proven classification method with the drawback of being slow. This suggests that it is desirable to use K-nearest neighbor as a comparison to classifiers based on the other methods. If real-time

Figure 15.13 An example plot from a PCA of road condition data for different road conditions. The different road conditions can be seen to be grouped, and by looking at more of the principal components the groupings for some of the road conditions can be seen more clearly.

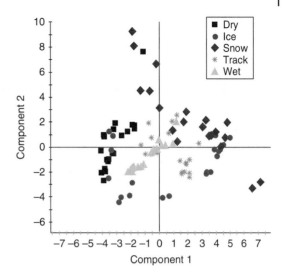

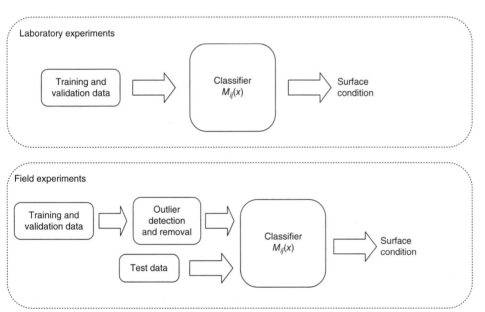

Figure 15.14 A method for developing road conditions classifiers. In the laboratory, a dataset is used to train and validate the classifiers. In the field, where the environment conditions are more varying, two datasets are used. One dataset is used for training and validation of the classifiers and another dataset is used to test the dataset. The test dataset from field experiments shows the performance of the classifiers.

calculations are required, methods such as partial least squares, NNs, and support vector machines should be considered as their resulting classification model is a mathematical polynomial. PLS is a method that is based on statistical methods, and therefore the performance gained from this classifier is sometimes easier to examine and understand.

The development of classifiers is preferably done in two steps: first a dataset is used to train and validate the classifiers; second, a different dataset is used to test the developed classifiers (see Figure 15.14).

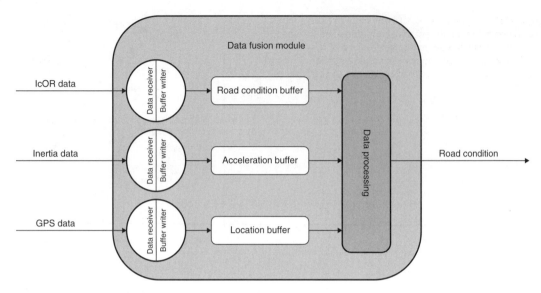

Figure 15.15 Sensor data fusion module for merging in-vehicle inertia and optical measurement data.

It was also a good idea to develop one classifier for each road condition. Then each classifier has a much simpler task as it only has to define if data belongs to a specific class or not. The drawback is that if road condition data is close to each other in the featured space, two or more classifiers can classify a pixel to belong to their road condition class. This is a multiclass problem that needs to be handled and one method is to use Bayesian networks. Such a Bayesian network would then use probabilities for selecting one of the conflicting road conditions; for example, it is more probable to have ice rather than being wet if the road temperature is −20°C.

Two sensors were used for road friction estimation: the IcOR camera system producing road friction classification information and the inertia unit complemented with the GPS data (see Figure 15.15). Data input structures from the IcOR, inertia, and GPS sensors were adapted from the specification provided by the German national C2X pilot project SimTD, using the standardized ASN.1 notation. This specification was selected since the protocol aligns with Car-2-Car communication consortium recommendations and is widely accepted by the European automotive industry. The architecture of the infrastructure sub-system is shown in Figure 15.16.

The IcOR system acquires images and executes analysis with 1 Hz frame rate. At 100 km/h speed, the vehicle will move 28 m during one measurement. To classify road condition in real time, the measurement area of the IcOR system was selected to be 30 m ahead of the vehicle. This measurement rate is sufficient because the road state is changing slowly and more important is the overall road condition.

15.5 Field Studies

In Sweden, field tests have been performed to evaluate the performance of classifiers using data from the visual and NIR spectra 0.4–0.9 μm and data from only the NIR spectra ranging from 1 to 2 μm.

By using a standard visual camera operating in the spectral region 0.4–0.9 μm together with RWIS data, it was found that it is possible to discriminate between different road conditions. The use of

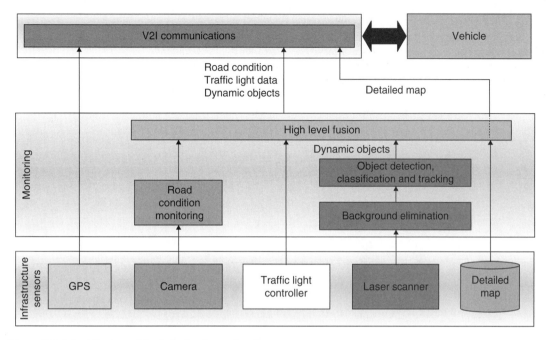

Figure 15.16 Architecture of the infrastructure subsystem.

Table 15.1 Success rates from road condition classification using 0.4–0.9 μm camera and RWIS data.

	Modeled road condition, percentage of correct classifications	
	PLS model (%)	NN model (%)
Observed road condition		
Dry	100	91
Ice	91	100
Snow	95	100
Wet	95	100

RWIS data together with image feature data makes it possible to more easily discriminate between a wet and icy surface because, as an example, temperatures well below 0°C implies an icy surface instead of a wet surface.

Results from field tests performed in Sweden showed successful results of classifying the overall road condition using PLS and NNs; see Table 15.1 [28, 29].

However, it was found that this method using standard cameras and RWIS data merely gives an overall classification of the road condition instead of details of individual parts of the retrieved image. By using the NIR spectra, 1–2 μm, it was found that it is possible to determine the road condition in each pixel of the retrieved images [22].

Table 15.2 Road condition differences in and between the wheel tracks at two sites.

Percentage dry or wet in wheel tracks	Percentage ice or snow between wheel tracks	Occasions at site 1	Occasions at site 2
70	40	69	86
70	50	50	62
70	60	35	47
70	70	22	33
70	80	14	24
70	90	2	12
80	40	46	75
80	50	33	52
80	60	21	38
80	70	15	27
80	80	11	19
80	90	1	11
90	40	23	60
90	50	17	41
90	60	10	30
90	70	7	23
90	80	5	16
90	90	0	8

Because each pixel is classified as a road condition, it is possible to find situations where the road condition differs in and between the wheel tracks. Such an analysis was done on the dataset, and it was found that there had been several occasions when the road condition were different in and between the wheel tracks; see Table 15.2 [30]. The data was collected during the winter season 2013–2014 close to the Sweden–Norway border.

Test runs were performed together with Finnish Meteorological Institute (FMI) and Foreca in the beginning of February 2009 [31]. The test route consisted of a wide variety of road types: interurban highways, rural roads, and urban roads in the Tampere region. The road conditions varied from salted, wet highway to snowy gravel road. The weather during the test drives was somewhat cloudy with temperatures in the neighborhood of −5°C.

In addition to the IcOR system, the test vehicle was also equipped with two of Vaisala's surface sensors [32]: remote road surface state sensor (DSC111) and remote surface temperature sensor (DST111). Vaisala sensors were placed on the roof, aiming at the rear of the vehicle (see Figure 15.17). The surface state sensor was measuring road surface directly behind the left back wheel. For the other sensor, the IcOR's camera was mounted on the windshield inside the vehicle, facing toward the road ahead.

The data consists of road parts giving a typical impression of the winter road condition. The reference road condition value is based on the measurements from the Vaisala sensor that provides ice, water, and snow layer thickness. The thickest layer was chosen as the road condition; for example, if the snow layer had the largest value of the three, the road would be considered snowy.

Figure 15.17 Vaisala's road weather sensors on the roof of the test vehicle.

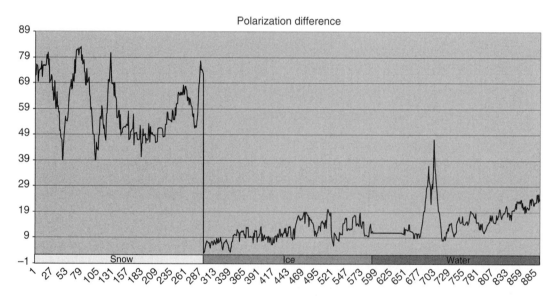

Figure 15.18 Polarization difference.

Figure 15.18 shows the polarization differences from certain road parts distinguishing the changing road conditions (snowy, icy, and wet from left to right). During a snowy route, the polarization values are quite large, creating a distinct gap between the snowy and icy/wet parts of the road. The difference between icy and wet road is small, but correlation with the reference can be seen.

Figure 15.19 contains graininess values of the same measurement points presented in Figure 15.18. There seems to be no significant differences between the values of three different road parts. The negative values in the snowy area are measurement errors probably due to overexposed images.

Earlier experiences have shown that the graininess analysis has the potential to analyze not just whether the road is covered by ice, water, or dry asphalt but also to recognize the type of the road. This would be interesting especially when the vehicle is navigating in an unstructured off-road

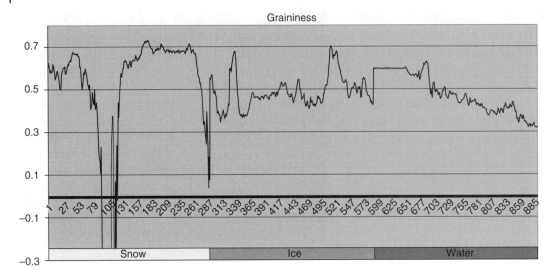

Figure 15.19 Graininess levels.

environment. Currently, control algorithms of the robots moving in forest roads, and so on, do not take into account influence of reduced friction.

15.6 Cooperative Road Weather Services

Vehicle environment perception today is primarily based on onboard units. However, reliability and versatility of perception can be improved by taking additional measures from other sources into account [33]. A cooperative safety system can fuse the data available from other vehicles and roadside units. This fusion utilizes, in many cases, maps of the region to compare in-vehicle and road weather service data.

The local dynamic map (LDM) concept has been introduced for detailed road- and traffic-related information [34]. In the LDM, the road geometry from a standard digital map is integrated with the information collected by vehicles and the infrastructure. When comparing this information with the LDM specifications, conflict areas and especially unique map lanes are not included. Lanes are not separate objects and their properties are linked to road elements. Other differences include, for example, the coordinate system in which paths are described.

The roadside equipment has been developed to offer processed data for vehicles entering a communication-equipped intersection. The infrastructure subsystem merges data of the traffic light controller, road condition information, road user locations and speed, and the static map information to the package which assists a vehicle to adapt speed and maneuvering to safely approach an intersection. The laser scanners of the infrastructure are used to detect pedestrians, other vehicles, and bicycles, whereas the camera system detects if the friction level of road surface is changing. The traffic light controller provides information of traffic signals and intelligently controls traffic flow by minimizing waiting times and vehicle accelerations. Static data (locations of curbs and poles) is available in the intersection model which is stored to the local dynamic map database.

Data input structures from the sensors, as well as the intersection topology data from the LDM, are described using the standardized ASN.1 metalanguage in accordance with the SIMTD specification

[35]. The high-level fusion (HLF) module receives data from sensors through data receiver processes. Data is written into the shared memory buffers that enable data transfer and communication across process boundaries. This approach also enables the attachment of timestamps to the data packets as they are written into the shared memory buffers. This feature is used in the fusion to associate the contents of the buffers with a certain moment of time. Incoming data examples are as follows:

- Traffic light controller data (signal state buffer) data contains periodic information about the state of the traffic light signal. Signal state buffer writer writes this data into the buffer approximately once per second.
- Road condition monitor data (road condition buffer) contains information about the road conditions in the intersection. Most critical pieces of information are the number of monitored areas, road condition type in the monitored areas, and the position of the monitored areas. A road condition buffer writer writes this data into the buffer approximately once per second.
- Laserscanner data (laserscanner buffer) contains information about the detected objects in the intersection. First of all, this data consists of information about age, time, classification, position, orientation, velocity, and size of the detected objects. A laserscanner buffer writer writes this data into the buffer within 12.5–25 Hz frequency.
- LDM data (intersection topology) provides a formal representation of the surrounding intersection topology. The HLF module reads the intersection topology stored in an XML file. This data has a more static character and is not read-in continually like the data to the other buffers.

15.7 Discussion and Future Work

In recent years, the automotive industry has introduced numerous new driver assistance systems into the markets. Measuring road friction is in the interests of original equipment manufacturers (OEMs) since future vehicles that adopt new autonomous features need predictive friction measurement for autonomy to be feasible in winter time in countries having snow and freezing conditions. The vehicles need to select an optimum speed and lanes in various road weather conditions for preventing accidents. The current problem is that the reliable sensors based on tire rotation speed differences, in practice, report data too late when considering, for instance, aquaplaning. On the other hand, in the markets where slippery roads due to snow and ice are not commonly encountered, this has not been a top priority for sensor development.

The optical sensing principle is the most promising to meet the automotive industry and road operator requirements for future systems. Still both software and, especially, hardware development is needed. Ice has certain optical properties which favor the use of the NIR band for detecting black ice on a road. The components unfortunately are still too expensive for moderately priced level cars, and hence this is one of the biggest future challenges—to design optics, sensor chips, and processing units to be at a sufficiently low cost (about €1000). The development of these systems will continue in the RobustSENSE-ECSEL-2015 project, which is co-funded by the European Commission.

Road operators need a system not only to monitor the actual road weather for informing vehicle drivers but also to be able to predict the changes in road weather-related conditions as early as possible. Road operators who are responsible for snowplowing/snowblowing and salting/sanding and maintenance are required to react before the road conditions change to a dangerous icy condition. The optical sensors both in-vehicle and infrastructure side are turning out to be the most promising option for sensing. However, an important consideration is the development of IT systems that are able to amalgamate and process the information and have a channel of communication to provide information to vehicle operators.

References

1 Wallman, C.-G. & Åström, H. 2001. Tema vintermodell—olycksrisker vid olika vinterväglag, Technical Report N60-2001, VTI, Linköping.

2 Annual Statistical Report 2007. 2007. Deliverable 1.16 of SafetyNet. Based on data Community database on Accidents on the Roads in Europe—CARE. Available in: http://ec.europa.eu/transport/road_safety/specialist/statistics/care_reports_graphics/index_en.htm, cited on July 29, 2010 (accessed on September 12, 2010).

3 SafetyNet. 2007. Annual Statistical Report 2007. Deliverable No: 1.16. Data based on Community database on Accidents on the Roads in Europe—CARE. Available in: http://ec.europa.eu/transport/road_safety/observatory/statistics/reports_graphics_en.htm, cited on February 10, 2010 (accessed on September 12, 2016).

4 Sihvola, N. & Rämä, P. 2008. Drivers' opinions about road weather conditions and road weather information—interviews during winter conditions. Helsinki 2008. *Finnish Road Administration 16/2008*, p. 62 + app. p. 10.

5 Yamada, M., Ueda, K., Horiba, I. & Sugie, N. 2001. Discrimination of the road condition toward understanding of vehicle driving environments. *IEEE Transactions on Intelligent Transportation Systems*, vol. 2, no. 1, pp. 26–31.

6 Gustafsson, F. 2009. Automotive safety systems. *IEEE Signal Processing Magazine*, vol. 26, no. 4, pp. 32–47.

7 Danoy, G., Boyvry, P., et al. 2011. Framework Definition. WiSafeCar EUREKA-CELTIC project deliverable 1.4.

8 Haas, T., Bian, N., Gamulescu, C., Köhler, M., Koskinen, S., Kutila, M., Jokela, M., Pesce, M., Hartweg, C. & Casselgrein, J. 2008. Fusion of vehicle and environment sensing in Friction project. In *Proceedings of the 17th Aachen Colloquium "Automobile and Engine Technology" Congress*, Aachen, Germany, October 6–8, 2008.

9 Koskinen, S. 2010. Sensor data fusion based estimation of tyre–road friction to enhance collision avoidance. Dissertation, VTT Publications 730, Espoo, Finland, p. 209.

10 Kutila, M., Jokela, M., Burgoa, J., Barsi, A., Lovas, T. & Zangherati, S. 2008. Optical road-state monitoring for infrastructure-side co-operative traffic safety systems. In *Proceedings of 2008 Proceedings of IEEE Intelligent Vehicles Symposium (IV'08)*, Eindhoven, The Netherlands, June 4–6, 2008.

11 Kutila, M. & Pyykönen, P. 2011. Prototype and demonstration of instrumented vehicle based data collection and visualisation in real-time. TIVIT Oy Cooperative Traffic–ICT subproject—Sensor Data Fusion and Applications. Deliverable 3.4.

12 Vaisala Oyj. 2010. Available in: www.vaisala.com, cited on February 10, 2010 (accessed September 12, 2016).

13 Holzwarth, F. & Eichhorn, U. 1993. Noncontact sensors for road conditions. *Sensors and Actuators, A: Physical*, vol. 37–38, pp. 121–127.

14 Casselgren, J. 2007. Road surface classification using near infrared spectroscopy. Licentiate thesis, Dept. Appl. Physics and Mech. Eng., Luleå Univ. of Tech, Luleå, Sweden.

15 Irvine, W. M. & Pollack, J. B. 1968. Infrared optical properties of water and ice spheres. *Icarus*, vol. 8, pp. 324–360.

16 Yuesheng, L. & Higgins-Luthman, M. 2007. Black ice detection and warning system. Patent application. Application number: US2007948086A.

17 Casselgren, J., Sjödahl, M. & LeBlanc, J. 2007. Angular spectral response from covered asphalt. *Applied Optics*, vol. 46, pp. 4277–4288.

18 White, S., Thornes, J. & Chapman, L. 2006. A Guide to Road Weather Systems. SIRWEC (online). Available in: http://www.sirwec.org/en/rwis_web_guide.pdf (accessed September 12, 2016).

19 Jonsson, P. 2009. Road status sensors—a comparison of active and passive sensors. In *16th ITS World Congress*, Stockholm, Sweden, September 21–25.

20 Kuehnle, A. & Burghout, W. 1998. Winter road condition recognition using video image classification. *Transportation Research Record*, vol. 1627, pp. 29–33.

21 Löfwing, S. 2007. Optisk metod att uppskatta egenskaper för ett is-eller vattenbelagt mätobjekt, Sweden SE 531949, Optical Sensors, PRV.

22 Jonsson, P., Casselgren, J. & Thörnberg, B. 2014. Road surface status classification using spectral analysis of NIR camera images. *IEEE Sensors Journal*, vol. 15, no. 3, pp. 1641–1656.

23 Saleh, B. E. A. & Teich, M. C. 2007. *Fundamentals of Photonics*, Second edition, John Wiley & Sons, Inc., New Jersey.

24 Löfwing, S. 2014. Road Eye. Available in: http://www.opticalsensors.se/Roadeye.html, cited on July 29, 2014 (accessed on September 12, 2016).

25 Born, M. & Wolf, E. 2003. *Principles of Optics*, 7th edition, Cambridge University Press, Cambridge.

26 Hecht, E. 1997. *Optics*, 3rd edition, Addison Wesley, Reading, MA.

27 Hubert, M. & Vanden Branden, K. 2005. Robust classification in high dimensions based on the SIMCA method. *Chemometrics and Intelligent Laboratory Systems*, vol. 79, no. 1–2, pp. 10–21.

28 Jonsson, P. 2011a. Classification of road conditions: from camera images and weather data. In *2011 IEEE International Conference on Computational Intelligence for Measurement Systems and Applications (CIMSA)*, September 19–21, 2011. http://ieeexplore.ieee.org/abstract/document/6059917/ (accessed on October 17, 2016).

29 Jonsson, P. 2011b. Road condition discrimination using weather data and camera images. In *2011 14th International IEEE Conference on Intelligent Transportation Systems (ITSC)*, October 5–7, 2011, pp. 1616–1621. http://ieeexplore.ieee.org/document/6082921/ (accessed on October 17, 2016).

30 Jonsson, P., Vaa, T., Dobslaw, F., et al. 2015. Road condition imaging—model development. In *Transportation Research Board 94th Annual Meeting*, Washington, DC, USA, January 11–15, 2015.

31 Jokela, M., Kutila, M. & Le, L. 2009. Road condition monitoring system based on a stereo camera. In *Proceedings of the 2009 IEEE 5th International Conference on Intelligent Computer Communication and Processing (IEEE ICCP)*, Cluj-Napoca, Romania, August 27–29, 2009, pp. 423–428.

32 Ewan, L., Al-Kaisy, A. & Veneziano, D. 2013. Remote sensing of weather and road surface conditions. *Transportation Research Record*, vol. 2329, no. 1, pp. 8–16.

33 Kutila, M., Pyykönen, P., Kauvo, K. & Eloranta, P. 2011. In-vehicle sensor data fusion for road friction monitoring. In *The 2011 IEEE 7th International Conference on Intelligent Computer Communication and Processing (2011 IEEE ICCP)*, Cluj-Napoca, Romania, August 25–27, 2011.

34 Yuen, A., Brown, C., Zott, C., Hiller, A., Ahlers, F., Wevers, K., Dreher, S., Schendzielorz, T., Bartels, C., Papp, Z. & Netten, B. 2008. Local dynamic map specification. SAFESPOT-EU-FP6 project Deliverable 3.3.3.

35 Hiller, A., Neumann, C., Matthess, M., Festag, A., Santos, H., Zhang, W., Sorge, C. & Wiecker, M. 2009. Spezifikation der Kommunikatiosprotokolle. simTD Deliverable 21.4.

Index

Computer Vision and Imaging in Intelligent Transportation Systems, First Edition.
Edited by Robert P. Loce, Raja Bala and Mohan Trivedi.
© 2017 John Wiley & Sons Ltd. Published 2017 by John Wiley & Sons Ltd.
Companion website: www.wiley.com/go/loce/ComputerVisionandImaginginITS